Manufacturing Systems: Design and Analysis

Manufacturing Systems: Design and Analysis

Editor:

Harinirina Randrianarisoa, Ph.D.

ArclerPress

www.arclerpress.com

Manufacturing Systems: Design and Analysis

Editor: Harinirina Randrianarisoa, Ph.D.

708 3rd Avenue, 6th Floor
New York
NY 10017
United States of America
www.arclerpress.com

ISBN: 978-1-68094-351-1

Library of Congress Control Number: 2016950936

About the Editor

Harinirina Randrianarisoa, Ph.D.

Hari Randrianarisoa obtained is Ph.D from Institute of Electronics, Microelectronics and Nanotechnology (IEMN) of the University of Science and Technology – Lille (France). His thesis focused for designing and manufacturing of thermal micro sensors. His interests span from on renewable energy especially solar energy. He has a strong knowledge in technology transfer field. He is currently working as a Business advisor for economic development organization in Québec City and as a consultant Scientist advisor.

Contents

List of Contributors

Antonio Gordillo
Tecnologías de la Información y Comunicación, Universidad Tecnológica del Suroeste de Guanajuato. Carr. Valle-Huanímaro km1.2, 38400 Valle de Santiago, Gto., Mexico

Adriana Giret
Departamento de Sistemas Informáticos y Computación, Universidad Politécnica de Valencia. Camino de Vera s/n 46022 Valencia, Spain

Xiaohui Lin
Department of Electrical and Computer Engineering, the University of Texas at Austin, Austin, TX 78758, USA

Harish Subbaraman
Omega Optics, Inc., 8500 Shoal Creek Blvd, Building 4, Suite 200, Austin, TX 78757, USA

Zeyu Pan
Department of Electrical and Computer Engineering, the University of Texas at Austin, Austin, TX 78758, USA

Amir Hosseini
Omega Optics, Inc., 8500 Shoal Creek Blvd, Building 4, Suite 200, Austin, TX 78757, USA

Chris Longe
Department of Electrical and Computer Engineering, the University of Texas at Austin, Austin, TX 78758, USA

Klay Kubena
Department of Electrical and Computer Engineering, the University of Texas at Austin, Austin, TX 78758, USA

Paul Schleicher
Department of Electrical and Computer Engineering, the University of Texas at Austin, Austin, TX 78758, USA

Phillip Foster
Department of Electrical and Computer Engineering, the University of Texas at Austin, Austin, TX 78758, USA

Sean Brickey
Department of Electrical and Computer Engineering, the University of Texas at Austin, Austin, TX 78758, USA

Ray T. Chen
Department of Electrical and Computer Engineering, the University of Texas at Austin, Austin, TX 78758, USA

Jiwei Huang
State Key Laboratory of Networking and Switching Technology, Beijing University of Posts and Telecommunications, Beijing 100876, China

Yeping Zhu
Agricultural Information Institute, Chinese Academy of Agricultural Sciences, Beijing 100081, China

Bo Cheng
State Key Laboratory of Networking and Switching Technology, Beijing University of Posts and Telecommunications, Beijing 100876, China

Chuang Lin
Department of Computer Science and Technology, Tsinghua University, Beijing 100084, China

Junliang Chen
State Key Laboratory of Networking and Switching Technology, Beijing University of Posts and Telecommunications, Beijing 100876, China

Z. Yang
Manufacturing Research Division, Faculty of Engineering, University of Nottingham, University Park, Nottingham NG7 2RD, UK

A.A. Popov
Manufacturing Research Division, Faculty of Engineering, University of Nottingham, University Park, Nottingham NG7 2RD, UK

S. McWilliam
Materials, Mechanics and Structures Research Division, Faculty of Engineering, University of Nottingham, University Park, Nottingham NG7 2RD, UK

Carlos Herrera
Departamento de Ingeniería Industrial, Universidad de Concepción, Edmundo Larenas, 215, Concepción, Chile

André Thomas
CRAN, Université de Lorraine, Nancy, France

Victor Parada
Departamento de Informática, Universidad de Santiago de Chile, Santiago de Chile, Chile

Muhammed Nafis Osman Zahid
Mechanical and Manufacturing Engineering, Loughborough University, Loughborough, Leicestershire LE11 3TU, UK

Keith Case
Mechanical and Manufacturing Engineering, Loughborough University, Loughborough, Leicestershire LE11 3TU, UK

Darren Watts
Mechanical and Manufacturing Engineering, Loughborough University, Loughborough, Leicestershire LE11 3TU, UK

Joakim Wikner
Department of Industrial Engineering and Management, School of Engineering, Jönköping University, Jönköping, Sweden

Eva Johansson
Department of Industrial Engineering and Management, School of Engineering, Jönköping University, Jönköping, Sweden

Jalel Ben Hmida
Systems Engineering, University of Louisiana at Lafayette, Rougeou, Hall Room # 244, Lafayette, LA 70504, USA

Jim Lee
Systems Engineering, University of Louisiana at Lafayette, Rougeou, Hall Room # 244, Lafayette, LA 70504, USA

Xinchun Wang
Systems Engineering, University of Louisiana at Lafayette, Rougeou, Hall Room # 244, Lafayette, LA 70504, USA

Fathi Boukadi
Department of Petroleum Engineering, University of Louisiana at Lafayette, Madison Hall, Room # 134, Lafayette, LA 70504, USA

Hossam Faris
King Abdullah II School for Information Technology, The University of Jordan, Amman, Jordan

Alaa F. Sheta
Computers and Systems Department, Electronics Research Institute (ERI), Giza, Egypt

Ertan Öznergiz
Marine Engineering Operations Department, Faculty of Naval Architecture and Maritime, Yildiz Technical University, Istanbul, Turkey

C.Q. Hou
School of Civil Engineering, Lanzhou University of Technology, Lanzhou 730050, People's Republic of China

S.Y. Li
School of Civil Engineering, Lanzhou University of Technology, Lanzhou 730050, People's Republic of China

Y. Cai
School of Civil Engineering, Lanzhou University of Technology, Lanzhou 730050, People's Republic of China

H.M. Wu
School of Science, Lanzhou University of Technology, Lanzhou 730050, People's Republic of China

A.M. An
College of Electrical and Information Engineering, Lanzhou University of Technology, Lanzhou 730050, People's Republic of China

Y. Wang
School of Civil Engineering, Lanzhou University of Technology, Lanzhou 730050, People's Republic of China

V. M. M. Thilak
Department of Production Engineering, PSG College of Technology, Coimbatore, Tamil Nadu 641004, India

S. R. Devadasan
Department of Production Engineering, PSG College of Technology, Coimbatore, Tamil Nadu 641004, India

N. M. Sivaram
Department of Mechanical Engineering, National Institute of Technology, Tiruchirappalli, Tamil Nadu 620015, India

Emad Abouel Nasr
Industrial Engineering Department, College of Engineering, King Saud University, Riyadh 11421, Saudi Arabia
Mechanical Engineering Department, Helwan University, Cairo 11792, Egypt

Abdulaziz M. El-Tamimi
Industrial Engineering Department, College of Engineering, King Saud University, Riyadh 11421, Saudi Arabia

Abdulrahman Al-Ahmari
Industrial Engineering Department, College of Engineering, King Saud University, Riyadh 11421, Saudi Arabia

Husam Kaid
Industrial Engineering Department, College of Engineering, King Saud University, Riyadh 11421, Saudi Arabia

Longfei He
College of Management and Economics, Tianjin University, Tianjin 300072, China

Huangli Peng
College of Management and Economics, Tianjin University, Tianjin 300072, China

Zhanwen Niu
College of Management and Economics, Tianjin University, Tianjin 300072, China

Haili Lu
Department of Logistics Management, Wuhan University of Technology, Hubei 430063, China

Xiangli Xie
College of Management and Economics, Tianjin University, Tianjin 300072, China

Guorong Chen
School of Electronic and Information Engineering, Chongqing University of Science and Technology, Chongqing 401331, China
CAIA, Dhurakij Pundit University, Bangkok 10210, Thailand
Faculty of Engineering, University of Ottawa, Ottawa, ON, Canada K1N 6M9

Jiangyuan Zhao
CAIA, Dhurakij Pundit University, Bangkok 10210, Thailand

Juli Deng
School of Electronic and Information Engineering, Chongqing University of Science and Technology, Chongqing 401331, China

Xun Gong
State Key Lab of Fluid Power Transmission and Control, Zhejiang University, Hangzhou 310027, China
Robotics and Microsystems Center, Soochow University, Suzhou 215006, China

Yixiong Feng
State Key Lab of Fluid Power Transmission and Control, Zhejiang University, Hangzhou 310027, China

Hao Zheng
State Key Lab of Fluid Power Transmission and Control, Zhejiang University, Hangzhou 310027, China

Jianrong Tan
State Key Lab of Fluid Power Transmission and Control, Zhejiang University, Hangzhou 310027, China

Jiafeng Zhang
School of Electro-Mechanical Engineering, Xidian University, Xi'an 710071, China
Chair of Automation and Energy Systems, Saarland University, 66123 Saarbrücken, Germany

Mohamed Khalgui
University of Carthage, 1054 Carthage, Tunisia

Wassim Mohamed Boussahel
Chair of Automation and Energy Systems, Saarland University, 66123 Saarbrücken, Germany
Zentrum für Mechatronik und Automatisierungstechnik, 66121 Saarbrücken, Germany

Georg Frey
Chair of Automation and Energy Systems, Saarland University, 66123 Saarbrücken, Germany

ChiTin Hon
Institute of Systems Engineering, Macau University of Science and Technology, Taipa, Macau

Naiqi Wu
Institute of Systems Engineering, Macau University of Science and Technology, Taipa, Macau

Zhiwu Li
Institute of Systems Engineering, Macau University of Science and Technology, Taipa, Macau
Faculty of Engineering, King Abdulaziz University, Jeddah 21589, Saudi Arabia

Fenglan He
Antai College of Economics & Management, Shanghai Jiao Tong University, 535 Fahua Zhen Road, Shanghai 200052, China

Ming Dong
Antai College of Economics & Management, Shanghai Jiao Tong University, 535 Fahua Zhen Road, Shanghai 200052, China

Dong Yang
College of Management, Donghua University, 1882 Yan-An Road, Shanghai 200051, China

Reza Raminfar
Department of Mechanical and Manufacturing Engineering, Faculty of Engineering, Universiti Putra Malaysia, 43400 Selangor, Selangor Darul Ehsan, Malaysia
Department of Industrial Engineering, Islamic Azad University, Lenjan Branch, isfahan, Iran

Norzima Zulkifli
Department of Mechanical and Manufacturing Engineering, Faculty of Engineering, Universiti Putra Malaysia, 43400 Selangor, Selangor Darul Ehsan, Malaysia

Mohammadreza Vasili
Department of Industrial Engineering, Islamic Azad University, Lenjan Branch, isfahan, Iran

Tang Sai Hong
Department of Mechanical and Manufacturing Engineering, Faculty of Engineering, Universiti Putra Malaysia, 43400 Selangor, Selangor Darul Ehsan, Malaysia

Preface

Manufacturing is the transformation of material into something useful and portable. A Manufacturing System is a set of machines, transportation elements, computers, storage buffers, people, and other items that are used together for manufacturing. This book brings together a wide range of published research and development studies under the manufacturing systems. The Book reviews the model control, production planning and tools for manufacturing. Furthermore, that studies a production approach to reduce cost. This study brings criteria for manufacturing. It embraces electronic, mechanical and automatic design field.

These studies demonstrate that whatever the nature of the manufacturing organization or the product being manufactured, all manufacturing systems have a number of common feature like to reduce time of manufacturing and to meet the market in the most cost-effective manner.

Editor
Harinirina Randrianarisoa, Ph.D.

CHAPTER 1

PERFORMANCE EVALUATION OF BIDDING-BASED MULTI-AGENT SCHEDULING ALGORITHMS FOR MANUFACTURING SYSTEMS

Antonio Gordillo [1] and Adriana Giret [2]

[1]Tecnologías de la Información y Comunicación, Universidad Tecnológica del Suroeste de Guanajuato. Carr. Valle-Huanímaro km1.2, 38400 Valle de Santiago, Gto., Mexico

[2]Departamento de Sistemas Informáticos y Computación, Universidad Politécnica de Valencia. Camino de Vera s/n 46022 Valencia, Spain

ABSTRACT

Artificial Intelligence techniques have being applied to many problems in manufacturing systems in recent years. In the specific field of manufacturing scheduling many studies have been published trying to cope with the complexity of the manufacturing environment. One of the most utilized approaches is (multi) agent-based scheduling. Nevertheless, despite the large list of studies reported in this field, there is no resource or scientific study on the performance measure of this type of approach under very common and critical execution situations. This paper focuses on multi-agent systems (MAS) based algorithms for task allocation, particularly in manufacturing applications. The goal is to provide a mechanism to measure the performance of agent-based scheduling approaches for manufacturing systems under key critical situations such as: dynamic environment, rescheduling, and priority change. With this mechanism it will be possible to simulate critical situations and to stress the system in order to measure the performance of a given agent-based scheduling method. The proposed mechanism is a pioneering approach for performance evaluation of bidding-based MAS approaches for manufacturing scheduling. The proposed method and evaluation methodology can be used to run tests in different manufacturing floors since it is independent of the workshop configuration. Moreover, the evaluation results presented in this paper show the key factors and scenarios that most affect the market-like MAS approaches for manufacturing scheduling.

INTRODUCTION

The manufacturers' success is no longer measured by their ability to cost-effectively produce a single product; success now seems to be measured in terms of flexibility, agility, and versatility [1]. In order to survive, manufacturing systems need to adapt at an ever-increasing pace to incorporate new technology, new products, new organizational structures, *etc.*

The above trends have motivated researchers in academia and industry to create and exploit new production paradigms on the basis of autonomy and co-operation because both concepts are necessary to create flexible behavior and thus to adapt to the changing production conditions. Such technologies provide a natural way to overcome such problems and to design and implement distributed intelligent manufacturing environments [1,2]. One of the complex problems in manufacturing systems is scheduling. Manufacturing scheduling is the process of assigning manufacturing resources over time to the set of manufacturing processes in the process plan. It determines the most appropriate time to execute each operation, taking into account the temporal relationships between

Citation: Antonio Gordillo and Adriana Giret, Performance Evaluation of Bidding-Based Multi-Agent Scheduling Algorithms for Manufacturing Systems, doi: 10.3390/machines2040233.

manufacturing processes and the capacity limitations of the shared manufacturing resources. The assignments also affect the optimality of a schedule with respect to criteria such as cost, tardiness, or throughput. In summary, scheduling is an optimization process where limited resources are allocated over time among both parallel and sequential activities. Such an optimization process is becoming increasingly important for manufacturing enterprises to increase their productivity and profitability through greater shop floor agility to survive in a globally competitive market [3].

In the past twenty years, researchers have been applying Artificial Intelligence (AI) techniques to many problems [4,5,6,7]. For a large literature review of AI applications for manufacturing systems see [8]. In the specific field of manufacturing scheduling many works are published that try to cope with the complexity of the manufacturing environment. One of the approaches most utilized in the specialized literature is (multi)agent-based scheduling. Nevertheless, despite the large list of studies reported in this field there is almost no resource or scientific study on the performance measure of this type of approaches under very common and critical execution situations. Even so, Bench4Start [9] must be pointed out as an interesting approach to define benchmarking solutions for manufacturing systems. Nevertheless, Bench4Start does not take into account multi-agent based approaches for manufacturing scheduling. The goal of the work presented in this paper is to provide a mechanism to measure the performance of agent-based scheduling approaches for manufacturing systems under key critical situations such as: dynamic environment, rescheduling, and priority change. With this mechanism it will be possible to simulate critical situations and to stress the system in order to measure a given agent-based scheduling method performance. In this work a specific method developed for bidding-based or market-like multi-agent approaches of manufacturing scheduling is presented.

TASK ALLOCATION IN INTELLIGENT MANUFACTURING SYSTEMS: STATE OF THE ART, BACKGROUND

Techniques from Artificial Intelligence have already been used in Intelligent Manufacturing for more than twenty years [1]. However, the recent developments in multi-agent systems in the domain of Distributed Artificial Intelligence have brought new and interesting possibilities. Distributed intelligent manufacturing systems are based on multi-agent system (MAS) technology [11]. MAS studies the coordination of intelligent behavior among a group of (possibly preexisting) agents. An agent is an autonomous and flexible computational system, which is able to act in the environment [11]. Today MAS is a very active area of research, which is beginning to see commercial and industrial applications [12,8,10,13].

The scheduling problem exists not only in manufacturing enterprises, but also in organizations like publishing houses, universities, hospitals, airports, and transportation companies. It is typically NP-hard, *i.e.*, it is impossible to find an optimal solution without the use of an essentially enumerative algorithm, with computation time increasing exponentially with problem size. However, the manufacturing scheduling problem is one of the most difficult of all scheduling problems.

A well-known manufacturing-scheduling problem is the classical job shop scheduling where a set of jobs and a set of machines are given. Each machine can handle at most one job at a time. Each job consists of a chain of operations, each of which needs to be processed during an uninterrupted time period of a given length on a given machine. The purpose is to find the best schedule, *i.e.*, an allocation of the operations to time intervals on the machines, which has the minimum total duration required completing all jobs. The total number of possible solutions for a classical job-shop scheduling problem with n jobs and m machines is $(n!)m$.

The problem becomes even more complex when unforeseen dynamic situations are considered. In a job shop manufacturing environment, things rarely go as expected. The system may be asked to include additional tasks that are not anticipated, or to adapt to changes to several tasks, or to neglect certain tasks. The resources available for performing tasks are subject to changes. Certain resources can become unavailable, and additional resources can be introduced. The beginning time and the processing time of a task are also subject to variations. A task can take more or less time than anticipated, and tasks can arrive early or late. Other uncertainties include power system failures, machine failures, operator absence, and unavailability of tools and materials. An optimal schedule, generated after considerable effort, may rapidly become unacceptable because of unforeseen dynamic situations on the shop floor and a new schedule may have to be generated. This kind of rescheduling problem is also called dynamic scheduling or real-time scheduling.

Agent-Based Scheduling for Manufacturing

In this paper the discussion on scheduling is restricted to (a) the allocation of production operations to specific resources and (b) the specification of timing (start, duration, and completion) for those operations. The key characteristics, which typify an intelligent manufacturing scheduling approach, are [14]:

- A local decision-making and computational capability associated with each agent.
- A cooperative interaction strategy that governs the way in which agents exchange information and determine mutually acceptable solutions.
- An interchange mechanism or protocol, which manages the exchange of message types, needed to execute the cooperative strategy.
- A means of ensuring that the global concerns of the factory are addressed.

A degree of central coordination (not present in all solutions).

Work on agent-based scheduling for manufacturing systems to date has predominantly experimented with different algorithms [15,16,17,18,19,20,21] and simulated testing although issues such as the desire for an emergent schedule *vs.* a fixed schedule structure, and the relationship between scheduling and execution has been examined also. Schiegg [22] compiled an extensive bibliography on multi-agent scheduling in manufacturing systems.

In agent-based manufacturing process planning and scheduling systems, bidding-based negotiations or market-like approaches are commonly used. In systems of this kind, the applied agent negotiation protocols require individual agents to reply to the incoming offers, to compete, and to negotiate or to bargain with other agents. As a result, rich knowledge bases and powerful learning and reasoning mechanisms are very important. Each agent should have at least knowledge of the capability, availability, and cost of the physical resource (e.g., a machine) represented by it. Some sophisticated agents need to have knowledge of other agents in the system, the products to be manufactured, and the know-how (historical experience, successful cases), *etc.*

The decision scheme of an individual agent depends primarily on two aspects [3]: coordination or negotiation mechanisms used by the multi-agent system and its local decision-making mechanisms based on knowledge. For example, a Contract Net protocol needs each individual agent to reply to an offer with requested information such as cost, starting time, processing time, *etc.* [23]. A game-theory-based multi-agent system needs agents to follow game rules [24]. While a multi-agent system implemented with a conversation scheme will need each agent to follow the conversation policies [25]. Local decision-making may use rule-based and case-based mechanisms reasoning on top of the knowledge the agent possesses. To update an agent's knowledge, learning mechanisms are needed. Such learning mechanisms may range from case-based reasoning to neural network and fuzzy logic-based reasoning.

Negotiation protocols are used in most agent-based manufacturing scheduling systems for resource allocation. The Contract Net Protocol or its modified versions are the most commonly utilized, although some other protocols such as the voting mechanism have also been considered. Although Contract Net and its variants are usually used as negotiation protocols in most agent-based scheduling systems, market-based approaches are becoming more and more popular. Market-based like protocols use the so-called bargaining process or auction process, which is also simple and easy to use. Market-based like approaches have recently been used in a number of agent-based scheduling systems [26,27,28,29].

Two types of agent-based distributed manufacturing scheduling systems can be distinguished according to the following characteristics.

- Scheduling is an incremental search process that can involve backtracking. Agents, responsible for scheduling orders, perform local incremental searches for their orders and may consider multiple resources. The global schedule is obtained through the merging of local schedules. This is very similar to centralized scheduling.
- An agent represents a single resource (e.g., a work cell, a machine, a tool, a fixture, a worker, *etc.*) and is responsible for scheduling this resource. This agent may negotiate with other agents to carry out the overall scheduling [27,28,30].

Most agent-based manufacturing scheduling systems proposed and developed in literature use the second approach.

The work presented in this paper is focused on the second approach for the following potential advantages. (I) Resource agents may be connected directly to physical devices they represented so as to realize real-time dynamic rescheduling (of course, not immediate rescheduling after any change in the working environment for the sake of system stability). It may therefore provide the manufacturing system with higher reliability and device fault tolerance. (II) Schedules are achieved by using mechanisms similar to those being used in manufacturing supply chains (*i.e.*, negotiation rather than search). This way, the manufacturing capabilities of manufacturers can be directly connected to each other and optimization is possible at the supply chain level in addition to the shop floor level and the enterprise level.

It is becoming clear that agent-based approaches offer many advantages for distributed manufacturing scheduling systems: modularity, reconfigurability, scalability, upgradeability, and robustness (including fault recovery). The results achieved so far in the agent research community provide excellent motivation for further development of solutions in this area. Moreover, at present, there are no other ways to solve these complex problems. However, whether the potential advantages of agent-based approaches can actually be realized in industrial systems will depend on the selection of a suitable system architecture for agent organization and an appropriate approach for agent encapsulation; on the design and implementation of effective mechanisms and protocols for communication, cooperation, coordination, and negotiation; and on the design and implementation of advanced internal architectures and efficient decision schemes of individual agents. Moreover, the key to success in any field is how well a given approach performs under very common and critical situations. Our goal is to assess this problem providing a mechanism that can help when evaluating the performance of a given agent-based scheduling approach for manufacturing.

A BIDDING-BASED MAS APPROACH FOR DYNAMIC SCHEDULING IN MANUFACTURING

In this work the main interest is the performance of bidding-based or market-like MAS approaches for dynamic scheduling of manufacturing systems under critical situations.

The bidding process implemented into a MAS is any protocol that allows an agent to focus on one or more resources to determine its final allocation as well as the involved transactions set in which the agents will be involved.

The bidding scheme has proven its efficiency as a general solution when applied to resource allocation integrated into a MAS [31]. Many variants of biddings have been developed, including single, combined, and multiple [32].

It is important to remark that the most common bidding scheme—ascendant bidding, known as English auction—represents just a small piece of a wide bidding spectrum. The fact that the bidding process is a relatively simple resource allocation mechanism, allows this working scheme to conform an infinite set.

The most common bidding model includes: one resource to be allocated, one seller, and many potential buyers. Each one of the buyers will set their own tag price for the item and want to acquire it at the lowest price as well.

Having multiple agents grouped at one side of the market environment makes them fall into the single-sided bidding category. The main goal defined for this classification is developing a specific protocol in order to solve a certain amount of defined global criteria, like creating an earn maximizing protocol, or establishing an economically efficient auction process, one that guarantees the potential buyer to acquire the selling item offering the maximum market value.

As previously stated, there is a wide range of working schemes for bidding processes. However, they all share a diverse structure, taxonomically speaking, common to all of them. For research purposes and simulation of real applications, they are to be considered as a structured framework for diverse negotiation threads. Each one of them includes certain rules, basically outlined by three main categories:

- Postulation rules: defining how, who, when, and what is to be contained for every auction bid.
- Clarification rules: defining when negotiations occur, what each one of them is related to in the bidding process (who obtains the resources and which bids are being modified)
- Information rules: related to bidders with active participation in the bidding process

The vast majority of the different bidding schemata include the above rules to a certain degree. However, more rules are likely to be added. However, aside from the bidding type and complexity aggregated to the industrial process, each design includes those three basic axes in its different implementations.

Given the vast amount of bidding schemes, the problem of how to select one of them may arise. With some considerations, the selection is not that relevant, as formalized in the following theorem:

Theorem 1. *Revenue Equivalence Theorem*. [33] Assume that each of n risk-neutral agents has an independent private valuation for a single good at auction, drawn from a common cumulative distribution $F(o)$ that is strictly increasing and atom less on [o,]. Then any auction mechanism in which the good will be allocated to the agent with the highest valuation, and any agent with valuation o has an expected utility of zero; this yields the same expected revenue, and, hence, results in any bidder with valuation o making the same expected payment.

From the above paragraph, it can be concluded that when bidders have independent and private bids, every auction scheme earns an equivalent. Aside from the theorem being useful to remark that the bidding type selection is not that relevant, it is a very useful analytical tool. In particular, it can be used to identify equilibrium strategies in that auction types fulfill the theorem conditions.

THE PROPOSED MODEL AND WORKING ENVIRONMENT

In this section the mechanism to measure the performance of bidding-based scheduling approaches for manufacturing systems is presented. It is defined as a MAS that can simulate critical situations in order to stress a manufacturing working environment and measure the performance of the scheduling approach (bidding-based) that is being used. It is important to point out that the work presented in this paper is focused on bidding-based mechanisms, nevertheless, other approaches for agent-based manufacturing scheduling are also important. The model described in this section can be modified in order to cope with other agent-based approaches. The final goal of the work, which is being developed by the authors, is to provide a complete package that the user can configure in order to test different agent-based approaches for manufacturing systems.

The proposed MAS architecture includes two levels. The first one is composed of those entities related to the real environment to be simulated. The second comprehends the responsible agents designed for controlling and handling the aforementioned entities. Furthermore, the MAS architecture requires more elements, such as information and communication management for each agent, as well as coordination and execution control for the assigned jobs.

The working environment in which the proposed approach will be tested consists of an assembly line system, with job stations receiving constant manufacturing assignments from different categories. The integrated agents are autonomous, each one representing a physical entity into the system.

To properly apply the bidding scheme, two agent categories are included: bidder and auctioneer agents, both autonomous and able to intercommunicate:

Auctioneer agent: coordinates the initial stage of the auction process. It sets the starting price of the job order. Moreover, it analyzes the generated bids and defines the best one.

Bidder agent: encapsulates the state of a given machine when a job order arrives. It also defines the bid during the entire process of the auction.

The implemented auction technique is the English auction. The auctioneer sets the assigned value for the job order as well as the prices randomly generated and offered by the bidder agents. The auctioneer agent generates and sends the reserved price for the resource to the system. A bidding agent is assigned for each assembly line. The individual bids are sequentially sent to the auctioneer when a new job unit requires processing.

In the following sections, a complete and detailed evaluation of the bidding-based MAS approach for task allocation in manufacturing environment is presented. The bidding-based approach is tested in order to measure its performance under key critical requirements of manufacturing systems. In order to simulate the situations for testing, dynamic environment and rescheduling, specific algorithms, which include these characteristics, are proposed. It is important to point out that if the concrete bidding mechanisms needs to be changed (for example, a Dutch auction, Japanese auction, *etc.*) it is only required to change the step in which the auctioneer evaluates the bids and the internal process by which every bidder agent defines their individual bid.

Dynamic Environment

One of the basic properties characterizing a modern manufacturing system is dynamism, defined as the set of changes in the structure and behavior during operation. This expresses different competencies, responsibilities, and relationships between entities [34].

The first scheduling requirement in the manufacturing field is for it to be dynamic [35]; in it, the job orders arrive at the assembly line with diverse characteristics like different priorities, manufacturing times, *etc.* In order to create a simulation environment for this requirement, the software elements depicted in Figure 1 were included.

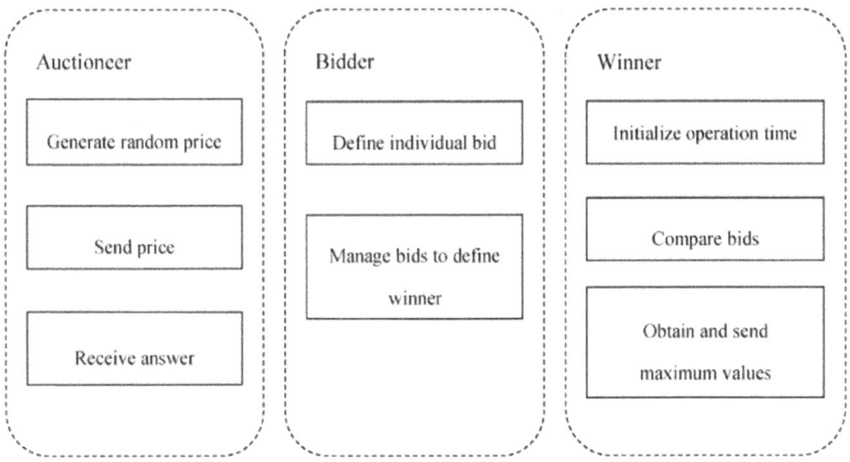

Figure 1: Conceptual classes for the multi-agent systems (MAS) Dynamic Environment.

For the proper execution of the Dynamic Environment simulation tests, the following conditions were considered:

Every agent can process only one job at a time.

A set of *n* operation jobs is available for processing at time zero.

Setup times for the operations are included in processing times.

m different machines are continuously available.

The evaluations for the bids, resource value assignations, and bid amount generations are effectuated randomly.

The priority values are randomly generated for every job order and classified as being of high, medium or low priority. Unless otherwise stated, the corresponding range values for all simulations are 0–3 for low, 4–7 for medium and 8–9 for high priorities, respectively.

Parameters evaluated: Allocation or rejection (when the maximum bid value is lower than the resource value), time for each resource, job order priority and occurrence of assignation with increased initial value

The operation flow related to the MAS agents for this test is in Table 1.

Table 1: Algorithm for job allocation in the Dynamic Environment test

(1) The auctioneer agent S_i sets the initial value V_i for the corresponding resource to every generated job assignment.

(2) Every registered bidder agent A_j presents one bid (ρ_j), constituting a set of eligible random options $C_{ij} = \{\rho_j, \rho_{j+1} \ldots \rho_{j+n}\}$ to allocate every job. S_i verifies every bid and once $\rho_j \geq V_i$ is accomplished; the corresponding resource is assigned to the Aj agent.

(3) $\rho_j \geq V_i + 1$ is evaluated, in order to maximize the allocation value of the corresponding resource. If the evaluation is valid, the resource will be allocated at $V_i + 1$ price, otherwise the allocation will take effect at V_i price.

(4) If $\rho_j < V_i$ is valid for every element of C_{ij}, the allocation will be considered empty. A new bidding stage takes effect.

(5) If $\rho_j = \ldots \rho_{j+1} = \rho_{j+n}$ is valid for every member of C_{ij}, ρ_j will be considered the maximum bid for the corresponding resource allocation.

Rescheduling

The Rescheduling process can be defined in general terms as a dynamic adjustment that updates the production scheme in response to sudden interruptions susceptible to occurrence in the manufacturing shop floor. There are diverse strategies that specify how and when to apply rescheduling to properly confront those occurrences. Within the scope of this work, the following strategies will be considered: High Priority and High Utilization [36].

The above strategies represent the combination of two elements: Rescheduling criteria and types of used algorithm for production rescheduling, both described in the following paragraph.

Rescheduling criteria: An interrupted job order by a failure in the manufacturing machines will be referenced as affected job; also, the amount of time for job completion will be the remaining time.

- Job priority: This criterion refers to the case of alternative machine susceptible of accepting the affected job, depending upon the priorities of the assigned job (actual job priority > minimum alternative job priority). If the affected job priority is higher than the equivalent in the processing job in the alternative machine, the latter is interrupted and the affected job starts to be processed.

- Machine utilization: Referred to when there are alternative machines able to receive the affected job depending upon its amount of utilization. When a failure in the machine affects a job, the availability of the alternatives is evaluated. If many options are presented, the machine with the lowest workload is selected.

Algorithm types

- High interference: Applied when a failure occurs. When activated, this algorithm will search alternative processing machines, ignoring the remaining time of the affected job process.
- Low interference: Consists of searching alternative machines only when failure time is higher than the affected job process remaining time.

In order to apply this test to the proposed algorithm, a set of software components was developed, as depicted in Figure 2.

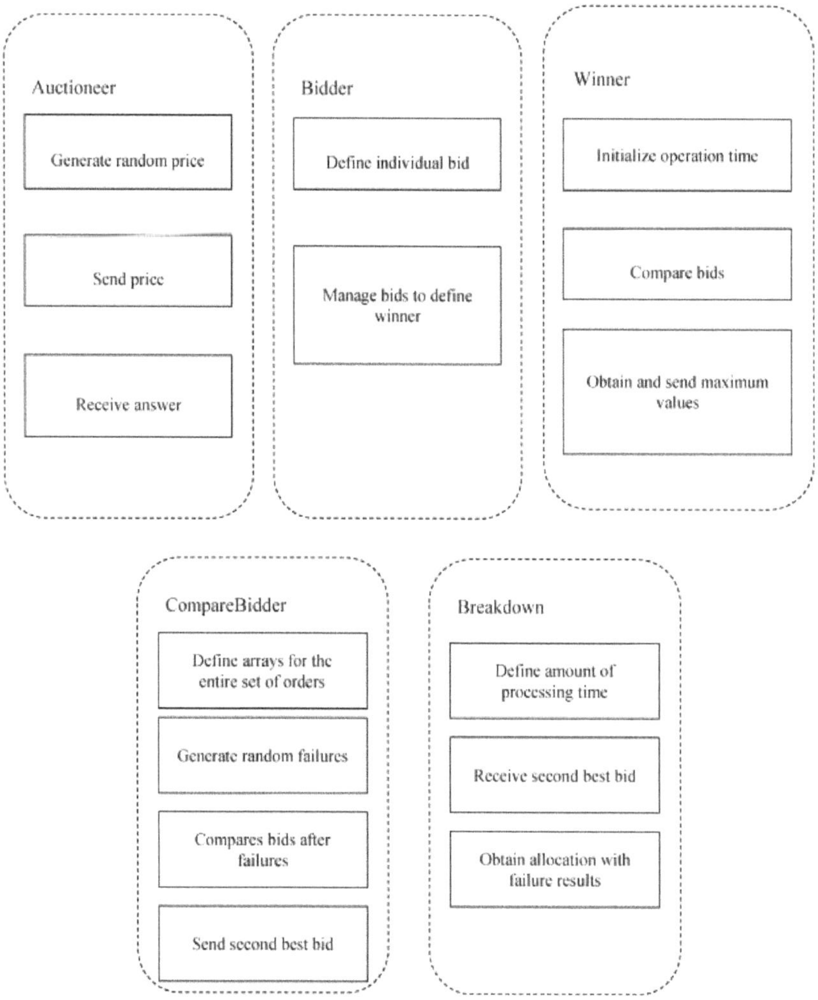

Figure 2: Conceptual classes for rescheduling.

In the following sections, the two types of simulated situations for rescheduling are implemented by the High Priority Algorithm and the High Utilization Algorithm. In this way the performance of the bidding-based MAS approach for scheduling can be validated when using the two mentioned techniques.

High Priority Algorithm

For simulation purposes of rescheduling requirements, the bidding-based algorithm is modified in order to use the high priority technique. In the first instance, when there is a breakdown occurrence affecting any of the scheduled jobs, the algorithm will select an alternative machine. If the priority of the affected job is higher than the available capacity of the remaining machines, the allocation is considered deserted. Otherwise, the job will be allocated to the machine with enough processing capacity.

In this test, the following conditions are considered:

- Every agent can process only one job at a time
- The random breakdown occurrence will always be applied on the machine with the best available capacity, 0 being for no machine failure and 1 for absence of breakdown. After the revision of this value for every scheduled job, the allocation process is initiated.
- A set of *n* operation jobs is available for processing at time zero.
- Setup times for the operations are included in processing times.
- *m* different machines are continuously available.

The characteristics of the High Priority Algorithm are defined in Table 2.

Table 2: High priority algorithm for rescheduling

(1) The auctioneer agent S_i sets the starting price (priority) V_i corresponding to every job order generated. Operation time t_i is initialized.
(2) Every registered bidding agent A_j offers a particular bid ρ_j, structuring one eligible random options set $C_{ij} = \{\rho_j, \rho_{j+1} \ldots \rho_{j+n}\}$ for assigning each job order. S_i verifies every bid and evaluates $\rho_j \geq V_i$, if valid, the resource is allocated to the A_j corresponding agent.
(3) A failure condition is generated randomly in the maximum bid agent max (ρ_j). As an alternative, the agent with the immediate lower bid is selected. The operation time increases due to the failure. The increment is defined by: $t_i = t_i + \Delta t_i$.
(4) $\rho_j \geq V_i + 1$, is evaluated in order to maximize the allocation value of the corresponding resource. When valid, the corresponding job will be definitely allocated at a price equal to $V_i + 1$, otherwise it will be allocated at a price equal to V_i.
(5) If $\rho_j < V_i$ is valid for every element of C_{ij}, the allocation will be considered deserted, and a new round of bids will take place. In a real environment, this is interpreted as a $\eth t$ waiting time, equivalent to the reincorporation of the failed machine.
(6) If $\rho_j = \rho_{j+1} = \ldots \rho_{j+n}$ is valid for every element of C_{ij}, ρ_j will be designated as the maximum bid for defining the resource allocation, evaluated for both failure or normal working processes.

High Utilization Algorithm

This algorithm includes basically the same processing rules as the High Priority Algorithm, with one difference: once the failure in the best machine is generated, a seek process is started in order to find the minimum bid presented in the system (the least capable agent) This algorithm is formalized in Table 3.

Priority Change Algorithm

The jobs that are under execution in a manufacturing system can experiment priority changes during exection due to different situations, *i.e.*, changes in the raw material bills, new job orders with different priority that enter the system affecting the already scheduled jobs, *etc.* In order to cope with this situation the Priority Change Algorithm is proposed to manage the system behavior when the priority level associated with the actual job rises at some point during execution. In order to simulate this situation, the proposed approach randomly generates two priority levels for the given job order, and defines a subsystem composed of the most and least occupied agents. Table 4 describes this algorithm in a more detailed fashion.

Table 3: High utilization algorithm for rescheduling

(1) The auctioneer agent S_i assigns the starting price (priority) V_i for the resource corresponding to every job order generated. Operation time t_i is initialized.
(2) Every registered bidder agent A_j generates a ρ_j bid, defining a random probable values set $C_{ij} = \{\rho_j, \rho_{j+1} \ldots \rho_{j+n}\}$ for assigning each job order. S_i verifies every bid and evaluates $\rho_j \geq V_i$, when valid, the resource is allocated to the A_j agent temporarily.
(3) A failure condition in the agent with the best offer: max (ρ_j), is generated. As an alternative, the agent accomplishing a min(C_{ij})—which represents the agent with the most inactive state in the set—is chosen. The total operation time reflects an increment due to the generated failure defined for $t_i = t_i + \Delta t_i$.
(4) $\rho_j \geq V_i + 1$ is evaluated, trying to maximize the allocation job value. When the comparison evaluates true, the job will be allocated at a $V_i + 1$ price tag, otherwise the final assignation price will be V_i.
(5) If $\rho_j < V_i$ is true for every element of C_{ij}, the allocation process is considered deserted. A new bidding round starts. This is considered as a $\eth t$ waiting time, equivalent to the reincorporation of the machine including a generated failure.
(6) If $\rho_j = \rho_{j+1} = \ldots \rho_{j+n}$ is true for every element of C_{ij}, ρ_j is designated as the highest bid for defining the resource allocation, and is evaluated for all system configurations.

Table 4: Priority change algorithm

(1)	The auctioneer agent S_i assigns the initial prices (priorities) V_i y H_i for the corresponding resource on each generated job. $H_i > V_i$. is true for all cases
(2)	From the priority ranges defined, the increment in priority will be classified as low to medium, medium to high or low to high, according to the generated values for V_i y H_i
(3)	Every registered bidder agent A_j presents a ρ_j bid, defining a set of random possible options $C_{ij} = \{\rho_j, \rho_{j+1} \dots \rho_{j+n}\}$ in order to allocate each job order. S_i verifies every bid and makes a selection on the best and worst offers $\max(\rho_j)$ and $\min(\rho_j)$, corresponding to the most and least occupied agents in the system, respectively.
(4)	$K = \max(\rho_j) + \min(\rho_j)$ is defined as the total capacity for the system to process the job order.
(5)	If $K < V_i$ evaluates true, the allocation process is considered deserted, and a new bidding round takes place.
(6)	If $K \geq V_i$ and $K < H_i$, the system is considered saturated when trying to process the actual job with a higher priority value. This result is added to the general evaluation of the system.
(7)	If $K \geq H_i$ is true, the system is considered not saturated and capable of processing the task at a higher priority. The result of the allocation is registered and incorporated to the general evaluation for the SMA on this specific test.

PERFORMANCE EVALUATION OF THE BIDDING-BASED SCHEDULING APPROACH USING THE PROPOSED MECHANISM

In this section the analysis of the execution of the bidding-based MAS approach for task allocation is described. The test where executed taking into account the performance criteria defined and modeled in the previous section: dynamic environment, rescheduling with high priority, rescheduling with high utilization, and priority change. The tests were executed running 100 simulations for the working environment (defined in Section 5) under the different complexity scenarios.

Dynamic Environment Test

The dynamic environment test was executed simulating the dynamic and random entrance of new job orders for one to 10 tasks. Moreover, every task was randomly labeled with a priority value: eight to nine for high priority, four to seven for medium priority, and zero to three for low priority. The results depicted in Figure 3 show that high priority allocation for jobs decreases when there are more job orders in the system. It varies, on average, from 24% with one job to 10% with ten jobs. On the other hand, the number of jobs does has almost no effect on the allocation of medium priority jobs in the system.

Finally, it can be noticed that when the number of jobs in the system increases, a very high amount of allocated jobs belongs to the low priority level, reaching up to 51.5% on average. Also, the unallocated job index with one job is very high (39%), while it remains almost stable (10.5% on average) when the number of jobs increases.

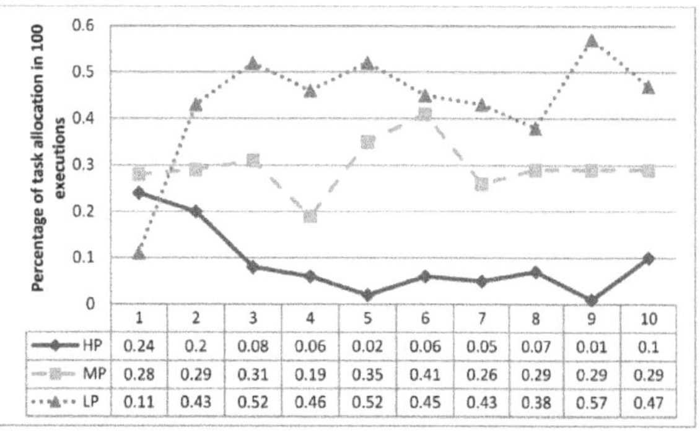

	1	2	3	4	5	6	7	8	9	10
HP	0.24	0.2	0.08	0.06	0.02	0.06	0.05	0.07	0.01	0.1
MP	0.28	0.29	0.31	0.19	0.35	0.41	0.26	0.29	0.29	0.29
LP	0.11	0.43	0.52	0.46	0.52	0.45	0.43	0.38	0.57	0.47

Figure 3: Dynamic Environment Test: Task allocation for one to 10 tasks order with high, medium, and low priority.

Rescheduling with High Priority Test

The rescheduling with High Priority test was executed with the following configuration: One hundred simulations; dynamic entrance of new job orders for one to 10 tasks, and priority values of eight to nine meaning high priority tasks, four to seven for medium, and zero to three for low. Random breakdown occurrences are simulated. The breakdown is always applied to the machine with the best available capacity, with zero being for no machine failure and one for absence of breakdown. After the revision of this value for every scheduled job, the allocation process is initiated.

Figure 4 shows that, despite the random failure generation in the most capable machine, the unassigned job levels maintain a low average for the medium and low priority scenarios. Although the high priority level scenario shows a decreasing performance when more job orders appear in the system, the percentage of allocated tasks is always between 30% and 50%. This result, in a real working environment scenario, indicates that the system is capable of processing a high percentage of jobs with a level of priority higher than the one initially required.

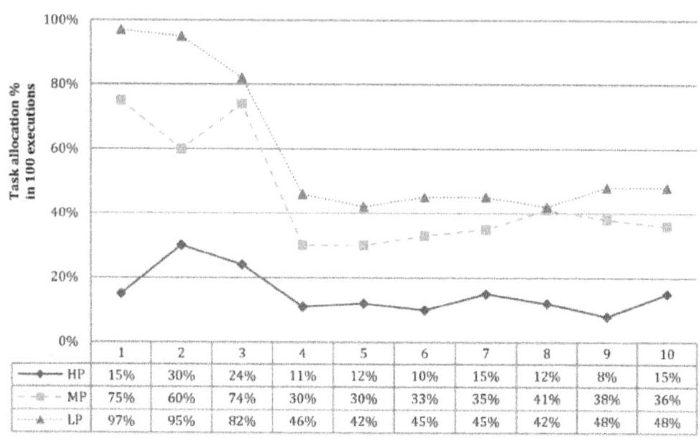

	1	2	3	4	5	6	7	8	9	10
HP	15%	30%	24%	11%	12%	10%	15%	12%	8%	15%
MP	75%	60%	74%	30%	30%	33%	35%	41%	38%	36%
LP	97%	95%	82%	46%	42%	45%	45%	42%	48%	48%

Figure 4: Rescheduling with High Priority Test: Task allocation for one to three task orders with high, medium, and low priority.

Once the testing process of the described software model finishes, having two generated job tasks, the unallocation index grows noticeably, moving from 9% to 20% for medium priority, and from 15% to 35% for high priority. This characteristic remains practically the same for low priorities. Despite that, the system behavior maintains acceptable levels of completed jobs, given that the unallocated job average level is lower than

10%. The observed general distribution is 25%, 30%, and 44% for high, medium, and low priority levels, respectively. The average completion time (50 ms) is slightly higher than the one observed in the previous test scenario (Dynamic Environment). This implies that the decision process before a machine failure generates an additional delay in the general allocation time.

Rescheduling with High Utilization Test

This test was executed under the same configuration of the working environment as for the previous tests (that is, 100 simulations; dynamic and random entrance of new job orders for one to 10 tasks; and, tasks' priority values: eight to nine for high priority, four to seven for medium priority, and zero to three for low priority). The processing rules for this test are similar to those of the High Priority test, with the following difference: once the failure in the best machine is generated, a seeking process is started to find the minimum bid presented in the system (the least capable agent).

Figure 5 shows the percentage of allocated tasks when random job orders enter the system with different priority levels. Analyzing these results again reveals the difficulty to allocate high priority level tasks in the system when the task orders increases.

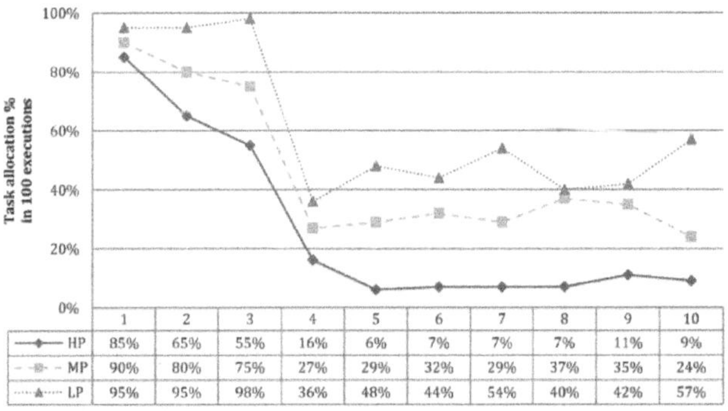

Figure 5: Rescheduling with High Utilization Test: Task allocation for zero to 10 task orders with high, medium, and low priority.

In the simulation using the High Utilization algorithm, the implemented Java code for the auctioneer seeks the lowest offer among the bidder agents and evaluates it against the randomly assigned value for the generated tasks. An independent routine is programmed in order to define a winner when the presented bids are equal; besides, three static functions use the bids as parameters to divide the information into decision blocks to allow the proper selection of the best bid for this particular case.

The allocation time using the High Utilization algorithm maintains practically the same observed levels in the application of the High Priority algorithm (arround 49 to 50 ms).

Priority Change Test

In this section the Priority Change test is decribed. In this test, the bidding-based MAS approach for task allocation is simulated with priority changes in the job orders that ramdonly enter the system. This way the average increment in the system saturation can be evaluated. For this test, the software simulator randomly generates the priority values as integer values in the range between zero and nine. To properly determine the coupled priority resulting from pairing two tasks of different values, the following conditions are considered: the total priority level is low if it falls in the range of zero to six, medium between seven and 12 and high between 13 and 18. For practical purposes, if the system is able to receive the job order at some initial priority but cannot complete it when its level is raised, it is considered oversaturated.

The high saturation in the system, when priority changes, is revealed in Figure 6 and in Table 5, with a 93% average for all the generated transitions, which leads to the conclusion that only with Low to Medium transitions the system has a good processing and allocating performance for the job tasks under this specific schema.

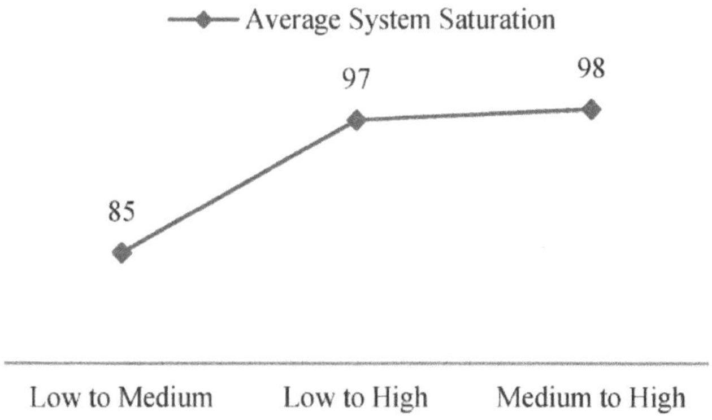

Figure 6: Average system saturation under simulations of the priority change test.

Table 5: Average system saturation levels

Total Experiments	Low to Medium Priority	Medium to High Priority	Low to High Priority
1–20	83%	99%	99%
21–40	81%	96%	97%
41–50	92%	96%	98%

PERFORMANCE EVALUATION ANALYSIS OF THE ALGORITHM

In this section, the overall performance of the bidding-based MAS approach when integrated by means of the algorithms presented in Section 5, to the diverse complexity scenarios is analyzed. The performance evaluation analysis considers the following: an optimum system performance is present when the processed orders are completed with a higher than initial priority (value), *i.e.*, the best bidder agent is able to accelerate the process; good performance is presen if the best bidder is able to allocate the job order at the initial price, and finally, a bad performance is present when the order cannot be allocated (Figure 7).

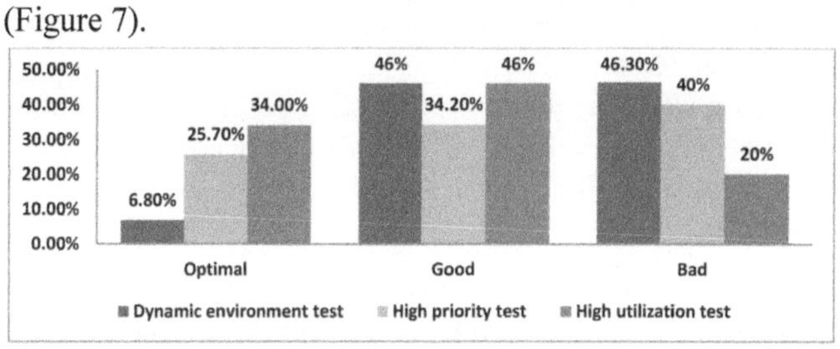

Figure 7: Performance evaluation for different tests.

Analyzing the results in Figure 7, it is clear that the overall behavior of the bidding-based MAS approach is good, on average, for those situations in which a high rescheduling requirement is present. For dynamic environments its behavior is not bad. Nevertheless, the number of unallocated tasks is higher than in the other tests.

In Figure 8, it can be noticed that the unallocated (NA) job level for the Dynamic environment Test is minimal (7%) and the entire allocation level is relatively high; however, a substantial amount of allocated jobs corresponds to low priority level orders, almost three times the result of high priority orders. Under the same optimum, good and bad classification scenarios, the High Priority test application results (Figure 8) show that despite the absence of the best bidder agent and the consequent utilization of an alternative, the accumulated amount of successfully allocated jobs is above 70%; however, categorizing the results, the high amount of unallocated jobs, almost equiparable to every allocated order category, is noticable.

For the High Utilization test, just as in the previous case, a failure occurrence in the best bidder agent is generated. Under these conditions, the alternative agent will be the least optimal. As expected, the amount of successfully completed orders decreases significantly (Figure 8). Furthermore, when the results for every priority level are distributed, the successfully completed high priority job order level is below 10%. Contrary to this, the low priority allocation level increases, reaching 32% on average.

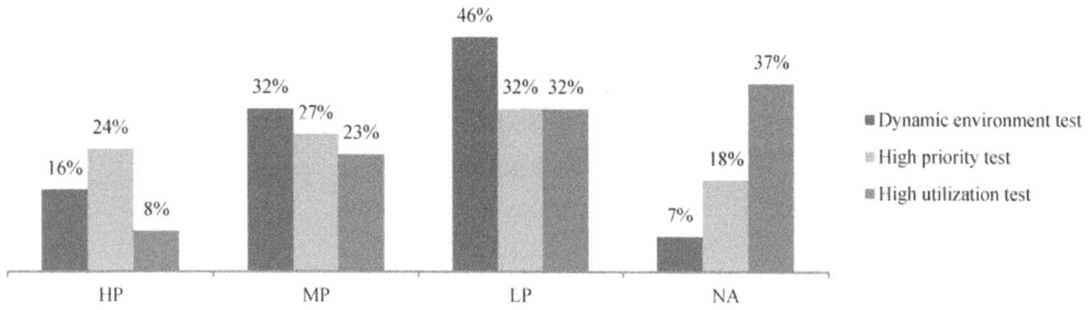

Figure 8: Task allocation distribution for different test and different priority values.

The final perspective is the Priority Change test analysis shown in Figure 6. In this, the average saturation level of the entire system is measured, *i.e.*, the system behavior when higher than initial priority levels are generated for all cases. For practical reasons, when reaching a 100% or higher saturation measure, the system is considered to surpass its capacity for processing incoming orders; for this reason, with the obtained values over 90% on average, it can be concluded that the system performs highly saturated under this conditions.

ROBUSTNESS MEASURE ANALYSIS OF THE BIDDING-BASED ALGORITHM

In this section, the robustness of the bidding-based MAS approach for task allocation is analyzed.

An optimized system, in a classical sense, can be very sensitive to small changes, without previous notification. It is preferable to design a system capable of delivering a high robustness level. Marczyk states that "optimization is precisely the opposite of robustness" [37] and, in this context, the system design is aimed at finding optimal solutions for the creation of a robust system. Precisely, the modeling process to attain this objective requires solving all related optimization problems, which are susceptible to demanding too much computing power [38].

The robustness of a schedule is not an easy concept to measure or even to define. A robustness measure can be calculated from the amount of lost time between the order finishing time and the order shipping time. In this case, a possible schedule robustness measure can be defined as:

$$R(S) = \frac{\sum_{j=1}^{n} \omega_j \left(d_j - C_j\right)}{\sum \omega_j d_j} \quad (1)$$

where:

dj, is the shipping time for job order *j*.

Cj, is the makespan for *j*.

ωj, is a weigh (priority) value assigned to the job order *j*.

This way the robustness measure becomes better as the value of *R*(*S*) increases.

Consider, for example, the processing time, shipping time, and priority values shown in Table 6 for three jobs.

Table 6: Initial values for simulation

Values	Job 1	Job 2	Job 3
p_j	10	10	10
d_j	10	22	34
ω_j	1	100	100

C_j, can be calculated from the cumulative processing time of the given job. In this case from Table 6 the following values are derived: $C_1 = 10$, $C_2 = 20$ and $C_3 = 30$.

The robustness measure using Equation (1) for schedule <1,2,3> is:

$$R(<1,2,3>) = \frac{600}{5610} = 0.11$$

while for the sequence <2,3,1>, the robustness measure is higher:

$$R(<2,3,1>) = \frac{2580}{5610} = 0.46$$

which means the second schedule is more robust than the first one. Moreover, this sequence has the higher value when analyzing all the possible sequences for this particular example.

In order to measure the robustness level of the generated schedules from the bidding-based MAS approach, a set of simulation tests taking into account the following issues, was executed:

- Oe the process is initialized, if the task is allocated at first bid, the p_j value remains without changes; however, if the capacity of the bidder allows the allocation at a higher value, the processing time is reduced and the p_j value is decreased in one time unit; however, in the case that the job could not be allocated, a system delay is considered and two time units are added for the processing time p_j
- Fty simulation tests were executed for this test, measuring the robustness index for each one and calculating their general average in order to obtain a comparative index against the optimal robustness level calculated initially.

As an example, when there is no adjudication of any task into the MAS, the robustness index level is defined as follows:

$$R(< 2,3,1 >) = \frac{100 \times 6 + 100 \times 8 + 1 \times (-26)}{5610} = 0.24$$

The general results from the robustness test reveals a 0.29 robustness unit for the average case (a 37% of the robustness measure for the most robust sequence) and a 0.38 robustness unit (17%) for the maximum capacity allocation in the system. Those values were calculated measuring the probability of interruption ocurrence, as well as the ability for reprogramming the assigned job tasks [39].

CONCLUSIONS

The research goal of the work presented in this paper is to provide a mechanism to measure the performance of agent-based scheduling approaches for manufacturing systems under key critical situations such as dynamic environment, rescheduling, and priority change. With this mechanism it will be possible to simulate critical situations and to stress the system in order to measure a given agent-based scheduling method performance. In this work a specific method developed for bidding-based or market-like multi-agent approaches for manufacturing scheduling was presented. The method presented and the evaluation test performed revealed that the bidding algorithm integrated into MAS is a useful model for effective utilization in real applications. However, some operating elements are in need of improvement, such as the system capacity for allocating job orders on failure occurrences into the best agents, which directly affects the robustness level. In addition, it is required that the methodology can include real measurement elements like operating times, incoming data with real variation, storing, and real time incorporation of historical data, among others.

In order to complete the mechanism for performance measure of agent-based scheduling approaches, more work is being developed in order to provide specific measurement methods for other kinds of agent-based scheduling approaches such as temperature equilibrium, swarm intelligence, *etc.* This will make it possible to have a performance measurement package that can be used to measure different agent-based approaches under different critical situations that affect the task allocation problem in manufacturing systems.

ACKNOWLEDGMENTS

A Marie Curie International Research Staff Exchange Scheme Fellowship within the 7th European Community Framework Program supported this research.

REFERENCES

1. Shen, W.; Norrie, D.H. Agent-Based Systems for Intelligent Manufacturing: A State-of-the-Art Survey. *Knowl. Inf. Syst.* 1999, *1*, 129–156.
2. Van Dyke Parunak, H.; Baker, A.D.; Clark, S.J. The AARIA agent architecture: From manufacturing requirements to agent-based system design. *Integr. Comput. Aided Eng.* 2001, *8*, 45–58.
3. Shen, W.; Wang, L.; Hao, Q. Agent-Based Distributed Manufacturing Process Planning and Scheduling: A State-of-the-Art Survey. *IEEE Trans. Syst. Man Cybern. Part C Appl. Rev.* 2006, *36*.
4. Marti, I.; Tomas, V.R.; Garcia, L.A.; Martinez, J.J. A Multi-agent system for managing adverse weather situations on the road network. *Integr. Comput. Aided Eng.* 2010, *17*, 145–155.
5. Bajo, J.; de Paz, J.F.; Rodríguez, S.; González, A. Multi-Agent System to Monitor Oceanic Environments. *Integr. Comput. Aided Eng.* 2010, *17*, 131–144.
6. Fernández de Alba, J.M.; Pavón, J. Talking Agents: A Distributed Architecture for Interactive Installation-Artworks. *Integr. Comput. Aided Eng.* 2010, *17*, 243–259.
7. López-París, D.; Brazález-Guerra, A. A new autonomous agent approach for the simulation of pedestrians in urban environments. *Integr. Comput. Aided Eng.* 2009, *16*, 283–297.
8. Badawy, R.; Yassine, A.; Heßler, A.; Hirsch, B.; Albayrak, S. A Novel Multi-Agent System Utilizing Quantum-Inspired Evolution for Demand Side Management in the Future Smart Grid. *Integr. Comput. Aided Eng.* 2013, *20*, 127–141.
9. Bench4star website. http://www.univ-valenciennes.fr/bench4star/.
10. Badawy, R.; Hirsch, B.; Albayrak, S. Agent-Based Coordination Techniques for Matching Supply and Demand in Energy Networks. *Integr. Comput. Aided Eng.* 2010, *17*, 373–382.
11. Wooldridge, M.; Jennings, N.R. *Intelligent Agents—Theories, Architectures, and Languages, Volume LNCS 890*; Springer-Verlag: Berlin, Germany, 1995.
12. Cristaldi, L.; Ponci, F.; Riva, M.; Faifer, M. Multi-agent Systems: An Example of Power System Dynamic Reconfiguration. *Integr. Comput. Aided Eng.* 2010, *17*, 359–372.
13. Pinto, T.; Praça, I.; Vale, Z.; Morais, H.; Sousa, T.M. Strategic Bidding in Electricity Markets: An Agent-Based Simulator with Game Theory for Scenario Analysis. *Integr. Comput. Aided Eng.* 2013, *20*, 335–346.
14. McFarlane, D.; Bussmann, S. Sate of the Art of Holonic Systems in Production Planning and Control. *Int. J. Prod. Plan. Control* 2000, *1*, 522–536.

15. Agre, J.; Elsley, G.; McFarlane, D.; Cheng, J.; Gunn, B. Holonic Control of Cooling Control Systems. In Proceedings of the Rensslaers Manufacturing Conference, New York, NY, USA, September 1994.
16. Brown, J.; Mccarragher, B. *Maintenance Resource Allocation Using Decentralized Co-Operative Control. Internal Report*; Australian National University: Camberra, Australia, 1998.
17. Gou, L.; Hasegawa, T.; Luh, P.; Tamura, S.; Oblack, J. Holonic Planning and Scheduling for a Robotic Assembly Testbed. In Proceedings of the 4th Rensselaer International Conference on Computer Integrated Manufacturing and Automation Technology, Rensselaer, NY, USA, 10–12 October 1994.
18. Gou, L.; Luh, P.; Kyoya, Y. Holonic Manufacturing Scheduling: Architecture, Cooperation, Mechanism, and Implementation. *Comput. Ind.* 1998, *37*, 213–231.
19. Heikkila, T.; Jarviluoma, M.; Juntunen, T. Holonic Control for Manufacturing Systems: Design of a Manufacturing Robot Cell. *Integr. Comput. Aided Eng.* 1997, *4*, 202–218.
20. Marcus, A.; Kis Vancza, T.; Monostori, L. A Market Approach to Holonic Manufacturing. *Ann. CIRP* 1996, *45*, 433–436.
21. Ramos, C. A Holonic Approach for Task Scheduling in Manufacturing Systems. In Proceedings of the IEEE Conference on Robotics and Automation, Minneapolis, MN, USA, 22–28 April 1996; pp. 2511–2516.
22. Schiegg, P. Bibliography on Multi-Agent Scheduling in Manufacturing Systems. Available online: http://farm.ecs.umass.edu/~pschiegg/bib/lit.html (accessed on September 2012).
23. Parunak, V.D. Manufacturing experience with the contract net. In *Distributed Artificial Intelligence*; Huhns, M.N., Ed.; Pitman: New York, NY, USA, 1987; pp. 285–310.
24. Babayan, A.; He, D. A distributed scheduling methodology for a two-machine flowshop using cooperative interaction via multiple coupling agents. *Int. J. Prod. Res.* 2004, *42*, 777–7104.
25. Lin, F.; Norrie, D.H.; Shen, W.; Kremer, R. Conversation specification—A schema-based approach to specifying conversation policies. In *Issues in Agent Communication, Lecture Notes in Computer Science*; Dignum, F., Greaves, M., Eds.; Springer-Verlag: New York, NY, USA, 2000; Volume 1916, pp. 193–204.
26. Goldberg, D. *Genetic Algorithms in Search, Optimization and Machine Learning*; Addison-Wesley: Reading, MA, USA, 1989.
27. Owliya, M.; Saadat, M.; Jules, G.G.; Goharian, G.; Anane, R. Agent-Based Interaction Protocols and Topologies for Manufacturing Task Allocation. *IEEE Trans. Syst. Man Cybern. Part A Syst. Hum.* 2011, *43*, 1, 38–52.
28. McDonnell, P.; Smith, G.; Joshi, S.; Kumara, S.R.T. A cascading auction protocol as a framework for integrating process planning and heterarchical shop floor control. *Int. J. Flex. Manuf. Syst.* 1999, *11*, 37–62.
29. Lee, Y.; Kumara, S.R.; Chatterjee, K. Multiagent based dynamic resource scheduling for distributed multiple projects using a market mechanism. *J. Intell. Manuf.* 2003, *14*, 471–484.
30. Wang, Y.H.; Yin, C.W.; Zhang, Y. A multi-agent and distributed ruler based approach to production scheduling of agile manufacturing systems. *Int. J. Comput. Integ. Manuf.* 2003, *16*, 81–92.
31. Boutilier, C. Sequential Auctions for the Allocation of Resources with Complementarities. In Proceedings of the International Joint Conference on Artificial Intelligence, Stockholm, Sweden, September 1999.
32. Shoham, Y. *Multiagent Systems, Algorithmic, Game-Theoretic, and Logical Foundations*; MAS Foundation: Miami, FL, USA, 2009.
33. Myerson, R. Optimal auction design. *Math. Oper. Res.* 1981, *6*, 58–73.
34. Caramia, M.; dell'olmo, P. *Effective Resource Management in Manufacturing Sytems*; Springer Series in Advanced Manufacturing; Springer: Berlin, Germany, 2006.
35. Wong, T.N.; Leung, C.W.; Mak, K.L.; Fung, R.Y.K. Dynamic shopfloor scheduling in multi-agent manufacturing systems. *Experts Syst. Appl.* 2006, *31*, 486–494.
36. Jain, A.K.; Elmaraghy, H.A. Production scheduling/rescheduling in flexible manufacturing. *Int. J. Prod. Res.* 1997, *35*, 281–289.
37. Marczyk, J. *Stochastic Multidisciplinary Improvement: Beyond Optimization, AIAA-2000–4929*; American Institute of Aeronautics and Astronautics: Madison Heights, MI, USA, 2000.
38. Huang, B.; Du, X. A robust design method using variable transformation and Gauss–Hermite integration. *Int. J. Numer. Methods Eng.* 2006, *66*, 1841–1858.
39. Pinedo, M.L. *Scheduling. Theory, Algorithms and Systems*, 3rd ed.; Springer: Berlin, Germany, 2008; pp. 480–485.

CHAPTER 2 TOWARDS REALIZING HIGH-THROUGHPUT, ROLL-TO-ROLL MANUFACTURING OF FLEXIBLE ELECTRONIC SYSTEMS

Xiaohui Lin [1], Harish Subbaraman [2], Zeyu Pan [1], Amir Hosseini [2], Chris Longe [1], Klay Kubena [1], Paul Schleicher [1], Phillip Foster [1], Sean Brickey [1] and Ray T. Chen [1]

[1]Department of Electrical and Computer Engineering, the University of Texas at Austin, Austin, TX 78758, USA

[2]Omega Optics, Inc., 8500 Shoal Creek Blvd, Building 4, Suite 200, Austin, TX 78757, USA

ABSTRACT

High-rate roll-to-roll (R2R) tracker systems are utilized for large volume flexible electronic device manufacturing, and the current alignment mechanism between layers is mainly achieved by relying on passive techniques. In this paper, we present a machine vision based alignment strategy that is used to achieve precise registration for stacking multilayers. Based on this strategy, we demonstrate two-layer printing with alignment accuracy better than 100 μm in web moving direction and 200 μm in lateral direction at a web rate of 5 m/min.

INTRODUCTION

Over the last decade, the printed electronics industry has been undergoing a significant growth in the market. Printed electronics refers to a set of printing methods, such as screen printing, flexography, gravure, offset lithography, inkjet printing, *etc.*, used to create electronic devices on various substrates. Among these methods, inkjet printing is an ideal solution for low cost, high speed printed electronics, due to the low material consumption, low equipment costs, and possibility for mass customization [1]. Some important application areas include RFID [2,3], solar cells [4,5], and displays [6,7], where device structures are mainly formed by stacking patterned layers, and where the critical dimension is in the tens of microns to millimeter range. The registration accuracy between layers is of great importance in determining the quality of the devices. For a tabletop printer, such as the ones printing on small area substrates, layer to layer registration is achieved via a microscopic system. The system determines the printing origin, and both the *X* and the *Y* direction offset values, with respect to previous layers. Therefore, one can precisely align two patterns with a maximum offset of the mechanical moving part accuracy, or to be more specific, the step motor accuracy.

However, the alignment strategy for a table-top printer is not applicable for a high-rate roll-to-roll system, where the registration needs to be performed while the substrate is moving at a high speed. There are several technologies that are used to realize roll-to-roll registration. For example, Krebs *et al.* punch a series of registration holes along the edge of the web so that the position of the web is kept relatively constant with respect to the printer head [8]. Jung *et al.* utilize no external setup for registration, but the mechanical system is capable of providing overlay printing registration accuracy of ±20 μm [2]. It is a practical way to make use of the precise tracking mechanism and the edge sensor to ensure minimum sideways wander of the web. However, such "passive" alignment strategies may

Citation: Xiaohui Lin, Harish Subbaraman, Zeyu Pan, Amir Hosseini, Chris Longe, Klay Kubena, Paul Schleicher, Phillip Foster, Sean Brickey and Ray T. Chen, Towards Realizing High-Throughput, Roll-to-Roll Manufacturing of Flexible Electronic Systems, doi:10.3390/electronics3040624.

encounter some off-registration problems if the original web has poor edge condition (e.g., edge flatness) that affects the width of the substrate. Due to the above-mentioned problems, in the actual industry environment, most ink-jet based roll-to-roll systems are only used to deposit the first layer, which does not require accurate alignment. For stacking of subsequent layers, screen printing is employed where the registration is performed when the web is stationery. This limits the final throughput of the manufacturing process.

In this research, we introduce an "active" alignment strategy using machine vision and demonstrate the entire system setup, including the software and hardware in order to realize multi-layer alignment. A feature of our method is that the active alignment is tolerant to the imperfections in the substrate condition, and does not pose demanding requirements on the web tracking system.

To better illustrate the system details, in Section 2, we first introduce the basic idea for realizing roll-to-roll registration. Based on this idea, the main system modules are determined. Next, based on the detailed system requirements, we specify the components in every module. A customized software implementation is described for synchronizing all the components for proper functioning in Section 3 Finally, multi-layer printing test is performed and the results are analyzed in Section 4.

SYSTEM COMPONENTS

Basic Idea

The Roll-to-Roll ink-jet printing system with alignment capability is a complex electro-mechanical system. In order to realize a high-speed, multi-layer printing system with good registration, an active alignment method should be considered. It should be noted that, in designing the system, the anisotropic mechanical properties of the flexible substrate also needs to be considered, since it plays an important role in achieving good alignment accuracy in a roll-to-roll process. In order to demonstrate the feasibility of our approach, and simplify the hardware, Kapton polyimide substrate is considered in our work due to its superior mechanical properties.

Alignment marks are utilized to determine the reference position, and all subsequent layer patterns are placed with respect to this reference position. Different from the table top printer used in our previous work [9,10,11], the printhead in the high-rate roll-to-roll system is stationary, which means that the alignment marks printed on first layer are not always exposed continuously to the same range of nozzles for the second layer printing due to sideways wander of the web.

A viable solution to this problem is to utilize a high speed imaging system to get the real-time alignment mark position, and adjust the printer nozzle actuation accordingly for each copy of the pattern. Additionally, the system should also be configured to provide correct timing to start the second layer printing by using a Print/Go device.

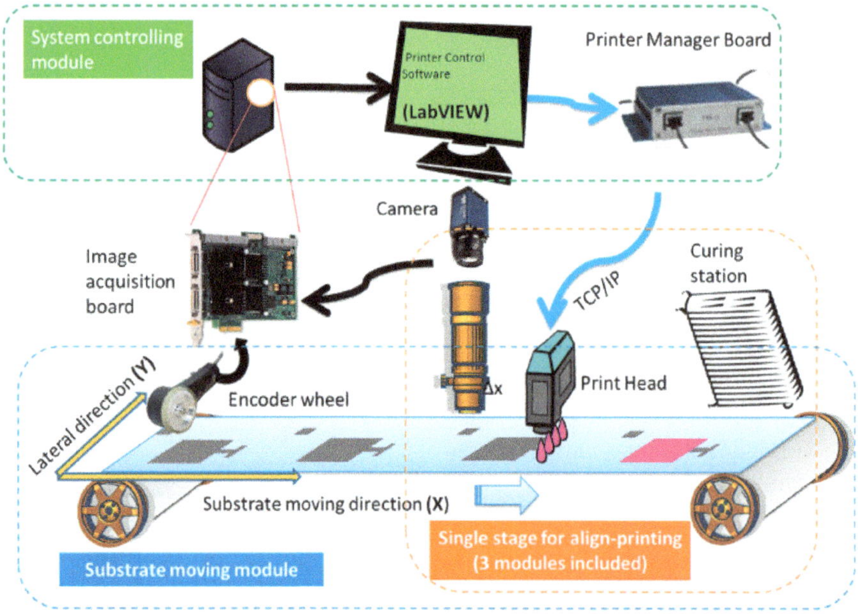

Figure 1: Schematic drawing of the main components in the roll-to-roll printing system with active alignment function. The main modules are (1) substrate moving module; (2) pattern detection module; (3) material ink-jet printing module; 4) fast curing module and (5) system control module. Modules (2) (3) and (4) form a standalone single stage system for aligned-printing, and can be configured in series to realize multi-stage continuous printing.

Main System Modules

Based on the above discussions, we determined the main modules for a single stage printer: pattern detection module, ink-jet printing module and fast curing module. To complete the web transfer and printing function, some other modules including substrate moving module (web tracker, encoder), system control module (desktop with the controlling software) are also needed. A cartoon briefly depicting all the components in a single stage system is shown in Figure 1.

The main functions of these modules are listed below:

- The substrate moving module delivers the web at a controlled speed.

- The pattern detection module captures the positions of the alignment and the Print/Go marks.
- The ink-jet printing module retrieves and prints patterns with the desired material.
- The curing system provides in-line fast curing of the ink.
- The system control module processes images from the image acquisition module, and controls the ink-jet printing module.

Web Tracking Module

The tracking system contains an edge detection sensor. The sensor detects the substrate edge and controls the tilting angle of a suspending shaft so that the web edge is aligned in real time to minimize the lateral direction (*Y*) offset and the sideways wander of the web. Although the edge detection sensor and the actuator system can ensure rough alignment of the web edge, the sideways wander is still inevitable. Moreover, since the width variations in the web can be large (which is common for commercial substrate material rolls), edge detection by itself is not sufficient to give satisfying registration results. In the current situation, the sideways wander is observed to be around 2 mm in each direction (4 mm total).

The maximum sideway wander determines the minimum required camera field of view (FOV). Generally speaking, reducing the FOV increases the physical resolution of the captured images. However, this requires the tracker to have less sideways wander, so that all the alignment marks can be captured. Another component in the web tracking module is the TR1 wheel encoder which provides a resolution of 15 μm per pulse. The purpose of this encoder is to feedback the real time web running speed for the ink-jet printing module to make sure the printed patterns have the correct aspect ratio.

Pattern Detection Module

The purpose of the pattern detection module is to capture the alignment and the Print/Go marks on the running web. We choose a line-scan camera and an optical system with coaxial illumination to build our image acquisition setup. With a single line of pixels, line-scan sensors build continuous images not limited in their vertical resolution. Vertical resolution is dependent on the web speed.

The line-scan camera has $n = 2048$ pixels, and the optical zoom lens is adjusted to set the FOV to 6 mm, which yields a camera system resolution of 3 μm/pixel. The image resolution can be increased further by reducing the FOV and/or increasing the line-scan camera pixels.

The Print/Go signal is generated by a photoelectric sensor. The signal is used to trigger the actuation of the printhead. When a Print/Go mark is detected, the sensor output produces a falling edge of a TTL signal, which in turn triggers the printing. As shown in Figure 2, the alignment mark first enters the FOV for determining the *Y* offset. Then, the photoelectric sensor sees the Print/Go mark in order to actuate the printing.

Material Ink-Jet Printing Module

The ink-jet printhead used in the current program is a high-speed head consisting of 1024 nozzles. The actuation of the cartridge is controlled by a "Print Manager Board" (PMB), which acts as the interface between the computer and the printhead. A "Print Server" software controls the PMB. It accepts external synchronizing signals from the encoder and routes the TCP/IP commands to the printhead. It also provides options for changing the waveform settings so that the jetting speed, jetting volume, and the transfer mechanism can be specified. In our system, the jetting speed is set at 5 m/s, providing a jetting volume of 14 pL per drop.

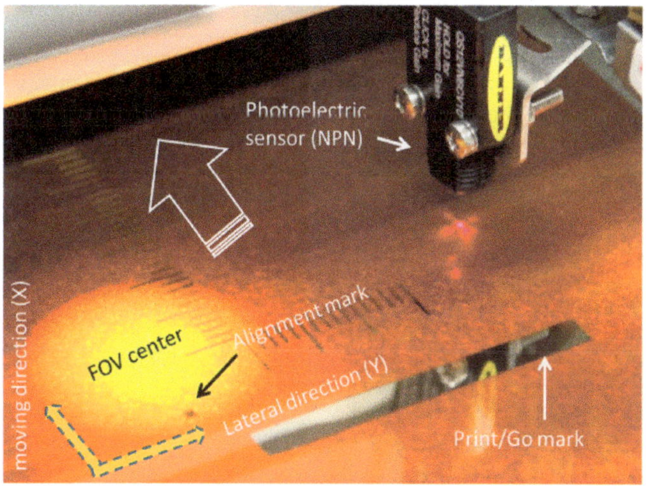

Figure 2: Alignment mark detection using line-scan camera, and Print/Go mark detection using photoelectric sensor.

Photonic Curing System

For the current roll-to-roll printing system, it is crucial to cure the printed ink quickly before it enters the next module. For inks and coatings containing nanoparticles, this process is called sintering. A conventional sintering process requires heating above 150 °C. However, it limits the applicable substrate, especially when the metal has a higher melting point than the flexible substrate, making it impossible to achieve sintering without causing subsequent damage to the substrate. Additionally, the curing speed of conventional systems is also relatively low for heating the ink above its melting points. To overcome these problems, a photonic curing/sintering method using high energy light pulse has been implemented.

Our photonic curing tool contains a 4.2" diameter spiral lamp (240 nm spectral cutoff). In our system, a 1.5 s exposure to 3 pulses, with a pulse energy 1500 Joules/pulse, completely sinters a printed silver nanoparticle film.

System Control Module

The system control module mainly refers to a PC equipped with the required hardware and software to perform the printing task, including acquiring images from the camera, processing the images, and sending the commands to the printhead.

The printer control is realized by TCP/IP commands. The client application initializes a command which is received by the server. The server processes the client command and starts the print engine operation. Once the operation is completed, the engine components respond to the server and the server feedbacks the client with related information. We set the printer to work in the queue print mode, thus enabling continuous addition of pre-offset images to the print queue.

SYSTEM INTEGRATION

With the basic functions realized by each individual module, system integration is realized by applying certain algorithm for aligning multi-layers. The detailed alignment strategy is discussed in the following sub-sections, along with software implementation using National Instruments NI 2011 LabVIEW software.

Alignment Strategy in the Y Direction (Lateral Direction)

The *Y* direction alignment is realized by the camera image processing. There are two steps involved: (a) determining the alignment mark offset; and (b) choosing the corresponding pre-offset second layer pattern to print.

By setting the correct FOV, one can make sure that whenever an alignment mark begins to pass under the camera, it appears in the FOV. The next step is to recognize the alignment mark and output its position. The alignment mark can be detected by utilizing pattern matching, which compares the captured image with a standard alignment mark template stored on the hard drive, and outputs the position error if they match to a certain degree. Alternatively, one can utilize an edge detection method as described below, which is preferred due to its lower processing time.

Edge detection is a set of mathematical methods which aims at identifying points in a digital image at which the image brightness changes sharply. It results in a highly reliable technique for detecting the alignment mark. The cartoon shown inFigure 3 presents the basic idea of detecting an alignment mark.

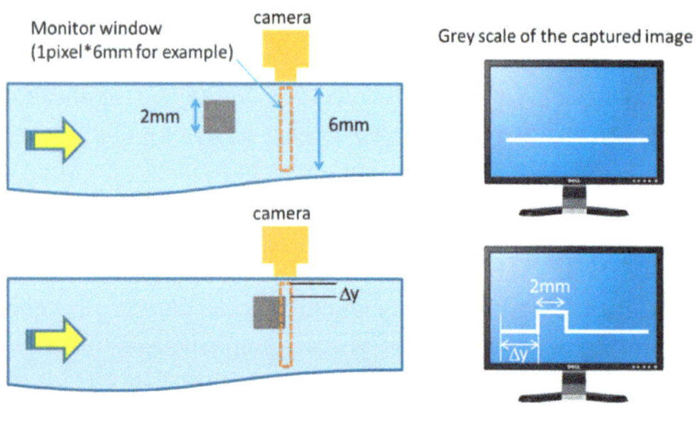

Figure 3: Edge detection scheme for *Y* direction alignment.

When an alignment mark passes under the camera, the abrupt greyscale image change is captured by the camera, and the first and last edge positions will be recorded. It should be noted, as mentioned above, that the line-scan camera has 2048 pixels, thus the direct output from the edge detection program is a value in pixel (0–2047). Here, in our application, we make use of the first edge to represent the position of the alignment mark. The last edge value is also recorded and used for checking the validity of the alignment mark.

The image processing speed should be able to match the image acquisition speed. All image processing applications are based on individual acquired images. When it comes to the case of line-scan camera, each image is formed by combining several line readout results together. For example, an image size of 2048 × 500 is formed by combining the results of 500 line readouts, with each line readout containing 2048 points. The line rate (number of line readouts per second) should be large enough in order to enable the capture of all the alignment marks. Therefore, the number of images, *N* that the system can acquire is determined using *N* = line rate/image height. A typical line rate is 20,000 Hz, and if we set the image height to be 200 pixels, it will generate *N* = 100 images per second for processing.

The image processing step should be efficient enough so that each acquired image can be processed. In our system, the line edge detection method takes around 0.2 ms, while the pattern matching method takes around 2.5 ms, which is about 10 times longer. During this processing time, the web is still being transferred, causing an uncontrollable registration error in the *X* direction. Thus, the image processing time should be as short as possible. This makes edge detection a natural choice for achieving aligned printing at high-rates.

After receiving the offset value, there are three methods to print the offset images. The first method involves physically moving the printhead, which requires a high cost, fast, and precise moving mechanism. For high-rate printing, the mechanical motion may not catch up with the web speed, thereby limiting the throughput. The second method involves offsetting the nozzles using a program. However, our current system does not provide this function. The third method involves loading the images with pre-offset patterns, and it turns out to be the most practical one. In this method, we build an array of bitmap (*.bmp) images with different pre-offset values and name them sequentially when storing in the hard disk. The images are formed by moving all "black pixels" down by 1 pixel (70.5 μm in physical offset) while maintaining the entire image size. We have developed a MATLAB code that offsets the input image by 1 pixel each time, and then stores the images in the hard drive. By mapping the offset value from edge detection to the image ID, the image with correct offset value can be selected and sent to print server for printing.

Alignment Strategy in the X Direction (Web Moving Direction)

After choosing the proper pre-offset image to print, we also need to determine the correct time to start printing. Therefore, we also need the *X* direction alignment, which is aimed at setting the right timing for the printhead to start printing the patterns. We use a photoelectric sensor with Print/Go mark to determine the printhead actuation moment. For example, as illustrated in Figure 4, we need to print a second layer pattern (UT AUSTIN filled in RED) on top of the first layer image. The alignment marks and Print/Go marks are placed at the ending positions of each copy of the first layer pattern (unfilled UT AUSTIN). The camera will first detect the alignment mark, and initiate the TCP/IP protocol. This triggers a chain of events, leading up to the availability of an error corrected print pattern ready for printing. When the Print/Go signal is detected by the photoelectric sensor, the print ready pattern is printed. The time delay between the detection of the alignment mark and the detection of the Print/Go signal on the same copy is called buffer time, and it should be set long enough to finish the processing, image loading and transfer tasks. In the actual setup, the buffer time can be adjusted by changing the value of B and D, which represent the distances between the photoelectric sensor and the camera to the Print/Go mark, respectively, as illustrated in Figure 4.

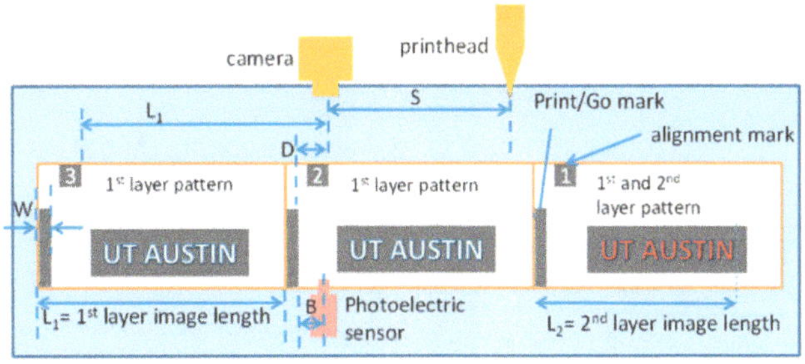

Figure 4: Alignment strategy in the *X* direction showing the relative positions of the printhead, camera, alignment mark, Print/Go mark and the photoelectric sensor.

Software Development Using NI LabVIEW

We use LabVIEW from National Instruments (NI) to integrate all the functional modules needed. Our software performs three major roles: image acquisition, image processing and printer control.

MODULE TESTING

Alignment Mark Detection

Whenever an alignment mark enters the field of view, it will be captured. The processed result is then used to select the corresponding pre-offset image, and the print command is sent out immediately. Figure 5 shows an offset testing result with 20 consecutive alignment marks captured by the camera. A non-flat behavior is typically expected due to sideways wander of the web. From the graph, the maximum sideways wander range of the web is calculated to be about 4 mm.

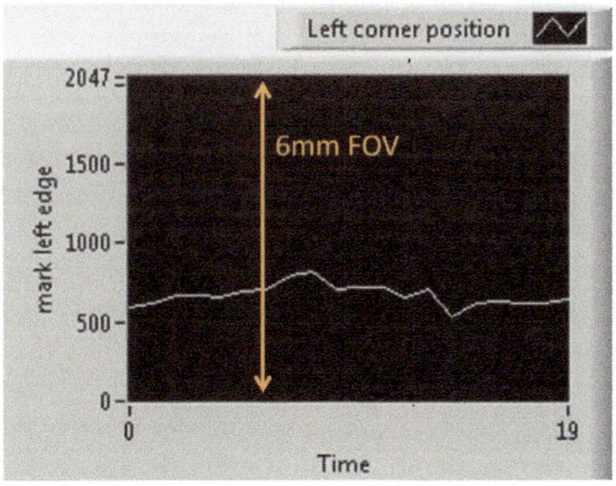

Figure 5: Detected positions of 20 consecutive alignment marks, indicating the web sideway wander of around 1.5–2 mm each way for this pass.

Registration for Multi-Layer Printing

We perform real-time aligned printing using a test pattern, as shown in Figure 6A. The first layer pattern contains a 2 mm × 2 mm square alignment mark, Print/Go mark and reference scales in *X* and *Y* directions. The second layer is a cross hair mark that is centered at $X = 40$ mm, $Y = 10$ mm location. In the experiment, we printed the first and the second layers using silver nanoparticle ink at a web rate of 5 m/min. The result is shown in Figure 6B, where the cross is printed at the desired location. We also perform continuous printing of 20 second-layer patterns to test its repeatability. The misalignment distribution with respect to the designed location results is shown in Figure 7, with inset pictures showing typical printed results on the web. As can be seen, for all copies, the off-registration errors are <100 μm in *X* and <200 μm in *Y* directions, which indicates the feasibility of our alignment strategy.

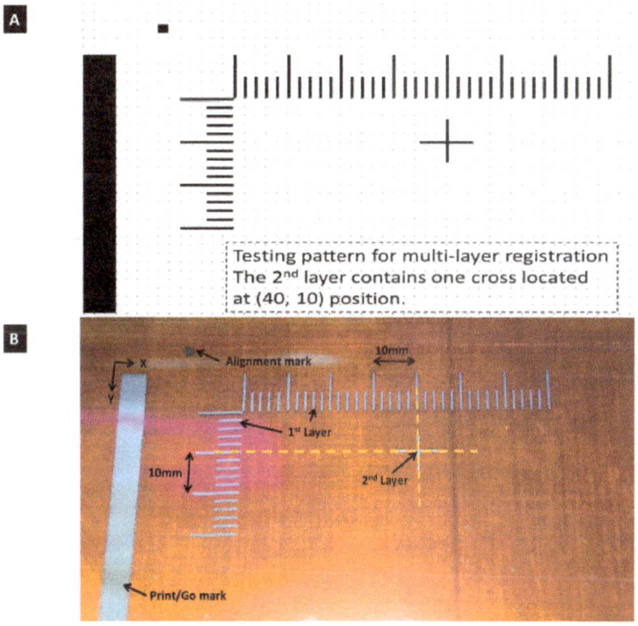

Figure 6: (**A**) Test pattern for aligned printing; (**B**) printed results showing registration of two layers with off-alignment less than 100 μm in both *X* and *Y* directions.

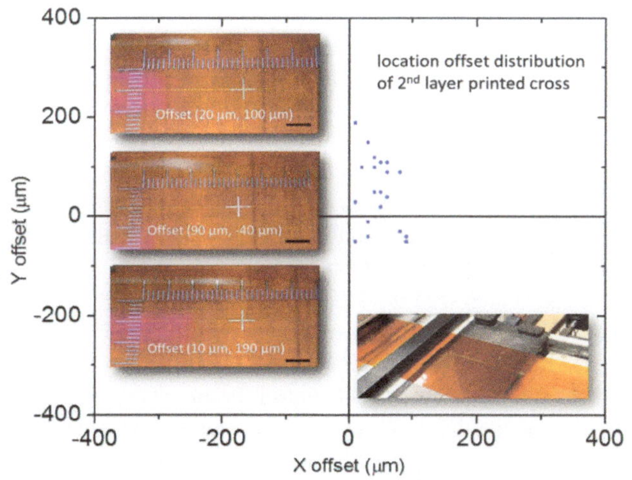

Figure 7: Registration offsets distribution showing continuous printing of 20 second-layer patterns with $\Delta X < 100$ μm and $\Delta Y < 200$ μm for all copies. The inset pictures have scale bars of 10 mm.

PERFORMANCE OVERVIEW

Although precise and effective algorithms are applied to ensure best registration accuracy, there are still several factors that need to be considered to minimize their effects to the system performance and auto registration accuracy. Factors that need further attention are:

- The resolution of the printhead.
- The time required to initialize the Print Server software.
- The time between the print command being sent and the Print Server being ready to start printing.
- Image acquisition and processing delays.
- The number of PMBs or printheads which can be connected to a single PC.

It is challenging to precisely predict or guarantee the performance of the software, as there are a number of different factors which contribute to it. These factors include:

- The operating system setup.
- The PC hardware.
- The other software or operations running on the PC.
- The Print Server configuration.

Table 1: Limiting factors for registration error

X Direction (absolute value)	
Limiting factors	Offset introduced
Encoder error due to slip when start/stop	100–200 μm
Printer delay (4 ms delay, web running is 5m/min)	333 μm during 4 ms
Image file transfer delay (uncertain)	Estimate ≈ 500 μm (500 KB file size) (*)
LabVIEW program image processing time (1 ms)	80 μm
Actual tested offset	<100 μm
Y Direction (absolute vale)	
Limiting factors	Offset introduced
Printer nozzle offset	70.5 μm
Sideway wander offset while transferring from FOV to printhead	Up to 2000 μm
Camera resolution	3 μm
Actual tested offset	<200 μm

(*) Including PMB-PC via USB (250 MB/s), thus 500 KB needs 2 ms Hard disk readout time: for 5400 rpm hard drive, 100 MB/s thus 500 KB needs 5 ms Total 7 ms delay (≈500 μm).

To be specific for our system, some limiting factors affecting the overall system registration resolution are shown in Table 1. In the *X* direction, the introduction of Print/Go marks gives the system extra buffering time for file loading/transferring and provides the exact timing to start printing. In the print runs, the buffer length (B) is set at around 10 mm (adjustable) which gives about 120 ms buffering time when the web is running at 5 m/min. Thus, many issues related to PC performance are eliminated, and the alignment errors are controlled well within 100 μm. This error is mainly due to the slip of the encoder. In the *Y* direction, the main limiting factors come from the printer nozzle offset and the web sideways wander. The printer nozzles have a pitch of 70.5 μm, which is larger than the camera resolution. Thus, the offset values will be rounded and mapped to the nearest integer when choosing the pre-offset images, as shown in Table 1. Besides, after a correct second layer image is selected, and until the moment printing starts, the web would have moved sideways due to web wander, which is beyond our current control. From the design point of view, reducing the camera/printhead separation and the second layer pattern width would be helpful in minimizing the effect of this uncertainty. In order to further improve the alignment accuracy, dynamic position control of printhead location can be utilized. Currently, the position of the printheads is fixed relative to the substrate. Mounting the printheads on precision motors will provide an additional degree of freedom for future systems in order to achieve sub 10 μm alignment accuracy, however, with a small throughput penalty. To ensure high yield, enough tolerance should be given for off-registration when designing the device pattern. In the future, we will also consider the effects of substrate shrinkage and buckling on the alignment accuracy.

CONCLUSIONS

In this paper, we demonstrated a customized, machine vision aided roll-to-roll ink-jet printing system for high rate, high yield device fabrication. Based on an active alignment technique, we built up all the necessary modules and developed the software to realize auto-registration for multi-layer stacking. The *X* direction registration was realized by in-line capturing the alignment mark and determining the onset moment of printing. The *Y* direction registration was achieved by real time image processing to get the alignment mark offset, and selecting the pre-offset images to compensate for the deviation. Several limiting factors, including the resolution of the printhead, the time required to initialize the Print Server software, the time between the print command being sent and the Print Server being ready to start printing, Image acquisition and processing delays, and the number of PMBs or printheads which can be connected to a single PC, affect the overall system performance. Specifically for registration, the printer nozzle offset, camera resolution, encoder error, print delay, image processing delay, leads to a current registration error of about 100 μm and 200 μm for the *X* and the *Y* directions, respectively.

ACKNOWLEDGMENTS

This work was supported by the Office of Naval Research (ONR) Small Business Technology Transfer (STTR) contract N00014-11-C-0704 monitored by Khershed Cooper. We would also like to acknowledge Ed Garcia from Industrial Inkjet Ltd and Kaj Larsen from KAJ's Technical services for their valuable suggestion in setting up the system.

AUTHOR CONTRIBUTIONS

Xiaohui Lin, Harish Subbaraman contributed to the system design, component selection, system setup, data analysis, and manuscript preparation. Zeyu Pan and Amir Hosseini contributed to data analysis and manuscript preparation. Chris Longe, Klay Kubena, Paul Schleicher, Phillip Foster, and Sean Brickey contributed to the alignment software development. Ray T. Chen contributed to overall system design, data analysis, and manuscript preparation.

REFERENCES

1. Parashkov, R.; Becker, E.; Riedl, T.; Johannes, H.; Kowalsky, W. Large area electronics using printing methods. *Proc. IEEE* 2005, *93*, 1321–1329.
2. Jung, M.; Kim, J.; Noh, J.; Lim, N.; Lim, C.; Lee, G.; Kim, J.; Kang, H.; Jung, K.; Leonard, A.D.; *et al.* All-Printed and Roll-to-Roll-Printable 13.56-MHz-Operated 1-bit RF Tag on Plastic Foils. *IEEE Trans. Electron Devices.* 2010, *57*, 571–580.
3. Noh, J.; Dongsun, Y.; Lim, C.; Cha, H.; Han, J.; Kim, J.; Park, Y.; Subramanian, V.; Cho, G. Scalability of Roll-to-Roll Gravure-Printed Electrodes on Plastic Foils. *IEEE Trans. Electron. Packag. Manuf.* 2010, *33*, 275–283.
4. Bundgaard, E.; Krebs, F.C. Low band gap polymers for organic photovoltaics. *Sol. Energy Mater. Sol. Cells.* 2007, *91*, 954–985.
5. Jørgensen, M.; Norrman, K.; Krebs, F.C. Stability/degradation of polymer solar cells. *Sol. Energy Mater. Sol. Cells.* 2008,*92*, 686–714.
6. Jacobson, J.; Comiskey, B.; Albert, J. Microencapsulated electrophoretic display. U.S. Patents US5961804 A, 5 October 1999.
7. Kim, C.A.; Myoung, H.J.; Kang, S.-Y.; Chung, H.-S.; Kim, G.H.; Ahn, S.D.; You, I.-K.; Oh, J.; Baek, K.H.; Soo, K. Towards a Colored Electronic Paper through a Fabrication of Color Microencapsulated Electrophoretic Display Panel. In Proceedings of Imid/Idmc 2006: The 6th International Meeting on Information Display/the 5th International Display Manufacturing Conference, Digest of Technical Papers, Daegu, Korea, 22–25 August 2006; pp. 1415–1418.
8. Krebs, F.C.; Tromholt, T.; Jørgensen, M. Upscaling of polymer solar cell fabrication using full roll-to-roll processing.*Nanoscale* 2010, *2*, 873–886.
9. Lin, X.; Ling, T.; Subbaraman, H.; Zhang, X.; Byun, K.; Guo, L.J.; Chen, R.T. Ultraviolet imprinting and aligned ink-jet printing for multilayer patterning of electro-optic polymer modulators. *Opt. Lett.* 2013, *38*, 1597–1599.
10. Lin, X.; Ling, T.; Subbaraman, H.; Guo, L.J.; Chen, R.T. Printable thermo-optic polymer switches utilizing imprinting and ink-jet printing. *Opt. Express* 2013, *21*, 2110–2117.
11. Subbaraman, H.; Pham, D.T.; Xu, X.; Chen, M.Y.; Hosseini, A.; Lu, X.; Chen, R.T. Ink-Jet Printed Two Dimensional Phased-Array Antenna on a Flexible Substrate. *IEEE Antennas Wirel. Propag. Lett.* 2013, *2*, 170–173.

CHAPTER 3 A PETRINET-BASED APPROACH FOR SUPPORTING TRACEABILITY IN CYBER-PHYSICAL MANUFACTURING SYSTEMS

Jiwei Huang [1], Yeping Zhu [2], Bo Cheng [1], Chuang Lin [3] and Junliang Chen [1]

[1]State Key Laboratory of Networking and Switching Technology, Beijing University of Posts and Telecommunications, Beijing 100876, China

[2]Agricultural Information Institute, Chinese Academy of Agricultural Sciences, Beijing 100081, China

[3]Department of Computer Science and Technology, Tsinghua University, Beijing 100084, China

ABSTRACT

With the growing popularity of complex dynamic activities in manufacturing processes, traceability of the entire life of every product has drawn significant attention especially for food, clinical materials, and similar items. This paper studies the traceability issue in cyber-physical manufacturing systems from a theoretical viewpoint. Petri net models are generalized for formulating dynamic manufacturing processes, based on which a detailed approach for enabling traceability analysis is presented. Models as well as algorithms are carefully designed, which can trace back the lifecycle of a possibly contaminated item. A practical prototype system for supporting traceability is designed, and a real-life case study of a quality control system for bee products is presented to validate the effectiveness of the approach.

INTRODUCTION

A cyber-physical system (CPS) is a system of collaborating computational elements controlling physical entities. It offers cyber capability in every physical component, a high degree of automation, and reorganizing/reconfiguring dynamics [1]. Due to its unique features, CPS has been widely used in manufacturing [2,3,4]. In a manufacturing system equipped with CPS techniques, which is called a cyber-physical manufacturing system, machines as well as human workers constitute the physical resources, and data is collected from the sensors of these resources which constitutes the cyber part of the system. Computations are carried out with the objective of maximizing the utilization of limited resources while guaranteeing the dependability of manufacturing processes, and suitable decisions are taken, based on which the physical resources are further controlled. Therefore, CPS makes a traditional fixed manufacturing process become a flexible dynamic one, which can significantly improve the efficiency of the manufactures. It has become to play an important role in the design and development of modern manufacturing systems.

With the growing number and high dynamics of complex procedures in a manufacturing activity [5], the monitoring of manufacturing processes has drawn more and more attention. Government institutions, companies, traders as well as researchers have stressed the importance of tracing the entire life of every product, especially for food, clinical materials, or similar items [6]. During the production processes, data from these processes has to be recorded; when an item were examined with any defect in the marketplace after production or in any process within production, it should be quite necessary to extract valuable information from such data supporting for back-tracing the sources of the defect. Such traceability

Citation: Jiwei Huang, Yeping Zhu, Bo Cheng, Chuang Lin and Junliang Chen, A PetriNet-Based Approach for Supporting Traceability in Cyber-Physical Manufacturing Systems, doi: 10.3390/s16030382.

problem has become one of the most important issues in the design, implement and management of manufacturing systems.

There have been several traceability solutions for manufactures in both industry and academia. Some of them started with the underlying hardware layer, by integrating radio frequency identification (RFID) tags and readers in different objects [7]. Furthermore, with advanced RFID techniques, networked software systems supporting traceability have been designed and implemented [8]. Meantime, others worked with the application of software engineering to industrial automation including manufacturing automation and process control, mostly aiming to reduce the cost as well as complexity of the systems [9]. Most of the existing works study the traceability problem from a technical point of view, aiming at advancing modern techniques to enable traceability of the productions. In this paper, however, such issue is studied from a very different theoretical angle. The objective is to provide modeling support for describing the detailed procedures of manufactures, based on which the entire life of every product can be fully monitored within production and precisely traced back after production. A generalized Petri net model is presented to formulate the manufacturing processes, and a traceability model is automatically built by model transformation. Model-based algorithms are proposed, enabling back-tracing to the source of an item with detailed productive data. A prototype system is designed which has been largely applied to bee products quality control in reality.

The contribution of this paper is threefold as follows.

- The traceability issue is studied from a modeling viewpoint, aiming at proposing precise modeling approaches for enabling traceability in fully automatic manufacturing systems. Different from most of the traditional approaches which trace an item by RFID tags, this paper digs into detailed production procedures and studies several basic patterns capturing elementary aspects of atomic manufacturing process.
- Petri net model is generalized for describing manufacturing processes, and an automatic methodology of traceability model generation is designed. Algorithms based on the model are proposed, with which all the manufacturing procedures related to the item for traceability analysis can be ordinally listed with detailed production data.
- A prototype system supporting traceability in CPS-based manufacturing is designed. Several techniques such as data entity generation, service provisioning, and service orchestration are applied, and a real-life case study for bee products quality control is presented.

The remainder of this paper is organized as follows. Section 2 discusses the related work most pertinent to this paper.Section 3 presents Petri net models for formulating manufacturing processes, and summarizes the basic patterns of them.Section 4 proposes detailed schemes of traceability analysis by designing corresponding models as well as algorithms from the original Petri net models. Section 5 presents the design of a prototype system, and introduces a case study of a quality control system for bee products in reality. Finally, the paper is concluded in Section 6.

RELATED WORK

Traceability in Manufacturing Systems

Traceability has been a critical concern for manufacturing systems, especially in food and clinical industries. It has been clearly demonstrated that the ability of systems to extract valuable data from manufacturing processes is of great importance [10], and thus advanced information technology (IT) has been widely adopted in modern manufacturing systems [11]. With the rapid growth of the complexity of manufacturing processes which lead to the costliness and inefficiency of monitoring [12], IT-based solutions show powerful strength of reducing the complexity associated with traceability by human operators [6].

Several traceability approaches started with applying RFID technology to industrial systems for tracking every individual item in a manufacturing process. Cao *et al.* [13] introduced RFID technology to automotive industry as an enabler of product lifecycle information management, and enhanced the traceability of the product throughout its value chain via automatic identification. Dai *et al.* [14] adopted RFID-enabled shop-floor manufacturing solutions across the whole operations in automotive small and medium sized enterprises, and extended the efforts of the companies in setting up and integrating manufacturing execution system and enterprise resource planning system. Furthermore, Fishkin *et al.* [7] equipped wearable gloves and bracelets with RFID techniques for detecting the interactions between users and unobtrusively tagged objects.

With RFID techniques, some approaches studied the traceability issue more systematically. With the emergence of cyber-physical systems, CPS has been widely applied for real-time processing in industrial activities [15]. Kelepouris *et al.* [8] proposed an information system architecture that supports efficient information management across the food supply chain utilizing RFID technology for small and medium enterprises. Sánchez *et al.* [6] designed a general framework for developing traceability solutions in small manufacturing companies based on cyber-physical systems. Mora-Mora *et al.* [16] proposed a distributed system equipped with RFID communication technologies for enabling traceability of the flow and movement of people in smart cities. Blackburn and Denno [17] presented a simulation platform for CPS-based industrial systems to support virtual design and verification of industrial process plant designs.

Most of the existing approaches dedicated to advancing practical techniques of sensing, data collection, data storage and database management in order to facilitate the traceability analysis in manufacturing processes. With such techniques, system engineers can manually design detailed schemes with few difficulty which have benefited a lot for the manageability of manufacturing activities and the safety of products. However, few of them have looked into the procedures of traceability analysis essentially by studying general methodologies that can be universally applied to different types of industrial systems. A general approach supporting traceability analysis in manufacturing systems is urgently required.

Theoretical Modeling in Cyber-Physical Manufacturing Systems

Besides technical solutions, there have been several research efforts that have extensively discussed the theoretical aspects in CPS. On one hand, some of them dedicated to design precise models in order to formulate basic activities in CPS especially for manufacturing [18]. Among them, Petri net, which allows the representation of very general discrete event systems whose operations depend on potentially complex control schemes, has been widely used in several aspects of manufacturing systems. Hu *et al.* [19] applied Petri nets to distributed control in large-scale manufacturing systems. Mitchell and Chen [20] used stochastic Petri nets for reliability analysis in CPS by formulating intrusion detection and response.

On the other hand, some researchers gave attention to some mathematical foundations of cyber-physical systems. Bogdan and Marculescu [5] studied the statistical characteristics of CPS workload, and highlighted the importance of stochastic control in CPS. Kolacinski and Loparo [21] designed a mathematical framework for stability and security analysis in cyber-physical power systems.

Many of the existing approaches have shown the advantages of theoretical research in cyber-physical manufacturing systems, such as its flexibility, wide applicability, high accuracy, *etc*. However, there are few efforts studying the traceability problem from a mathematical aspect. Theoretical frameworks as well as quantitative schemes of traceability analysis in manufacturing systems remain largely unexplored. To fill this gap, this paper starts with formulating manufacturing processes using mathematical models and designs a model-based quantitative approach to support traceability of the lifecycle for every industrial product. Considering the basic requirements of industrial manufacturing, the objective of the approach is to enable precise and efficient traceability analysis especially for cyber-physical systems in which complex non-stationary physical processes like in hydraulics systems are not involved, and theoretical optimal (stochastic) control of the system is out of the scope of this paper.

PETRI NET MODEL FOR MANUFACTURING PROCESSES

Basic Concepts of Process Petri Net

The traditional Petri net mostly focuses on the control flow of dynamic processes, studying explicit conditions under which events can be enabled. In a traceability solution for manufacturing activities, besides explicit control flow, some implicit data that closely related to the manufacturing processes has to be precisely recorded and displayed, such as the weight and volume of the items in each atomic process. Therefore, this paper generalizes Petri net with quantitative data related to the transitions.

Firstly, the basic definition of a generalized Petri net, which is called *Process Petri Net (PPN)*, is presented as follows.

Definition 1 (Process Petri Net).

A Process Petri Net (PPN) Σ for formulating a manufacturing process is defined by a 5-tuple

$$\Sigma = (P, T, A, w, m_0)$$

where

- *P is the finite set of places.*
- *T is the finite set of transitions.*
- $A \subseteq (P\times T) \cup (T\times P)$ *is the set of arcs from places to transitions and from transitions to places.*
- $w: A \rightarrow \mathbb{R}^+ \cup \{0\}$ *is the weight function of the arcs.*
- $m_0: P \rightarrow \mathbb{R}^+ \cup \{0\}$ *is the initial state of the net.*

A manufacturing process can be formulated by a PPN, where the *transitions* represent the atomic manufacturing activities, while the *places* denote the products (either raw materials, work-in-progress items, or products ready for sale). The *arcs*connect places to transitions and transitions to arcs, bridging the gap between activities and their related items. The *weight*related to an arc is a nonnegative real number indicating the implicit data that closely related to the manufacturing process, which can represent either weight, volume, amount, *etc*.

In a PPN, *tokens* are assigned to places, where a token essentially indicates the fact that the product described by that place has been finished by its precursor activity. Different from traditional Petri net, data of items is also attached with tokens in the PPN model. The way in which tokens are assigned to a PPN defines a *marking*, which is formally defined as follows.

Definition 2 (marking).

A marking m of a PPN $\Sigma=(P,T,A,w,m_0)$ *is a function* $m: P \rightarrow \mathbb{R}^+ \cup \{0\}$

Marking *m* defines a row vector $\mathbf{m} = [m(p_1), m(p_2), \ldots, m(p_n)]$, where the *i*-th entry indicates the data attached to the tokens in place pipi. Such marking row vector defines the *state* of a PPN at certain time point. In Definition 1, m0m0 is a marking indicating the state of a PPN when it is initialized.

When drawing a PPN, the convention is to use circles to represent places and bars to represent transitions. Directed arrows connecting places and transitions represent elements of the arc set, and their weights are written accompanied by the arcs. In a PPN graph, given a certain marking *m*, tokens are indicated by dark dots in the appropriate places where $m(p_i) \neq 0$. Example 1 illustrates the PPN model of a typical manufacturing process.

Example 1.

Figure 1 shows a PPN describing a typical manufacturing process, which is defined by $\Sigma=(P,T,A,w,m_0)$*.Specifically, such process starts with some amount of raw material ready to proceed, whose volume is indicated by* w_1*, i.e.,* $m_0(p_1) = w_1$*, and* $m_0(p_i) = 0$ *for* $\forall i \neq 1$. *Firstly, raw material is divided into several packages for further processing in a parallel way. On finishing all the processing, some of the intermediate products are packed together for final processing. Next, the final products are delivered to markets for sale.*

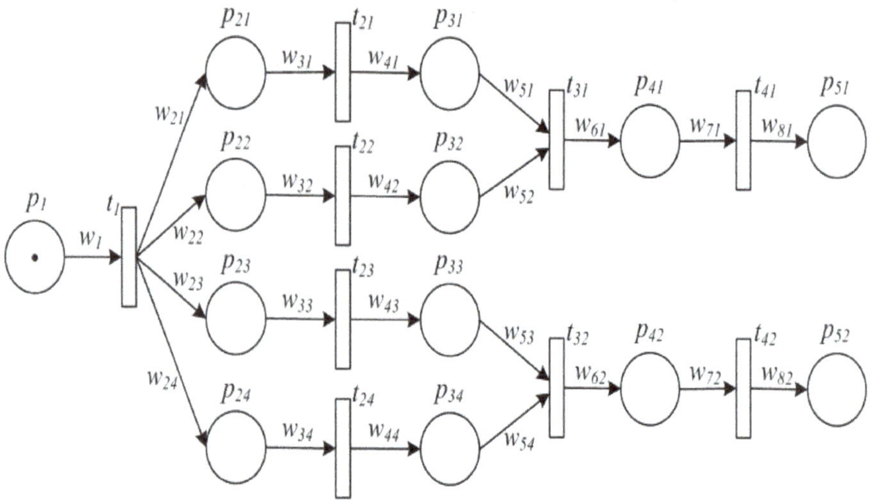

Figure 1: An example of PPN model for a typical manufacturing process.

In order to use PPN to model dynamic manufacturing processes, the state transition mechanism has to be defined. Basically, for a transition t∈Tt∈T to "happen", it is required that there should be sufficient tokens in each place which is input to the transition. Formally, the notion of *enabled transition* is introduced at first as follows.

Definition 3 (enabled transition).

A transition tj∈T *in a PPN* Σ=(P,T,A,w,m_0) *is enabled under marking m if and only if (iff)*

$$m(p_i) \geq w(p_i, t_j) > 0, \quad \forall p_i \in P$$

When a transition is enabled, we say that it can *fire*. Its formal definition can be found in Definition 4.

Definition 4 (firing).

When a transition tj∈T *in a PPN* Σ=(P,T,A,w,m0) *is enabled under marking m, the firing of transition t is denoted by* $m \xrightarrow{t_j} m'$ *where*

$$m'(p_i) = m(p_i) - w(p_i, t_j) + w(t_j, p_i), \quad \forall p_i \in P$$

An observation about the dynamic behavior of PPN is that not all states in $\mathbb{R}^n$ can necessarily be reached from a given initial state m0m0. All the states that can be reached form the set of *reachable states*, whose definition is as follows.

Definition 5 (reachable states).

The set of reachable states of PPN Σ=(P,T,A,w,m_0), *represented by* R[Σ], *is the minimal set that satisfies:*

- m_0∈R[Σ];
- *If* m1∈R[Σ] *and there exists* $t_j \in T$ *s.t.* $m_1 \xrightarrow{t_j} m_2$, *then* $m_2 \in R[\Sigma]$.

With all the definitions presented above, basic procedures as well as dynamic behaviors of a manufacturing process can be formulated by a PPN. Besides formulating explicit control flow like traditional Petri net model, the generalized PPN model is able to describe detailed implicit data bonded with each sub-process. Such ability is important for both manufacturing process modeling and traceability implementation, and readers will appreciate this advantage later when the traceability solution is described in Section 4 and Section 5.

Basic Patterns of Manufacturing Processes

For complex manufacturing processes in reality, there are several workflow patterns. Most of them have been well summarized in [22], and it has been universally acknowledged that five basic patterns are the most popular and common among them. According to the characteristics of manufacturing systems and the traceability requirement, PPN models of these five basic patterns are presented in this part. Most of the manufacturing processes can be formulated using these five patterns, and other more complex processes could be modeled with these basic blocks using similar methodology.

Sequence Pattern

Sequence pattern means that an activity in a manufacturing process is enabled after the completion of another activity in the same process. Such pattern is the most basic, and examples include production and processing on a single ingredient or intermediate product. A sequence pattern is a "single-input single-output" process, whose PPN model is shown by Figure 2. The implicit data related to the activity—the amount/volume of the item before and after such activity—is indicated by w1w1and w2w2, respectively. We should note that, it is possible that w1≠w2w1≠w2 due to the processing on the material/item.

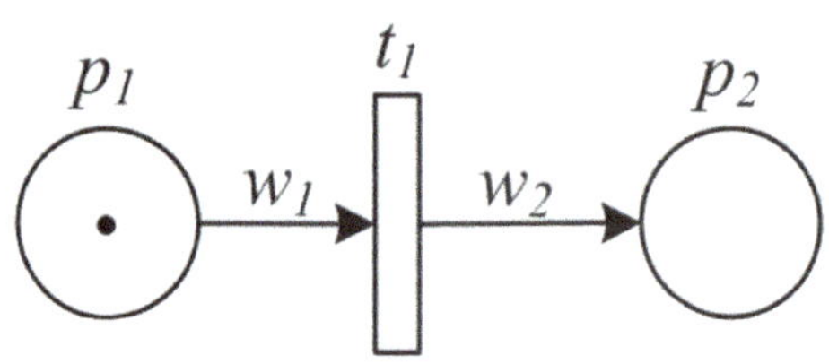

Figure 2: PPN model of sequence pattern.

Split Pattern

Split pattern is a point in the manufacturing process where a single item splits into multiple ones which can be processed in parallel, thus allowing further activities to be executed simultaneously or in any order. An example of split pattern is packaging products into small bags before delivery to the market, and another example is to split a large amount of raw material into small units each of which will be further processed by multiple pipelines. The PPN model for a split pattern is illuminated by Figure 3.

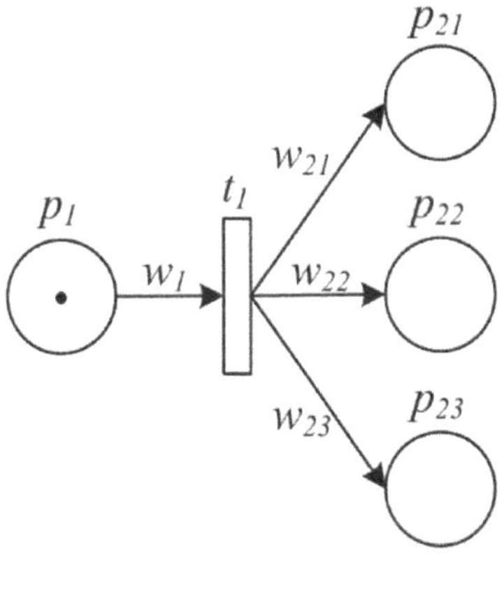

Figure 3: PPN model of split pattern.

Synchronization Pattern

In synchronization pattern, multiple items converge into a single one, thus synchronizing multiple activities. One example of such pattern is that middlemen collect materials/goods from small-sized producers, and converge them into a big package for centralized processing. Another example is to integrate multiple ingredients for synthesizing a new item. Figure 4 shows the synchronization pattern formulated by PPN model.

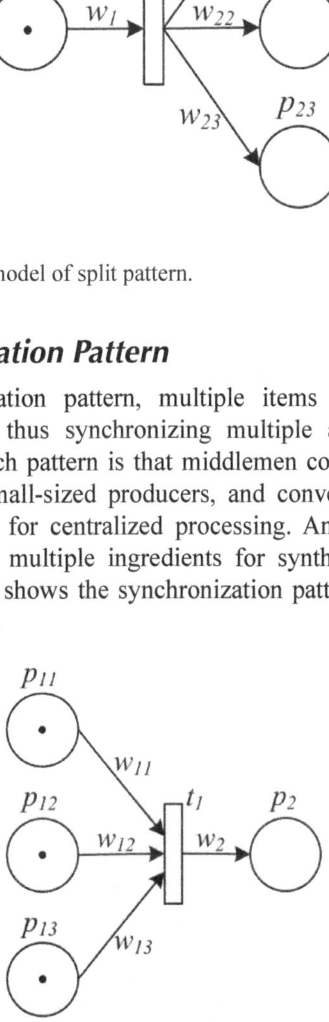

Figure 4: PPN model of synchronization pattern.

Exclusive Choice Pattern

Exclusive choice pattern indicates the divergence of a branch into two or more branches such that when the incoming task is enabled, the item is immediately passed to precisely one of the outgoing branches based on a mechanism that can select one of the outgoing branches. In reality, the activity of choosing one of the OEMs to produce the required item is a typical example of exclusive choice pattern. In addition, delivering each item to one of the pending markets is another exclusive choice activity in the lifecycle of the products. An intuitive PPN model of exclusive choice pattern is given by Figure 5.

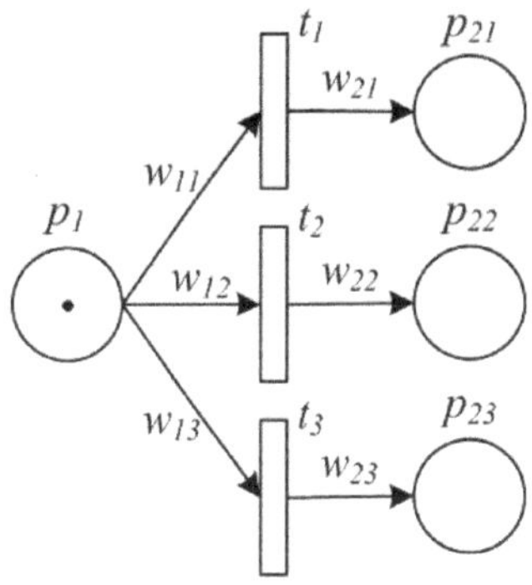

Figure 5: PPN model of exclusive choice pattern.

In a traceability solution for manufacturing systems, much more attention has been paid to tracing the liftcycle of the products than the control automata. In order to facilitate traceability analysis which will be introduced in detail in Section 4, the PPN model of exclusive choice pattern is transformed into a split one. With proper data recording in manufacturing process, it can be theoretically proved that the dynamics remain equivalent to the original model after model transformation. Formally, the transformation and its related proofs are given in Theorem 1.

Theorem 1.

Once exclusive choice is determined with definite outcome, its PPN model $\Sigma=(P,T,A,w,m_0)$ *can be transformed into a split model* $\tilde{\Sigma} = (P, \tilde{T}, \tilde{A}, \tilde{w}, m_0)$ *with the same places and initial state (marking). Suppose the i-th activity is chosen,* $\tilde{w}_1$ *in the split PPN model is set to be* w_{1i} *of the exclusive choice model,* $\tilde{w}_{2i}$ *in the split one to be* w_{2i} *of the exclusive choice pattern while* $\tilde{w}_{2j} = 0$ *for* $\forall j \neq i$.

Proof (Proof of Theorem 1).

In order to prove the soundness of the model transformation, the following two points have to be guaranteed: (1) the dynamics of the state transition remains equivalent; (2) the recorded data related to the activity should be correct.

On one hand, the dynamics issue can be proved by analyzing the state (marking) of the PPN model after the transition. In the original exclusive choice model, given the *i*-th activity is chosen, one can obtain that transition t_i fires. According to Definition 4, the marking m′ of the PPN after t_i fires can be expressed as follows.

$$m'(p_1) = m(p_1) - w_{1i} \tag{1}$$

$$m'(p_{2i}) = m(p_{2i}) + w_{2i} \tag{2}$$

$$m'(p_{2j}) = m(p_{2j}),\ \forall j \neq i \tag{3}$$

In the transformed split PPN model, $\tilde{m}$ and $\tilde{m}'$ are supposed to be the markings before and after the transition respectively, and all the model parameters are marked with tildes in order to distinguish the ones in the original exclusive choice model. Since t_i is enabled in the original model, we have $\tilde{m}(p_1) = m(p_1) \geq w_{1i}$, and thus $\tilde{t}_1$ of the split model is enabled. After the firing of $\tilde{t}_1$, one can obtain that,

$$\tilde{m}'(p_1) = \tilde{m}(p_1) - \tilde{w}_1 = m(p_1) - w_{1i} \quad (4)$$

$$\tilde{m}'(p_{2i}) = \tilde{m}(p_{2i}) + \tilde{w}_{2i} = m(p_{2i}) + w_{2i} \quad (5)$$

$$\tilde{m}'(p_{2j}) = \tilde{m}(p_{2j}) + \tilde{w}_{2j} = m(p_{2j}) + 0 = m(p_{2j}) \ \forall j \neq i \quad (6)$$

Therefore, it can be concluded that $\tilde{m}' = m'$.

On the other hand, as t_i fires in the original exclusive choice PPN model, the data related to such transition is w_{1i} and w_{2i}. In the transformed split model, the nonzero (meaningful) data related to t_i is $\tilde{w}_1 = w_{1i}$ and $\tilde{w}_{2i} = w_{2i}$. Hence, the correctness of data recording has also been proved.

Simple Merge Pattern

For simple merge pattern, two or more alternative branches come together without synchronization, and any one of them can trigger the successor activity. For instance, when there are multiple suppliers for the same raw materials, the provisioning from one of them will immediately start further production activities. Such process pattern can be modeled by the PPN as Figure 6.

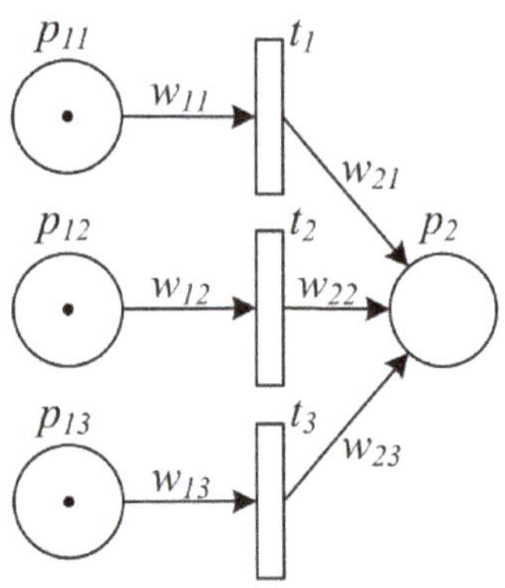

Figure 6: PPN model of simple merge pattern.

Also, a simple merge pattern can be transformed into a synchronization one in a traceability solution. The methodology of model transformation can be found in Theorem 2.

Theorem 2.

Given transition t_i fires in a simple merge pattern, its PPN model Σ=(P,T,A,w,m_0) can be transformed into a synchronization model $\tilde{\Sigma} = (P, \tilde{T}, \tilde{A}, \tilde{w}, m_0)$ with the same places and initial state (marking). $\tilde{w}_{1i}$ in the synchronization PPN model is set to be w_{1i} of the simple merge model, and $\tilde{w}_{1j}$ is set to be 0 for $\forall$j≠i. $\tilde{w}_2$ in $\tilde{\Sigma}$ is set to be w_{2i} of the original PPN model.

Proof (Proof of Theorem 2).

Similarly, we firstly study the dynamics of the state transitions. Given transition titi fires, we obtain the markings of the original PPN model Σ before and after the firing denoted by *m* and m'm', respectively.

$$m'(p_1 i) = m(p_{1i}) - w_{1i} \quad (7)$$

$$m'(p_{1j}) = m(p_{1j}), \ \forall j \neq i \quad (8)$$

$$m'(p_2) = m(p_2) + w_{2i} \quad (9)$$

In the transformed synchronization PPN model $\tilde{\Sigma}$, the markings before and after the transition are denoted by $\tilde{m}$ and $\tilde{m}'$, respectively. Given t_i is enabled before transition, we have $m(p_{1i}) \geq w_{1i}$, and thus we know that $\tilde{m}(p_{1i}) = m(p_{1i}) \geq w_{1i} = \tilde{w}_{1i}$. Therefore, $\tilde{t}_1$ in the transformed model is enabled. After the firing of transition $\tilde{t}_1$, one can obtain that,

$$\tilde{m}'(p_{1i}) = \tilde{m}(p_{1i}) - \tilde{w}_{1i} = m(p_{1i}) - w_{1i} \quad (10)$$

$$\tilde{m}'(p_{1j}) = \tilde{m}(p_{1j}) - \tilde{w}_{1j} = m(p_{1j}) - 0 = m(p_{1j}), \ \forall j \neq i \quad (11)$$

$$\tilde{m}'(p_2) = \tilde{m}(p_2) + \tilde{w}_2 = m(p_2) + w_{2i} \quad (12)$$

Therefore, we have $\tilde{m}' = m'$, and thus prove the equivalence of their dynamics.

Secondly, for all nonzero data related to the manufacturing process, w_{1i} and w_{2i} are recorded in the original PPN model while $\tilde{w}_{1i} = w_{1i}$ and $\tilde{w}_2 = w_{2i}$ are recorded in the transformed model. It is clear that the data recorded remains same after the model transformation.

TRACEABILITY MODEL AND ALGORITHMS

The previous section introduces a PetriNet-based approach for process modeling in an automated manufacturing system. With typical patterns, manufacturing processes can be formulated using uniformed bases. Furthermore, such model can be transformed into a traceability model automatically, which can provide powerful support for tracing the entire life of a product or intermediate item.

Traceability Model

The PPN model formulates the manufacturing process in a forward order from raw materials to final products. For a traceability solution, however, tracing backward from a retail product or an intermediate item to its origin is another important requirement, in order to identify sources of contamination and other potential causes of accidents. To facilitate automatic traceability analysis, *Traceability Petri Net (TPN)* is proposed to formulate a traceability process.

Definition 6 (Traceability Petri Net).

A Traceability Petri Net (PPN) $\hat{\Sigma}$ for a manufacturing process Σ=(P,T,A,w,m0) *is defined by a 5-tuple*

$$\hat{\Sigma} = (\hat{P}, \hat{T}, \hat{A}, \hat{w}, \hat{m}_0)$$

where

- $\hat{P}$=T *is the finite set of places.*
- $\hat{T}$ =P *is the finite set of transitions.*

- $\hat{A} = \{(p,t)|(t,p) \in \dot{A}, t \in T, p \in P\} \cup \{(t,p)|(p,t) \in A, t \in T, p \in P\}$ *is the set of arcs from places to transitions and from transitions to places.*
- $\hat{w} : A \to \mathbb{R}^+ \cup \{0\}$ *is the weight function of the arcs, where* w(p,t)=w(t,p) *for* ∀(t,p)∈A, t∈T, p∈P *and* $w(t,p) = w(p,t)$ *for* $\forall(p,t) \in A, t \in T, p \in P$.
- $\hat{m}_0 : \hat{P} \cup \hat{T} \to \mathbb{R}^+ \cup \{0\}$ *is the initial state of the net, indicating the starting point of the tracing process.*

The basic idea of building the transformed TPN model from a PPN is as follows. All the transitions in the original PPN model are transformed into places, while places of PPN are transformed into transitions. The connection relationship of arcs in Σ remains the same, however, their direction are turned into their opposite. The weights are basically unchanged, and the initial state of $\hat{\Sigma}$ indicates starting point where the tracing is originated. Here, the starting point can be either some places of $\hat{\Sigma}$ indicating that the tracing can be started from a manufacturing activities, or some transitions of $\hat{\Sigma}$ which means that one can also start to trace the lifecycle of a intermediate item during the process or a retail product from the market.

The basic process of building the TPN model is to traverse all the places, transitions and arcs of the original PPN, and thus one can obtain that the computational complexity of constructing TPN is $O(|P| + |T| + |A|)$. Therefore, the time-complexity is proportional to the scale of the original PPN.

With Definition 6, the corresponding TPN model of the manufacturing process shown in Example 1 can be constructed. Details of the TPN model can be found as follows.

Example 2.

Figure 7 shows a TPN model of the PPN in Example 1.

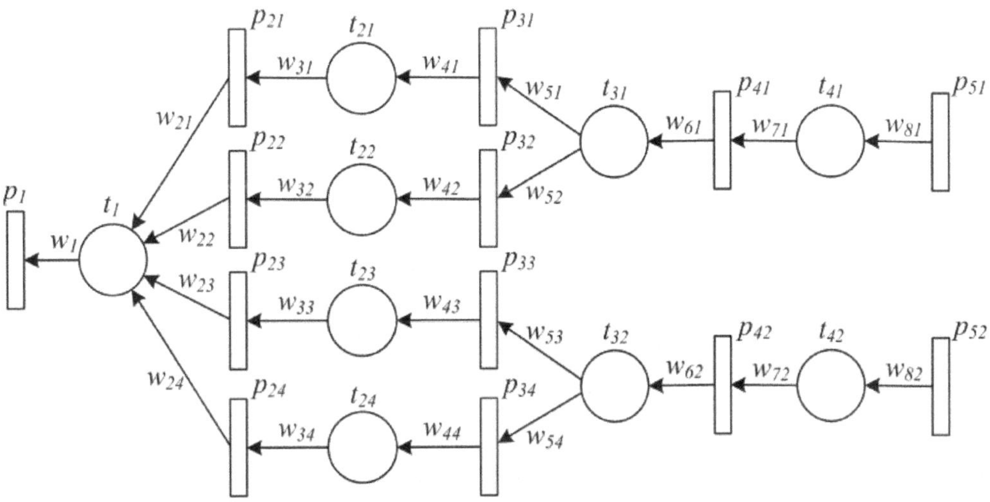

Figure 7: An example of TPN model for a typical manufacturing process.

The definition of *marking* is exactly the same as the one of PPN. Readers may refer to Definition 2 for details. However, each value in a marking has different meaning from PPN, which indicates the proportion of an ingredient in the final product, or the probability that each item has been contaminated. In most cases, the values should be in [0,1][0,1]. On the other hand, it should be noted that the definition of the initial state in TPN is different from PPN. Both places and transitions can be defined as one of the initial states, and thus $\hat{m}_0$ may not be a marking of the TPN. Such different definition provides the TPN model with the ability of tracing backward the manufacturing process from both items and atomic activities. We defer readers to the next subsection for detailed means of dealing with such differences.

In order to facilitate the traceability analysis, we present a different definition of enabled transition in a TPN from that of PPN, shown by Definition 7.

Definition 7 (enabled transition of TPN).

A transition $t_j \in \hat{T}$ *in a TPN* $\hat{\Sigma} = (\hat{P}, \hat{T}, \hat{A}, \hat{w}, \hat{m}_0)$ *is enabled under marking m iff*

$$m(p_i) > 0, \quad \text{for } \forall p_i \in \hat{P}, w(p_i, t_j) > 0.$$

The formal definition of firing of an enabled transition in a TPN can be found in Definition 8.

Definition 8 (firing of TPN).

When a transition $t_j \in \hat{T}$ *in a TPN* $\hat{\Sigma} = (\hat{P}, \hat{T}, \hat{A}, \hat{w}, \hat{m}_0)$ *is enabled under marking m, the firing of transition* t_j *is denoted by* $m \xrightarrow{t_j} m'$ *where*

$$m'(p_i) = \begin{cases} 0 & w(p_i, t_j) > 0 \\ m(p_i) + \dfrac{w(t_j, p_i)}{\sum_k w(t_j, p_k)} \cdot \sum_l \dfrac{w(p_l, t_j)}{\sum_k w(p_l, t_k)} m(p_l) & w(t_j, p_i) > 0 \\ m(p_i) & \text{otherwise} \end{cases}$$

Algorithm for Traceability Analysis

The TPN model presented in the above subsection is able to facilitate the traceability analysis in manufacturing processes. The basic idea is to construct a reachability tree of the TPN. In order to introduce the approach in detail, we first present the basic concept of *reachability tree* of a TPN.

Definition 9 (reachability tree).

The reachability tree of a traceability Petri net $\hat{\Sigma} = (\hat{P}, \hat{T}, \hat{A}, \hat{w}, \hat{m}_0)$ *is a tree that describes the dynamics of* $\hat{\Sigma}$*, where tree nodes are Petri net states and arcs represent transitions.*

A finite representation of reachability tree can be constructed by an algorithm which will be presented later. To do so, we next introduce some notation as follows.

- *Root node*. Root node is the first node of the reachability tree, obtained from the initial state of the given TPN.
- *Terminal node*. This is any node from which no transition of the TPN can fire.
- *Duplicate node*. This is a node that is identical to a node already in the reachability tree.
- *Node dominance*. Let m_1 and m_2 be two markings (states), *i.e.*, nodes in the reachability tree. We say that "m_1 dominates m_2", denoted by $m_1 >_d m_2$, if the following two conditions hold:

  $$m_1(p_i) \geq m_2(p_i), \text{ for all } i = 1, 2, \ldots, n;$$
  $$m_1(p_i) > m_2(p_i), \text{ for at least some } i = 1, 2, \ldots, n.$$

- *Symbol* ω. The symbol w in a reachability tree means "infinity" in representing the marking of an unbounded place. For $\forall n \in \mathbb{N}$, we specify $n < \omega$ and $\omega + n = \omega - n = \omega$. It should be noted that, although common in traditional Petri nets, such symbol is rare in TPN, for most of the manufacturing processes are finite and non-iterative. For preciseness, however, such symbol is still introduced in this approach.

With all the notations presented above, the reachability tree of a TPN can be constructed by the traceability algorithm, which is formally presented by Algorithm 1.

The construction procedures of the reachability tree meanwhile are tracing backward from the manufacturing process from the initial state of the TPN. Firstly, all the transitions in m^0m^0 are fired, and the initial marking can be obtained which forms the root note of the reachability tree. Next, the algorithm iteratively fires all enabled transitions in a breadth-first order and recursively constructs the reachability tree. The information of the fired transition is pushed into a queue which gathers all the information of semi-finished products that may be contaminated and need to be further examined. Besides the basic information, a weight is pushed into the queue as well, which indicates the probability that each semi-finished product in the manufacturing process being flawed. Finally, once completing the construction process, we may either directly dequeue traceability information in the (reversed) order of the original manufacturing process, or sort the items according to their corresponding weights putting first things first.

The construction of reachability tree has been an open problem in Petri net research for decades. Recently, existing literatures on reachability problem of Petri net have proved that the time complexity of the algorithm can be bounded by $O(\mathbb{N}^{|\hat{P}|})$ in worst case [23]. However, in the traceability analysis of a manufacturing process where the activities have been determinate resulting in the finiteness of the manufacturing process, Algorithm 1 presented in this paper can be normally completed in $O(n \cdot |\hat{T}|)$ time where *n* is the number of activities in the process. It should be noted that, although very rarely, it might also be exponential to the number of places of the TPN model in the worst cases.

Algorithm 1: Traceability Algorithm on TPN

Input: TPN $\hat{\Sigma} = (\hat{P}, \hat{T}, \hat{A}, \hat{w}, \hat{m}_0)$
Output: traceability information Q

Let $m(p_i) \leftarrow \hat{m}_0(p_i)$ for all $p_i \in \hat{P}$.
for all $t_j \in \hat{T}$ and $\hat{m}_0(t_j) > 0$ **do**
  Let $m'(p_i) \leftarrow \begin{cases} m(p_i) + \dfrac{w(t_j, p_i)}{\sum_k w(t_j, p_k)} \hat{m}_0(t_j) & w(t_j, p_i) > 0; \\ m(p_i) & \text{otherwise.} \end{cases}$
  Let $m \leftarrow m'$.
end for
Initialize m as the root node of reachability tree. Let $\Psi \leftarrow \{m\}$.
for all $\psi \in \Psi$ **do**
  if no transition is enabled at state (under marking) ψ **then**
    ψ is a terminal node.
  else
    Create a new node ψ' s.t. $\psi \xrightarrow{t_j} \psi'$ for some $t_j \in \hat{T}$.
    Q.enqueue$(t_j, \sum_l \frac{w(p_l, t_j)}{\sum_k w(p_l, t_k)} m(p_l))$.
    if $\psi(p_i) = \omega$ for some $p_i \in \hat{P}$ **then**
      Set $\psi'(p_i) \leftarrow \omega$.
    end if
    if there exists a node θ such that $\psi' >_d \theta$ **then**
      Set $\psi'(p_i) \leftarrow \omega$ for all $p_i \in \hat{P}$ s.t. $\psi'(p_i) > \theta(p_i)$.
    end if
    Let $\Psi \leftarrow \Psi \cup \{\psi'\}$
  end if
  Let $\Psi \leftarrow \Psi - \{\psi\}$.
end for

PROTOTYPE SYSTEM

In this section, a practical approach of traceability solution in manufacturing systems is introduced. Detailed designs, techniques, as well as a case study in reality are presented.

Framework

Figure 8 shows an overall framework of the approach for supporting traceability in cyber-physical manufacturing systems. At the very beginning, a manufacturing process is formulated by a PPN describing detailed activities in the whole lifecycle of the product from raw materials to final market, and thus the manufacturing process can be guided by the model. During the productions, atomic activities are encapsulated by web services, which provide precise interfaces for collecting data that enable the traceability process. Once a request for traceability analysis is submitted by either a consumer or a producer, the corresponding TPN model is build and traceability algorithm is conducted.

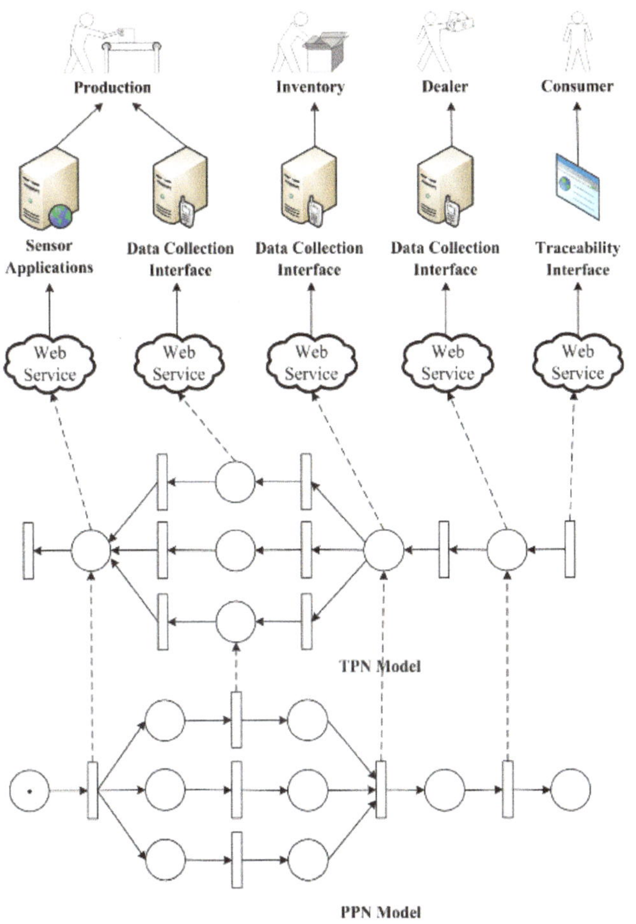

Figure 8: Framework of a traceability solution for cyber-physical manufacturing systems.

The approach presented in this paper is able to trace the whole life of a product. Besides traditional manufacturing process, transportation, inventory as well as delivery can be tracked, since most of them follow the basic patterns of manufacturing processes and thus can be well modeled by PPN. With modern CPS techniques, some of the activities can be accomplished fully automatically, and data can be collected by sensors attached to the manufacturing equipments. For others which involve human activities and lack of sensors, data collection interfaces are provided via web services for manual input. In specific, such data input procedure can be accomplished with either manually keyboard typing, speech recognition, barcode scanners, or RFID readers.

Technical Implementation

With the rapid development of services computing technology [24], web service has been widely applied in many IT-based systems. It provides programmable modules with standard interface descriptions that enable universal accessibility through standard communication protocols. It encapsulates heterogeneous functionalities of services, and can be universally monitored, discovered, invoked, and controlled through Uniform Resource Identifier (URI). Due to its powerful ability in dealing with heterogeneity, web service technology is applied into this approach to encapsulate atomic activities in manufacturing processes.

Figure 9 shows a basic framework of a web service based system. Devices especially sensors in manufacturing processes are encapsulated by a uniform devices abstraction approach based on Open Service Gateway Initiative (OSGi) technology [25], and heterogeneous data and device operations can be accessible with a uniform web service interface. It should be noted that, the details about the control on manufacturing devices are out of the scope of this paper which mainly seeks to solve the traceability problem. However, if the device control provides a web service interface, it can be easily embedded into the manufacturing process; otherwise, it may be treated as a legacy system and wrapped into a uniform service interface with some middleware adaptor techniques such as IBM WebSphere Business Integration Adaptor [26] and Jave EE Connector.

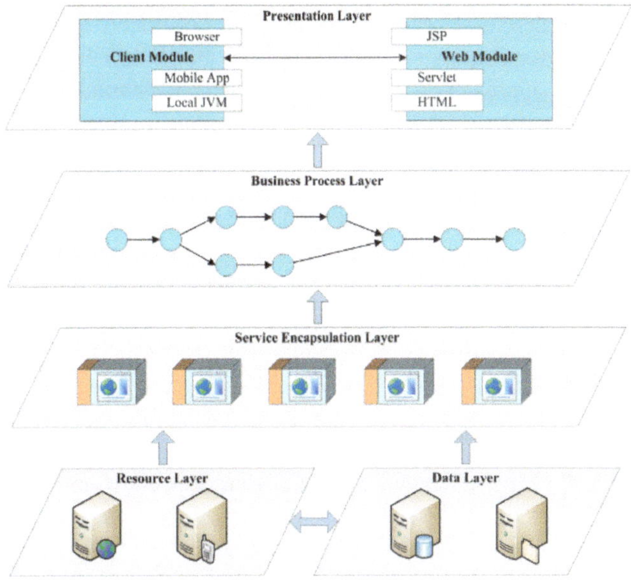

Figure 9: Framework of web service encapsulation.

In order to facilitate traceability analysis, database servers are deployed in this prototype system to record information related to manufacturing processes. According to the characteristics of basic patterns of manufacturing activity that have been presented in Section 3, an automatic data entity generation scheme is designed in order to improve the efficiency of system development and avoid human errors in service design and programming. In specific, basic data should include IDs and weights of items before and after the activities, and some of them can be obtained or generated automatically by the system. For example, the IDs of items before an activity can be obtained by barcode or RFID readers, and the weights/amounts before and after the activity can be obtained by measurement equipments. Furthermore, the item IDs after the activity can be automatically generated following certain encoding rules, and RFID tags or barcode labels can be produced with generated information. More specifically, the data entities for different manufacturing process patterns are formulated according to their traceability requirements, shown as Table 1.

Table 1: Data entities of basic manufacturing process patterns

	Sequence Pattern	Split Pattern	Synchronization Pattern
Data obtained by reader/scaner	id_1	id_1	$id_{11}, id_{12}, \ldots, id_{1n}$
Data obtained by measurement	w_1, w_2	$w_1, w_{21}, w_{22}, \ldots, w_{2n}$	$w_{11}, w_{12}, \ldots, w_{1n}, w_2$
Automatically generated data	id_2	$id_{21}, id_{22}, \ldots, id_{2n}$	id_2

Besides the basic traceability information presented above, other data related to each manufacturing activity throughout the supply chain should also be recorded. Since such data is closely related to different activities and varies significantly, we leave it for domain developers to implement data collection interfaces and design database tables when realizing specific web services for manufacturing activities. This variety can be partly illustrated later when we describe a case study in Section 5.3.

A manufacturing process may consist of a collection of atomic activities, each of which is encapsulated by a web service for automatic control and traceability data collection. In this approach, the process is practically formulated by Web Services Business Process Execution Language (WS-BPEL), which is an OASIS standard executable language for specifying actions within business processes with web services. Using such formulation, composite services can be executed and managed by BPEL-supported workflow engines. The transformation between Petri Net and WS-BPEL has been studied for years, which has been presented in several existing approaches for service composition analysis [27].

Finally, the services are provisioned based on a browser/server (B/S) architecture, and users can invoke the services through JSP or HTML web pages on their browsers. Also, in order to support various mobile devices especially embedded CPS equipments, customized mobile applications can also invoke the services for supporting traceability of manufacturing processes.

Case Study

The traceability approach presented in this paper has been applied to a quality control system for bee products (including honey, propolis, royal jelly, *etc.*). Such system monitors the whole manufacturing process and traces all the information during the lifecycle of every product, in order to improve the safety of bee products. It has been widely deployed in China, covering more than 113,600 bee colonies in 13 provinces.

The manufacturing process of bee products basically consists of the 5 following steps. (1) Raw materials are raised and harvested by farmers from farms, and they are labeled according to their beehives or colonies; (2) The brokers then evaluate the quality of the raw materials, purchase them from farmers, and resell them to the factories. In such step, some materials form different beehives/colonies might be merged together into bigger bottles; (3) Once finishing collecting raw materials, factories process them by melting, purification, and sterilization; (4) After the processing, edible products are packaged into small bottles/boxes for sale; (5) Finally, they are delivered to retail on the market.

Figure 10 shows both theoretical and practical frameworks of the system. According to the procedures of the bee product manufacture presented above, the PPN model of the manufacturing process is first constructed. Raw materials are the starting points of the process, regarded as initial states of the PPN model. Brokers collect raw materials from different origins and put them together into bigger bottles, and thus this activity can be modeled following the typical synchronization pattern. The processing and delivery are simply two sequence activities, and the packaging conforms to a split pattern.

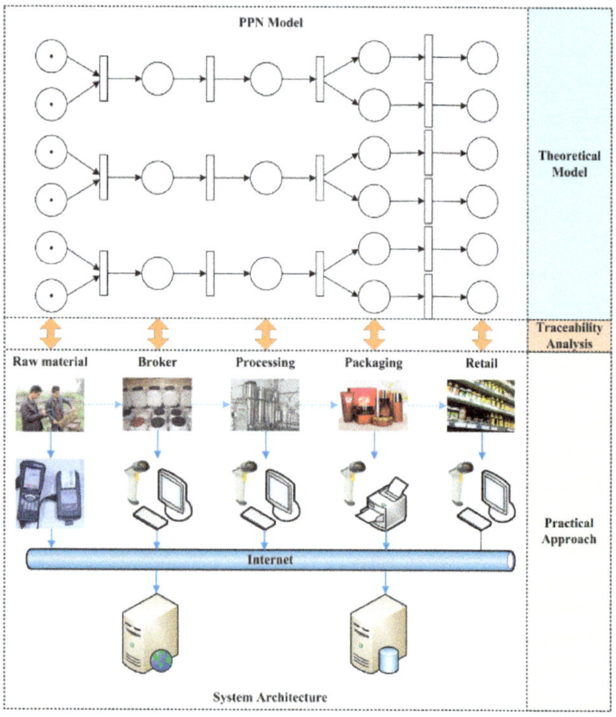

Figure 10: Framework of quality control system for bee products.

For the practical part of the system, we design and develop applications on both mobile device and PC for automatically collecting manufacturing data to facilitate traceability analysis. Each farmer is equipped with a PDA and a portable barcode printer when harvesting raw materials. Basic information on the materials and farms are collected by either voice input or manual typewriting, and traceable barcodes are automatically generated using the symbology of GS1-128 (formerly known as UCC/EAN-128). Once the information has been confirmed, an adhesive barcode label is automatically printed by the portable printer attached with the PDA (see Figure 10), and is stickered on the bottle of raw materials. On the other hand, most of the other activities are processed indoor, and PCs with Internet connection are used for data collection. Each PC is equipped with a USB barcode scanner which enables fast label reading, and a label printer which produces barcode labels adhered to semi-finished products.

All the data collection activities are handled through a uniform web service interface, and their underlying data is stored in a database server. In Figure 10, the solid arrows represent the data flows in the manufacturing process, while the dashed ones show the control flows.

More specifically, the underlying data entities are shown by Figure 11. Blue arrows indicate the control flow of the manufacturing process, while the black ones represent the data flow. According to the PPN model, some of the data entities can be automatically generated, indicated by colored ones in the figure. They include the barcodes of the items before and after each activity which are blue-colored, and their corresponding amounts shown in green. Other data entities are designed by system developers according to the characteristics of the manufacturing activities, such as the ID and address of each participant, the date and time of each activity, and the category of the raw materials harvested by farmers (*i.e.*, Chinese date honey, sunflower honey, *etc.*).

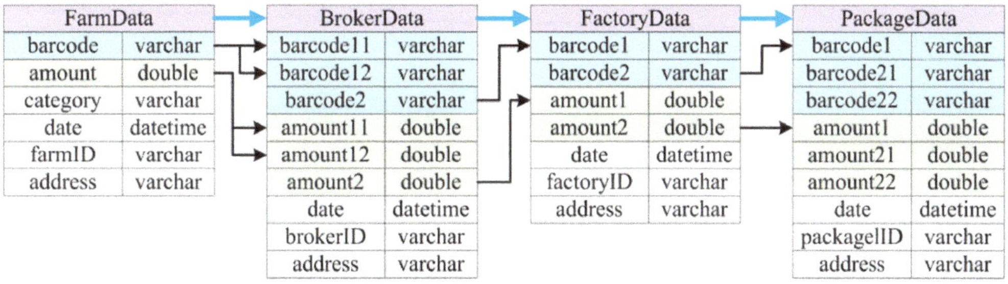

Figure 11: Data entities of quality control system for bee products.

In order to illuminate the effectiveness of the approach, an example from reality is presented, showing the traceability process of an orange honey item. The raw materials are collected from two different farms in Sichuan Province (in China) before being merged together for processing, and they are finally divided into two different sized bottles for retail. All the data related to each of the activity in such manufacturing process has been precisely recorded, and traceability analysis from a retail item is conducted using the approach proposed in this paper. Figure 12 shows the original PPN model of the manufacturing process, and hence Figure 13 and Figure 14 show the original recorded data and the result of the analysis, respectively.

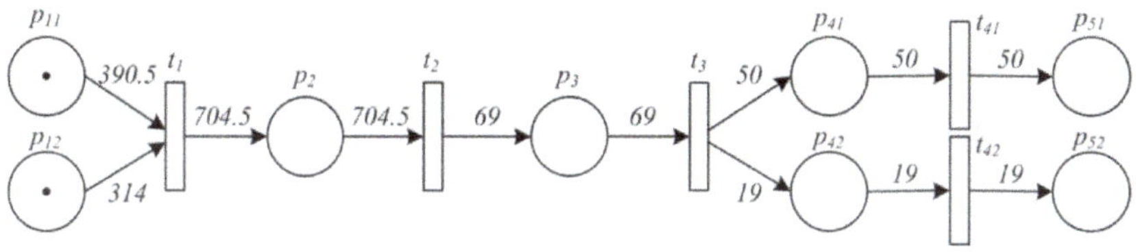

Figure 12: PPN model of the manufacturing process of an orange honey product.

FarmData					
barcode	amount	category	date	farmID	address
70301565101310100313120503100 01	390.5	orange honey	2012/5/3 0:00:00	51013101003	Guangming Village, Sichuan
7030156511424010011312050210004	314	orange honey	2012/5/2 0:00:00	51142401001	Wansheng Village, Sichuan

BrokerData					
	barcode	amount	date	brokerID	address
11	7030156510131010031312050310001	390.5	2012/11/21 0:00:00	21010001005	Meishan, Sichuan
12	7030156511424010011312050210004	314			
2	7030156210100010051312112110001	704.5			

FactoryData					
	barcode	amount	date	factoryID	address
1	7030156210100010051312112110001	704.5	2013/8/20 16:39:40	51013103001	Pujiang, Sichuan
2	7030156510131030011313082010001	69			

PackageData					
	barcode	amount	date	packageID	address
1	7030156510131030011313082010001	69	2013/9/25 0:00:00	61126301002	Pujiang, Sichuan
21	510131030011308200 01	50			
22	51013103001130820002	19			

RetailData					
	barcode	amount	date	marketID	address
1	51013103001130820001	50	2013/10/9 0:00:00	11010801001	Haidian, Beijing
2	51013103001130820001	50			
1	51013103001130820002	19	2013/10/9 0:00:00	11010801001	Haidian, Beijing
2	51013103001130820002	19			

Figure 13: Original data of the manufacturing process of an orange honey product.

Starting Point		
TPN Model Parameter	place	value
	p_{S1}	1.0

↓

Traceability Results					
TPN Model Parameter	place		value		
	p_{41}		1.0		
Corresponding Data	barcode	amount	date	marketID	address
	51013103001130820001	50	2013/10/9 0:00:00	11010801001	Haidian, Beijing
↓					
TPN Model Parameter	place		value		
	p_3		1.0		
Corresponding Data	barcode	amount	date	packageID	address
	7030156510131030011313082010001	69	2013/9/25 0:00:00	61126301002	Pujiang, Sichuan
↓					
TPN Model Parameter	place		value		
	p_2		1.0		
Corresponding Data	barcode	amount	date	factoryID	address
	7030156210100010051312112110001	704.5	2013/8/20 16:39:40	51013103001	Pujiang, Sichuan
↓					
TPN Model Parameter	place		value		
	p_{11}		0.5543		
Corresponding Data	barcode	amount	date	brokerID	address
	7030156510131010031312050310001	390.5	2012/11/21 0:00:00	21010001005	Meishan, Sichuan
TPN Model Parameter	place		value		
	p_{12}		0.4457		
Corresponding Data	barcode	amount	date	brokerID	address
	7030156511424010011312050210004	314	2012/11/21 0:00:00	21010001005	Meishan, Sichuan

↓

Raw Material Information						
Corresponding Data	barcode	category	amount	date	farmID	address
	7030156510131010031312050310001	orange honey	390.5	2012/5/3 0:00:00	51013101003	Guangming Village, Sichuan
	7030156511424010011312050210004	orange honey	314	2012/5/2 0:00:00	51142401001	Wansheng Village, Sichuan

Figure 14: Results of traceability analysis of an orange honey product.

Specifically, the analytical results obtained by the algorithm are shown in red, indicating every place and its corresponding value (proportion of ingredient or probability of contamination) in the process of traceability analysis. Based on them, production data can be found from the database according to the barcodes, and detailed information is displayed to users which is shown by the blue tables in Figure 14.

CONCLUSIONS

Traceability has always been an important issue in manufacturing, especially for food and clinical materials. This paper proposes a theoretical approach for traceability analysis of CPS-based manufacturing systems. The Petri net model is generalized to formulate a manufacturing process, and a scheme for obtaining its corresponding traceability model is presented. An algorithm is proposed for traceability analysis, based on which a practical prototype system is designed. The approach presented in this paper has been successfully applied to a real-life quality control system for bee products, which has been widely deployed in China. It is expected to provide both theoretical and practical reference value for the design and development of modern automatic manufacturing systems.

There are several avenues for future work. Firstly, the traceability problem is important not only to manufacturing but also to other systems like healthcare and transportation where CPS technologies have also been widely applied for enhanced robustness and efficiency [28]. The basic idea of the modeling approach proposed in this paper may have potential contribution to other domains by theoretically formulating related fundamental activities and conducting traceability analyses. Secondly, time is an essential issue in CPS. Although the time points at which manufacturing activities are enabled are recorded in the database for potential traceability analysis in this approach, such issue can be further studied theoretically. Time can be precisely modeled and formulated by some generalized Petri net models, e.g., Timed Petri Net, Stochastic Petri Net, *etc.*, by further equipping the Petri Net model with a clock structure, which can be valuable especially for time series analysis and dynamic optimal control. Thirdly, the performance and trustworthiness issues can be further addressed in the future. Proper trust management, fault tolerance and performance optimization on system components, sensing data and underlying networking infrastructure are of value for the traceability analysis or even the whole cyber-physical manufacturing system.

ACKNOWLEDGMENTS

The authors would like to appreciate the anonymous reviewers for their extensive and constructive comments for the improvement of this paper. This work is supported by the National Natural Science Foundation of China (No. 61502043, 61132001, and 61271364), Beijing Natural Science Foundation (No. 4162042), BeiJing Talents Fund (No. 2015000020124G082), and the Fundamental Research Funds for the Central Universities (No. 2015RC22).

AUTHOR CONTRIBUTIONS

Jiwei Huang designed the modeling approach and the algorithm, and wrote the paper. Yeping Zhu designed the prototype system and implement the bee products quality control system. Bo Cheng and Junliang Chen proposed the schemes of service encapsulation and data entity generation. Chuang Lin designed the scheme of quantitative analysis based on Petri nets.

REFERENCES

1. Khaitan, S.; McCalley, J. Design Techniques and Applications of Cyberphysical Systems: A Survey. *IEEE Syst. J.* 2015,*9*, 350–365.
2. Lee, E.A. The Past, Present and Future of Cyber-Physical Systems: A Focus on Models. *Sensors* 2015, *15*, 4837–4869.
3. Saldivar, A.A.F.; Li, Y.; Chen, W.N.; Zhan, Z.H.; Zhang, J.; Chen, L.Y. Industry 4.0 with Cyber-Physical Integration: A Design and Manufacture Perspective. In Proceedings of the 21st International Conference on Automation and Computing (ICAC 2015), Glasgow, UK, 11–12 September 2015; pp. 1–6.
4. Neves, P.; Ribeiro, L.; Dias-Ferreira, J.; Onori, M.; Barata, J. Exploring Reconfiguration Alternatives in Self-Organising Evolvable Production Systems Through Simulation. In Proceedings of the 12th IEEE International Conference on Industrial Informatics (INDIN 2014), Porto Alegre, Brazil, 27–30 July 2014; pp. 511–518.
5. Bogdan, P.; Marculescu, R. Towards a Science of Cyber-Physical Systems Design. In Proceedings of the 2011 IEEE/ACM International Conference on Cyber-Physical Systems (ICCPS 2011), Chicago, IL, USA, 12–14 April 2011; pp. 99–108.
6. Bordel Sánchez, B.; Alcarria, R.; Martín, D.; Robles, T. TF4SM: A Framework for Developing Traceability Solutions in Small Manufacturing Companies. *Sensors* 2015, *15*, 29478–29510.
7. Fishkin, K.P.; Philipose, M.; Rea, A. Hands-on RFID: Wireless Wearables for Detecting Use of Objects. In Proceedings of the 9th International Symposium on Wearable Computers (ISWC 2005), Osaka, Japan, 18–21 October 2005; pp. 38–41.
8. Kelepouris, T.; Pramatari, K.; Doukidis, G. RFID-Enabled Traceability in the Food Supply Chain. *Ind. Manag. Data Syst.* 2007, *107*, 183–200.
9. Vyatkin, V. Software Engineering in Industrial Automation: State-of-the-Art Review. *IEEE Trans. Ind. Inform.* 2013, *9*, 1234–1249.
10. Woodall, W.H.; Montgomery, D.C. Some Current Directions in the Theory and Application of Statistical Process Monitoring. *J. Qual. Technol.* 2014, *46*, 78–94.
11. Maturana, F.; Mann, L.; Asenjo, J.; Chatrola, S.; Staron, R. Information Framework for Energy Systems Using Agent-Based Cloud Computing Technology. In Proceedings of the 11th International Conference on Informatics in Control, Automation and Robotics (ICINCO 2014), Vienna, Austria, 1–3 September 2014; Volume 2, pp. 338–345.
12. Condea, C.; Thiesse, F.; Fleisch, E. RFID-Enabled Shelf Replenishment with Backroom Monitoring in Retail Stores.*Decis. Support Syst.* 2012, *52*, 839–849.
13. Cao, H.; Folan, P.; Mascolo, J.; Browne, J. RFID in Product Lifecycle Management: A Case in the Automotive Industry.*Int. J. Comput. Integr. Manuf.* 2009, *22*, 616–637.
14. Dai, Q.; Zhong, R.; Huang, G.Q.; Qu, T.; Zhang, T.; Luo, T. Radio Frequency Identification-Enabled Real-Time Manufacturing Execution System: A Case Study in an Automotive Part Manufacturer. *Int. J. Comput. Integr. Manuf.*2012, *25*, 51–65.
15. Wang, Y.; Vuran, M.C.; Goddard, S. Cyber-Physical Systems in Industrial Process Control. *SIGBED Rev.* 2008, *5*, 12:1–12:2.
16. Mora-Mora, H.; Gilart-Iglesias, V.; Gil, D.; Sirvent-Llamas, A. A Computational Architecture Based on RFID Sensors for Traceability in Smart Cities. *Sensors* 2015, *15*, 13591–13626.
17. Blackburn, M.; Denno, P. Virtual Design and Verification of Cyber-physical Systems: Industrial Process Plant Design.*Procedia Comput. Sci.* 2014, *28*, 883–890.
18. Sharma, A.B.; Ivancic, F.; Niculescu-Mizil, A.; Chen, H.; Jiang, G. Modeling and Analytics for Cyber-Physical Systems in the Age of Big Data. *Perform. Eval. Rev.* 2014, *41*, 74–77.
19. Hu, H.; Liu, Y.; Zhou, M. Maximally Permissive Distributed Control of Large Scale Automated Manufacturing Systems Modeled With Petri Nets. *IEEE Trans. Control Syst. Technol.* 2015, *23*, 2026–2034.
20. Mitchell, R.; Chen, I.R. Effect of Intrusion Detection and Response on Reliability of Cyber Physical Systems. *IEEE Trans. Reliab.* 2013, *62*, 199–210.
21. Kolacinski, R.M.; Loparo, K.A. A Mathematic Framework for Analysis of Complex Cyber-Physical Power Systems. In Proceedings of the 2012 IEEE Power and Energy Society General Meeting, San Diego, CA, USA, 22–26 July 2012; pp. 1–8.
22. Van der Aalst, W.M.P.; Ter Hofstede, A.H.M.; Kiepuszewski, B.; Barros, A.P. Workflow Patterns. *Distrib. Parallel Databases* 2003, *14*, 5–51.

23. Finkel, A.; Leroux, J. Recent and Simple Algorithms for Petri Nets. *Softw. Syst. Model.* 2014, *14*, 719–725.

24. Zhang, L.J.; Zhang, J.; Cai, H. *Services Computing*; Springer Berlin Heidelberg: Heidelberg, Germany, 2007.

25. Cheng, B.; Wei, Z. RESTful M2M Gateway for Remote Wireless Monitoring for District Central Heating Networks.*Sensors* 2014, *14*, 22447–22470.

26. Budinsky, F.; Decandio, G.; Earle, R.; Francis, T.; Jones, J.; Li, J.; Nally, M.; Nelin, C.; Popescu, V.; Rich, S. WebSphere Studio Overview. *IBM Syst. J.* 2004, *43*, 384–419.

27. Tan, W.; Fan, Y.; Zhou, M.; Zhou, M. A Petri Net-Based Method for Compatibility Analysis and Composition of Web Services in Business Process Execution Language. *IEEE Trans. Autom. Sci. Eng.* 2009, *6*, 94–106.

28. Ghorbani, M.; Bogdan, P. A Cyber-Physical System Approach to Artificial Pancreas Design. In Proceedings of the Ninth IEEE/ACM/IFIP International Conference on Hardware/Software Codesign and System Synthesis (CODES+ISSS 2013), Montreal, QC, Canada, 29 September–4 October 2013; pp. 1–10.

CHAPTER 4

VARIATION PROPAGATION CONTROL IN MECHANICAL ASSEMBLY OF CYLINDRICAL COMPONENTS

Z. Yang[a], A.A. Popov[a], S. McWilliam[b]

[a] Manufacturing Research Division, Faculty of Engineering, University of Nottingham, University Park, Nottingham NG7 2RD, UK

[b] Materials, Mechanics and Structures Research Division, Faculty of Engineering, University of Nottingham, University Park, Nottingham NG7 2RD, UK

ABSTRACT

Variation propagation control is one of the procedures used to improve product quality in the manufacturing assembly process. The quality of a product assembly depends on the product type and the optimization criteria employed in the assembly. This paper presents two assembly procedures of component stacks by controlling variation propagation. The procedures considered are: (i) straight-build assembly by minimizing the distances from the centres of components to table axis; (ii) parallelism-build assembly by minimizing the angular errors between actual and nominal planes. Simulation results are presented for the assembly of four cylindrical components. The results indicate that the variation can be reduced significantly by using these procedures, compared to that without minimization. The results also indicate that the variation not only greatly relies on the assembly procedures, but also on the number of available orientations at the assembly stage. The radial variation increases with the stage for the straight-build assembly, while the angular error decreases with the stage for the parallelism-build assembly. The assembly quality for the two assembly procedures can be improved by increasing the number of orientations. The variation decreases exponentially and monotonically with the number of orientations. The information obtained is useful for manufacturing processes and the assembly modeling.

INTRODUCTION

Variations always exist due to imperfections in the manufacturing processes and materials, and various other random errors. They cause small deviations in parts from the nominal geometry. These deviations propagate and accumulate as parts are assembled and can quickly drive assembly dimensions out of specification [1], [2], [3],[4] and [5]. Tolerance assignment in mechanical engineering product design and manufacturing is critical both for product quality and performance, as well as for its manufacturing cost. A tighter tolerance normally requires more extensive manufacturing effort, resulting in a higher manufacturing cost [6].

Improving quality and reducing cycle time and cost are the main objectives for competitive manufacturing today. These objectives can be achieved partially by effectively controlling the variation propagation in manufacturing [7]. The traditional methods for studying assembly tolerance stackup are usually based on engineering experience, worst on worst tolerance analysis (WOW) method [8], [9], [10] and [11], or the root-sum-square tolerance analysis (RSS) method [6], [12] and [13]. These methods are used frequently in the

Citation: Z. Yang, A.A. Popov, S. McWilliam, Variation propagation control in mechanical assembly of cylindrical components, Journal of Manufacturing Systems, Volume 31, Issue 2, April 2012, Pages 162-176, ISSN 0278-6125, http://dx.doi.org/10.1016/j.jmsy.2011.09.003.

analysis of a single-dimensional chain, and are not suitable for the analysis of geometrical tolerances in three-dimensional space. Furthermore, these methods do not take into account the practical assembly procedures, resulting in conservative and expensive solutions.

To reduce the costs and minimize the variation propagation, in recent years, some researchers have explored new methods for the variation control. Mease et al. suggest[14] a cost-effective method, selective assembly. This method focuses on the fit between mating parts rather than the absolute dimension of each component. It can be used to achieve high-precision assembly from relatively low-precision components. In this approach, the mating components are measured, sorted by dimension and binned into groups prior to the assembly process. The final product is assembled by selecting the components from appropriate bins to achieve optimal assembly dimensions. There are costs associated with using a larger number of bins. If there are no components of one type in a particular bin at a particular time, the mating components in the corresponding bin cannot be used. Mantripragada and Whitney [1] suggest that the variation propagation is controlled by designing contact features to make adjustments along the assembly process. However, in practice, the contact features are hardly changed in the assembly process. Thus, this paper proposes that the variation is controlled by an effective selection of assembly options. For example, for any rotationally symmetric component assembly, the assembly can be changed by rotating the component about the symmetry axis, and the assembly quality can be improved by a suitable selection of orientations. This would result in either improving the product quality or reducing the manufacturing costs. Therefore, it would be valuable to find out how the available options influence assembly quality. The information obtained would provide significant insight into assembly for manufacturing processes and be useful for the modeling in practical applications. For instance, Rachuri et al. [15] propose a model to capture product assembly information for developing representational methodologies for the next generation of assembly-related standards. The assembly information model requires information regarding parts and their assembly relationships. Shi et al. [16], [17],[18] and [19] recommend a systematic variation reducing strategy by monitoring process consistency and identifying the variation sources based on available online measurements. This systematic variation reducing strategy involves very complex modeling work, such as components deviation modeling, assembly process modeling, and measurement deviation modeling. This systematic variation reducing strategy also requires information about components deviation and their inter-correlation among different assembly stages. Furthermore, the dimension of systematic variation models increases dramatically with the number of process routes and stages, resulting in heavy computation.

The quality of a product assembly is also greatly dependent on the product type and the optimization techniques employed in the assembly [20] and [21]. Due to the complexity of a manufacturing process, an effective method for the assembly would be highly desirable[20]. For example, in a multistage radial flow submersible pump assembly, it requires a parallelism-build assembly, and axial play between the impeller and the volute casing is required to have a tolerance of ±0.5 mm, to prevent rubbing action [21]. For the assembly of aero-engine components, such as compressor stages, Module 04 shown in Fig. 1[22], a precise alignment and clamping device of rotationally symmetric parts is required [23]. The concentricity deviation of the part must be within a 2.5 μm tolerance to meet the strict vibration requirements of the engine [24]. It requires a straight-build assembly where a key characteristic here is to give a 'straight line' between the centres of the parts for the assembly process.

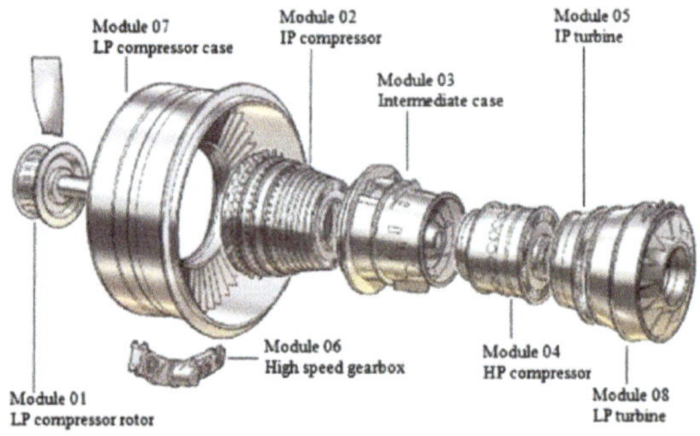

Figure 1: The modular breakdown of a Trent family engine [22].

Furthermore, the practical assembly mainly involves cylindrical or/and polyhedral components [25], [26], [27], [28] and [29]. For example, in the assembly of aero-engine components, all of the parts are cylindrical components, as shown in Fig. 1[22]. The cylindrical and polyhedral components belong to the same family [30]; once the kinematic parameters within a tolerance range are identified, the results are readily usable for any tolerance specification in the same family. Thus, this paper considers a practical cylindrical-component assembly. For the assembly considered, the variation control usually focuses on minimizing the concentricity deviation and/or axial runout, and these types of variation control will be considered here.

In summary, this paper presents two assembly procedures for controlling variation propagation in the assembly of cylindrical component stacks, as follows:

- Straight-build assembly by minimizing the distances from the centres of components to table axis to control the radial variation, where the radial variation is used as a measure of the quality of straight-build assembly.
- Parallelism-build assembly by minimizing the angular errors between actual and nominal planes to control axial runout, where the angular error is used as a measure of the quality of the parallelism-build assembly.

The minimal variation propagation can be achieved by minimizing the variation stage-by-stage according the assembly process, or by minimizing the sum of squared variations based on all the information for all the assembly stages. The latter is based on ideal conditions without process errors, disturbances etc. in manufacturing processes. However, due to the complexity and randomness of uncertainties and disturbances in manufacturing processes, the minimizing the sum of squared variations for the all assembly stages cannot guarantee the delivery of desired product quality. Thus the paper considers the practical assembly operations to provide an economical and effective solution to analyze tolerance build up. During the assembly, variation propagation is minimized stage-by-stage by rotating the component about its symmetry axis, according to the available number of practical orientations. To validate the two assembly methods, Monte Carlo simulations are performed. The relationship between variation propagation and the number of orientations is also investigated.

Details of the two assembly methods are given in Section 2. A four-cylindrical-component assembly is described in Section 3. The results obtained using the two assembly methods are presented in Section 4, the relationship between variation propagation and the number of orientations is also given in Section 4. The conclusions from the study are presented in Section 5.

TWO ASSEMBLY TECHNIQUES

The proposed assembly techniques are: (i) straight-build assembly and (ii) parallelism-build assembly. Here, the paper uses two cylindrical components to demonstrate the straight-build assembly and parallelism-build assembly, as shown in Fig. 2 and Fig. 3, respectively, where C_1 and C_2 are the two component centres at their top ends. The nominal plane is defined by the base of the first component. V_1 and V_2 are normal vectors to the nominal and actual uppermost planes, respectively. The axis normal to the nominal base plane is called "table axis". The techniques are described as follows:

- Straight-build assembly: minimize radial variation from the component centre to table axis stage-by-stage, as shown in Fig. 2.
- Parallelism-build assembly: minimize angular error between actual and nominal planes stage-by-stage, as shown in Fig. 3.

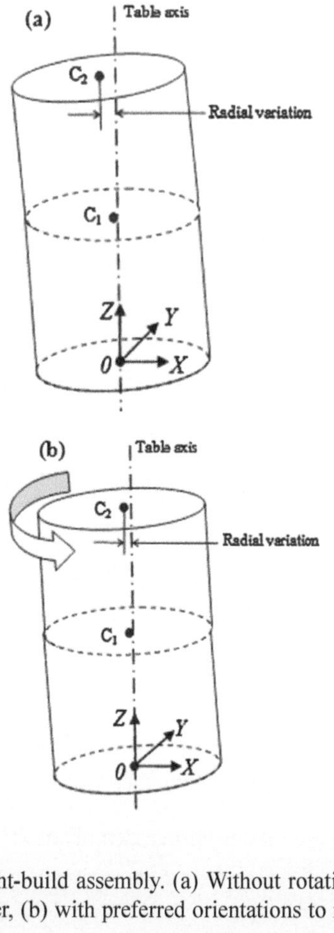

Figure 2: Straight-build assembly. (a) Without rotating components against each other, (b) with preferred orientations to minimize radial variation.

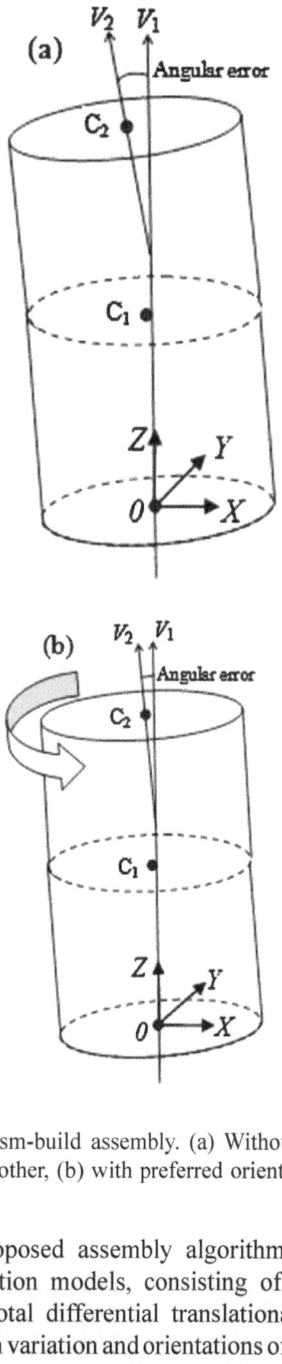

Figure 3: Parallelism-build assembly. (a) Without rotating components against each other, (b) with preferred orientations to minimize angular error.

The two proposed assembly algorithms are developed from state transition models, consisting of three steps: (1) calculating the total differential translational and rotational vectors for a given variation and orientations of the components; (2) calculating of the translational error vector and rotational errors from the differential translational and rotational vectors; and (3) minimizing radial variation or angular errors for the straight-build assembly or parallelism-build assembly from the translational error vector and rotational errors. The calculations for the first two steps are given by the developed state transition models in Appendix A. The minimization of radial variation or angular errors for the straight-build assembly or parallelism-build assembly is presented as follows.

Straight-build assembly algorithm

To minimize radial variation, the second (top) component is rotated about its symmetry axis, and then the translational error vector $(dp_2^x(i), dp_2^y(i), dp_2^z(i))$ is calculated from the state transition models described in Appendix A, according to the available number of practical orientations. The number of orientations is described by rotational matrix R in Eq. (A-3) in Appendix A.

The radial variation for orientation i at Stage 2 is calculated as follows:

$$\textit{Radial}\ \ \mathrm{var}(2, i) = \sqrt{(dp_2^x(i))^2 + (dp_2^y(i))^2} \tag{1}$$

where Stage 2 corresponds to the assembly when the second component is added.

The minimum radial variation for all possible orientations gives the best straight-build assembly at Stage 2 and can be calculated as follows:

$$\textit{Radial}\ \ \mathrm{var}(2) = \min\left\{\sqrt{(dp_2^x(i))^2 + (dp_2^y(i))^2}\right\} \tag{2}$$

where i = 1,2, … , N, and N is the maximum number of orientations.

In a similar way, the minimum radial variation at Stage k can be calculated by:

$$\textit{Radial}\ \ \mathrm{var}(k) = \min\left\{\sqrt{(dp_k^x(i))^2 + (dp_k^y(i))^2}\right\} \tag{3}$$

where Stage k corresponds to the assembly when the kth component is added.

Therefore, the minimum radial variation can be obtained at each stage.

Parallelism-build assembly algorithm

In a similar way, using the state tranistion models described in Appendix A, the rotational errors $(\delta_k^x(i), \delta_k^y(i), \delta_k^z(i))$ can be calculated at each stage for the available orientation i by rotating the component about its symmetry axis, as shown in Fig. 3. The nominal uppermost plane at Stage k is:

$$z = (H_1 + H_2 + \cdots + H_k) \tag{4}$$

where Hi (i = 1,2, … , k) is a height of the ith cylindrical component, and k corresponds to the assembly when the kth component is added.

The normal vector to the nominal uppermost plane at Stage k is:

$$V_{k1} = \begin{bmatrix} 0 \\ 0 \\ 1 \end{bmatrix} \tag{5}$$

The normal vector to the actual uppermost plane which has rotational errors $(\delta_k^x(i), \delta_k^y(i), \delta_k^z(i))$ at Stage k is:

$$V_{k2} = \begin{bmatrix} x \\ y \\ z \end{bmatrix} = R_x(\delta_k^x(i))R_y(\delta_k^y(i))R_z(\delta_k^z(i))V_{k1} \tag{6}$$

where

$$R_x(\delta_k^x(i)) = \begin{bmatrix} 1 & 0 & 0 \\ 0 & \cos(\delta_k^x(i)) & -\sin(\delta_k^x(i)) \\ 0 & \sin(\delta_k^x(i)) & \cos(\delta_k^x(i)) \end{bmatrix}, \tag{7}$$

$$R_y(\delta_k^y(i)) = \begin{bmatrix} \cos(\delta_k^y(i)) & 0 & \sin(\delta_k^y(i)) \\ 0 & 1 & 0 \\ -\sin(\delta_k^y(i0) & 0 & \cos(\delta_k^y(i)) \end{bmatrix}, \tag{8}$$

$$R_z(\delta_k^z(i)) = \begin{bmatrix} \cos(\delta_k^z(i)) & -\sin(\delta_k^z(i)) & 0 \\ \sin(\delta_k^z(i)) & \cos(\delta_k^z(i)) & 0 \\ 0 & 0 & 1 \end{bmatrix} \tag{9}$$

Assuming that the rotation angle errors are small, first order small angle approximations have been used such that: $\cos(\delta) \approx 1$, $\sin(\delta) \approx \delta$. Eqs. (7), (8) and (9)can be approximated as follows:

$$R_x(\delta_k^x(i)) \cong \begin{bmatrix} 1 & 0 & 0 \\ 0 & 1 & -\delta_k^x(i) \\ 0 & \delta_k^x(i) & 1 \end{bmatrix}, \tag{10}$$

$$R_y(\delta_k^y(i)) \cong \begin{bmatrix} 1 & 0 & \delta_k^y(i) \\ 0 & 1 & 0 \\ -\delta_k^y(i) & 0 & 1 \end{bmatrix}, \tag{11}$$

$$R_z(\delta_k^z(i)) \cong \begin{bmatrix} 1 & -\delta_k^z(i) & 0 \\ \delta_k^z(i) & 1 & 0 \\ 0 & 0 & 1 \end{bmatrix} \tag{12}$$

Substituting Eqs. (5), (10), (11) and (12) into Eq. (6) gives:

$$V_{k2} \cong \begin{bmatrix} \delta_k^y(i) \\ -\delta_k^x(i) \\ 1 \end{bmatrix} \tag{13}$$

The angle $d\theta(k, i)$ between the nominal and actual planes is calculated from the two normal vectors Vk_1 and Vk_2, such that:

$$\cos(d\theta(k, i)) = \frac{|V_{k1} \cdot V_{k2}|}{|V_{k1}||V_{k2}|} = \frac{1}{\sqrt{(\delta_k^x(i))^2 + (\delta_k^y(i))^2 + 1}} \tag{14}$$

$$d\theta(k, i) = \cos^{-1}\left[\frac{1}{\sqrt{(\delta_k^x(i))^2 + (\delta_k^x(i))^2 + 1}}\right] = \sin^{-1}\left\{\sqrt{\frac{(\delta_k^x(i))^2 + (\delta_k^y(i))^2}{(\delta_k^x(i))^2 + (\delta_k^y(i))^2 + 1}}\right\} \tag{15}$$

where the rotation angle errors (δ_k^x and δ_k^y) are very small compared to 1 and negligible, such that Eq. (15) can be rewritten as

$$d\theta(k, i) \cong \sin^{-1}\left\{\sqrt{(\delta_k^x(i))^2 + (\delta_k^y(i))^2}\right\} \cong \sqrt{(\delta_k^x(i))^2 + (\delta_k^y(i))^2} \tag{16}$$

The minimum angular error for all possible orientations gives the best parallelism-build assembly at Stage k and can be calculated as follows:

$$d\theta(k) = \min\left\{\sqrt{(\delta_k^x(i))^2 + (\delta_k^y(i))^2}\right\} \quad (17)$$

where i = 1,2, ... , N and N is the maximum number of orientations.

AN ASSEMBLY EXAMPLE OF FOUR CYLINDRICAL COMPONENTS

To illustrate the two optimization techniques for minimizing uncertainty propagation, a practical four-cylindrical-component assembly is considered. The cylindrical component has a height of H (70 mm) with tolerance ±h, and a diameter of Φ (100 mm) with tolerance ±φ, as shown in Fig. 4, where h = φ = 0.1 mm. In practical, each component has limited orientations, here each component is assumed to have 4, 8, 16 or 20 possible uniformly arranged orientations. Fig. 4 shows the situation when the component has 4 orientations. To analyze the variation propagation statistically, the errors for the diameter and height of the cylindrical component are assumed to be normally distributed, where the standard deviation (σ) is taken to be one sixth of the tolerance range, i.e. one third of 0.1 mm. Each assembly procedure is simulated 10,000 times using standard Monte Carlo methods. For simplicity, the following assumptions are made for the component with 4 possible orientations.

- The top and base surfaces are flat.
- Frames F1 and F1′ are attached to the assembly features.
- Base centre (0,0,0) is the origin in the world coordinate system.
- Points 1, 2, 3 and 4 are uniformly separated by 90° angle in the XOY plane with centre O, and lie on the circumference of the cylinder on the OX and OY axes.
- Frame F1′ related to Frame F1 has a translation of (0,0,H) with a translation error of (dx,dy,dz) and an angular error of (dθx, dθy, dθz).
- The top centre O′ lies at (dx,dy,dz + H) relative to O.
- Points 1′, 2′, 3′ and 4′ are uniformly separated by 90° angle on the top surface with centre O′, and lie on the circumference of the cylinder on the O′X′ and O′Y′ axes.

As the cylindrical component has a height of H with tolerance ±h, and a diameter of Φwith tolerance ±φ, the size tolerance zone at the component top is enclosed by two cylinders with a height of 2h having a diameter of Φ − φ and a diameter of Φ + φ, respectively, as shown in Fig. 5(a). In the size tolerance zone, the following inequities can be obtained.

$$|dx| \le \varphi, \quad (18)$$

$$|dy| \le \varphi, \quad (19)$$

$$\sqrt{dx^2 + dy^2} \le \varphi, \quad (20)$$

$$|dz| \le h, \quad (21)$$

$$|d\theta_x| \le \tan^{-1}\left(\frac{h - |dz|}{0.5\Phi}\right), \quad (22)$$

$$|d\theta_y| \le \tan^{-1}\left(\frac{h - |dz|}{0.5\Phi}\right), \quad (23)$$

$$|d\theta_z| \le \tan^{-1}\left(\frac{\varphi - |dy|}{0.5\Phi}\right) \quad (24)$$

Since (h−|dz|)/0.5Φ and (φ−|dy|)/0.5Φ are very small, the right sides of inequities (22), (23) and (24) can approximately be (h−|dz|)/0.5Φ, (h−|dz|)/0.5Φ and (φ−|dy|)/0.5Φ, respectively. Inequities (22), (23) and (24) can be written as:

$$|d\theta_x| \le \frac{h - |dz|}{0.5\Phi}, \quad (25)$$

$$|d\theta_y| \le \frac{h - |dz|}{0.5\Phi}, \quad (26)$$

$$|d\theta_z| \le \frac{\varphi - |dy|}{0.5\Phi} \quad (27)$$

From inequities (21) and (25), dθx and dz are constrained to lie within the shaded "bounded" region shown in Fig. 5(b). Similarly, from inequities (21) and (26), dθy anddz are constrained to lie within the same shaded "bounded" region as dθx and dz; from inequities (19) and (27), the constrained bounds for dθz and dy can be constructed as shown in Fig. 5(c); and from inequities (18), (19) and (20), the constrained bounds fordx and dy can be constructed as shown in Fig. 5(d). The dx and dz are normally distributed. Each of the variables is randomly and normally generated, and then the second variable generated is limited to the range left by the first one. For example, inFig. 5(d), if the variable dx is generated first, then the variable dy will be limited to the shaded circled region left by the dx. Any instances generated outside [−3σ, +3σ] are disregarded to respect the shaded "bounded" regions, where the standard deviation is described by σ. Details of the generated variables can be found in [31].

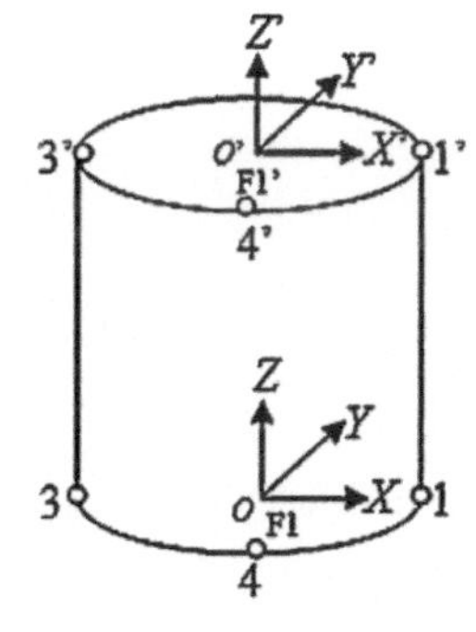

Figure 4: A cylindrical component.

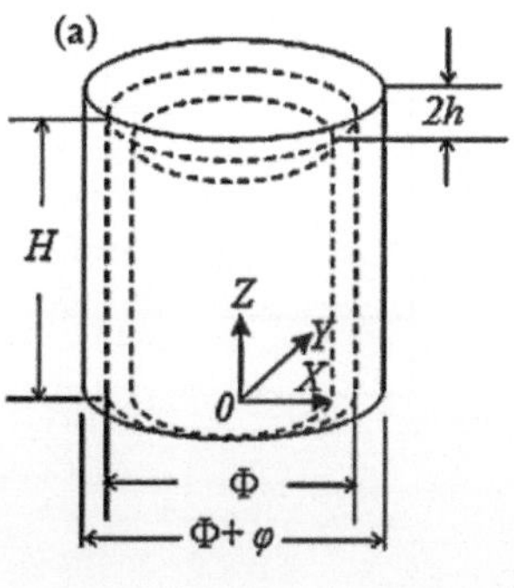

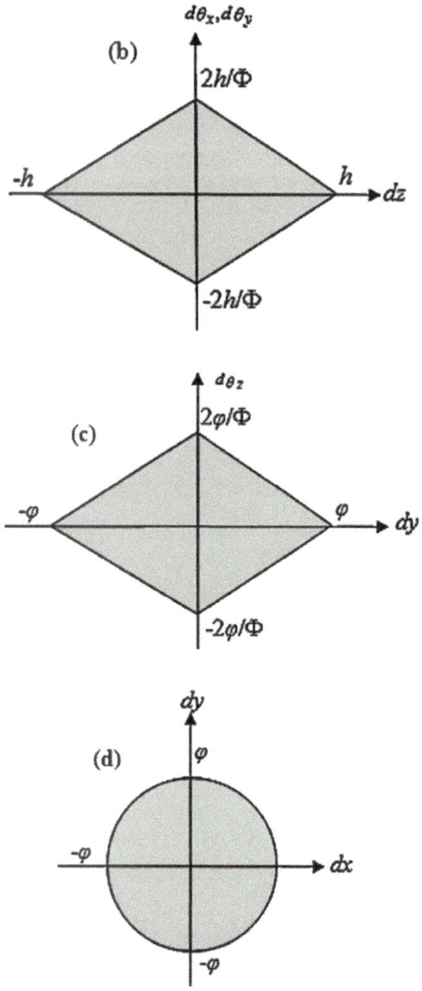

Figure 5: A size tolerance zone and the bounds used to specify the kinematic parameter space. (a) A size tolerance zone, (b)–(d) define the constrained regions.

In a similar way, for component with different orientations, similar assumptions as described above can be made and will not be reiterated here.

RESULTS AND DISCUSSION

Influence of orientations on variation propagation

For the straight-build assembly, the radial variation propagation from the table axis is calculated using Eq. (3). This paper only gives details for the orientations 0, 4, 8 and 16, since the results for 16 orientations are very similar to those for 20 orientations. The average stage-by-stage radial variation against stage from the table axis is shown in Fig. 6(a). The average stage-by-stage radial variation against the number of available orientations from the table axis for Stages 2–4 is shown in Fig. 6(b). The mean and standard deviations of the variations are provided in Table 1, together with the maximum and minimum values. The histograms of the radial variations for 0, 4, 8 and 16 orientations are shown in Fig. 7.

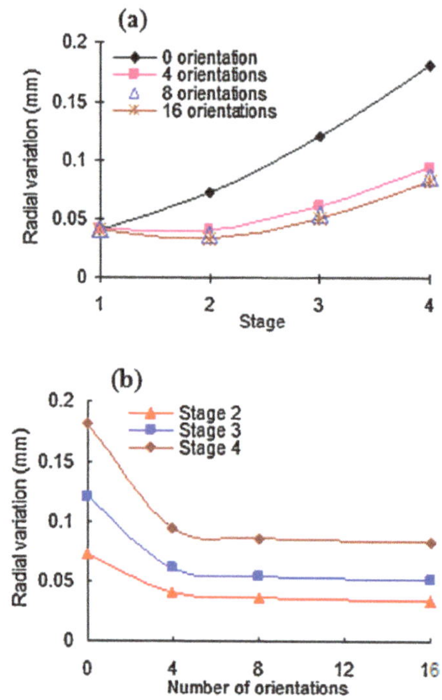

Figure 6: Variation propagation from the table axis against stage and the number of orientations, respectively. (a) Variation against stage, (b) variation against the number of orientations.

Table 1: Radial variation propagation from the table axis for straight-build assembly

Number of orientations	Stage	Max variation (mm)	Min variation (mm)	Average variation (mm)	Standard deviation of variation (mm)
0	1	0.0999	4.0078×10^{-4}	0.0412	0.0208
	2	0.2481	7.7686×10^{-4}	0.0727	0.0380
	3	0.4434	0.0018	0.1198	0.0643
	4	0.6599	5.9583×10^{-4}	0.1813	0.0978

4	1	0.0999	4.0078×10^{-4}	0.0412	0.0208
	2	0.1782	3.3417×10^{-4}	0.0405	0.0249
	3	0.3069	6.1712×10^{-4}	0.0605	0.0439
	4	0.5117	3.4981×10^{-4}	0.0946	0.0726
8	1	0.0999	4.0078×10^{-4}	0.0412	0.0208
	2	0.1782	3.3417×10^{-4}	0.0356	0.0253
	3	0.3069	2.6078×10^{-4}	0.0535	0.0436
	4	0.5220	7.0305×10^{-4}	0.0856	0.0721
16	1	0.0999	4.0078×10^{-4}	0.0412	0.0208
	2	0.1782	3.3417×10^{-4}	0.0338	0.0259
	3	0.2981	3.8648×10^{-4}	0.0513	0.0440
	4	0.5120	2.1112×10^{-4}	0.0829	0.0723
20	1	0.0999	4.0078×10^{-4}	0.0412	0.0208
	2	0.1782	2.7312×10^{-4}	0.0335	0.0260
	3	0.3000	1.0063×10^{-4}	0.0510	0.0441
	4	0.5130	2.8536×10^{-4}	0.0825	0.0724

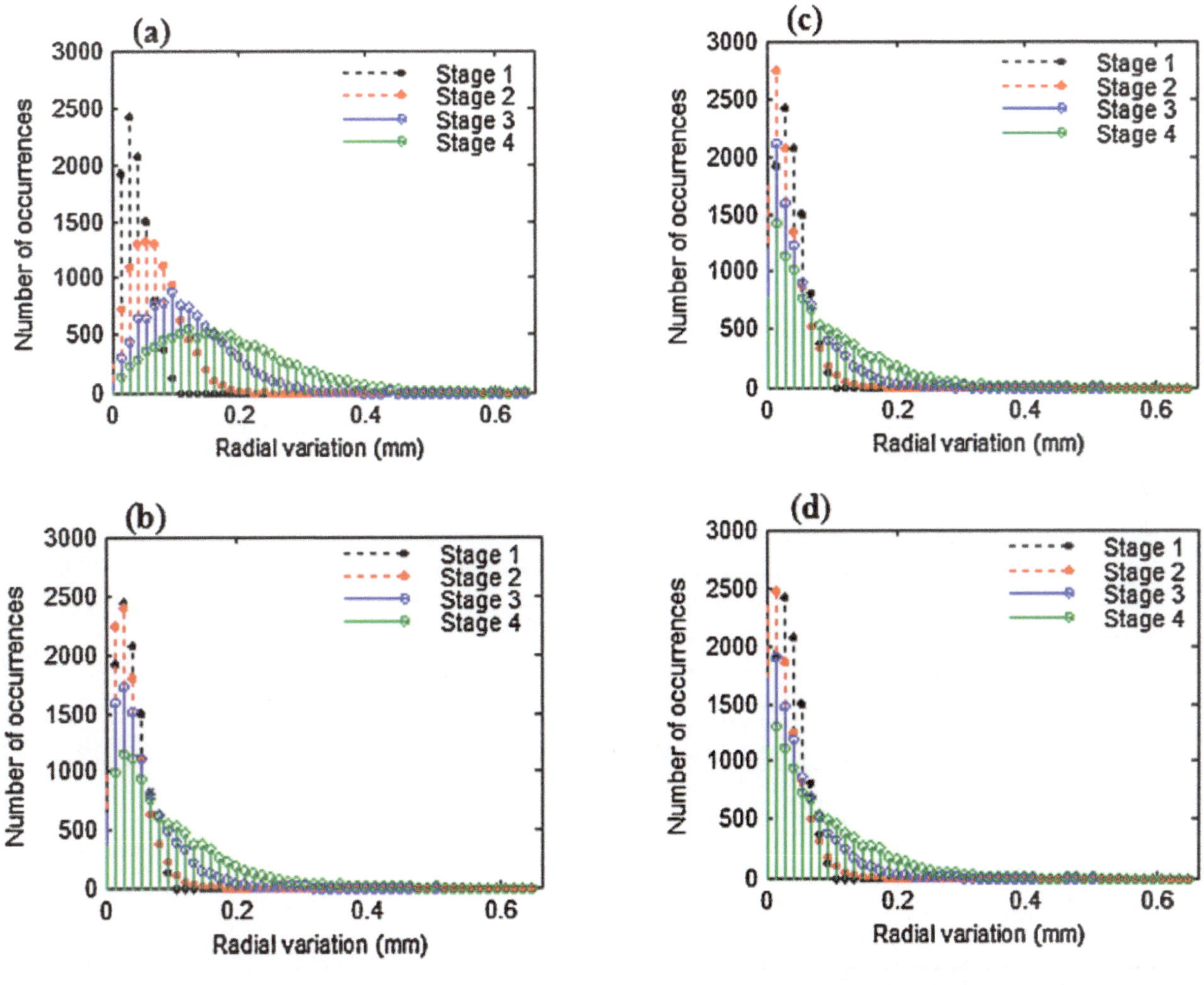

Figure 7: Histogram of variation propagation from the table axis. (a) 0 orientation, (b) 4 orientations, (c) 8 orientations, (d) 16 orientations.

For the parallelism-build assembly, the angular error is calculated using Eq. (17). In a similar way, the paper only gives the results for 0, 4, 8 and 16 orientations. The average angular error is shown in Fig. 8. The mean and standard deviations of the errors are provided in Table 2, together with the maximum and minimum values. The histograms of the angular errors for each procedure are shown in Fig. 9.

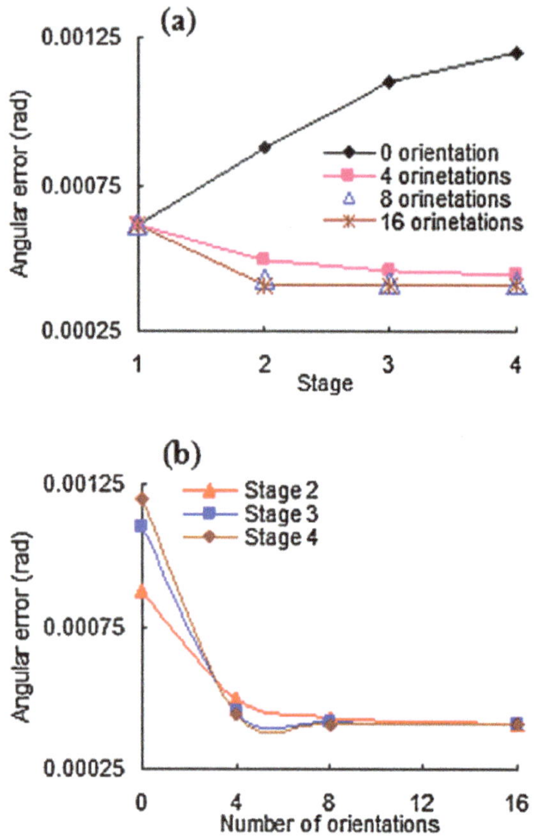

Figure 8: Angular error against stage and the number of orientations, respectively. (a) Variation against stage, (b) angular error against the number of orientations.

Table 2: Angular error for parallelism-build assembly

Number of orienta-tions	Stage	Max error (rad)	Min error (rad)	Average error (rad)	Standard deviation of error (rad)
0	1	0.0020	3.6311×10^{-7}	6.0901×10^{-4}	3.5427×10^{-4}
	2	0.0030	4.2030×10^{-6}	8.7466×10^{-4}	4.7417×10^{-4}
	3	0.0044	1.8471×10^{-5}	0.0011	5.7669×10^{-4}
	4	0.0045	1.0520×10^{-5}	0.0012	6.6198×10^{-4}
4	1	0.0020	3.6311×10^{-7}	6.0901×10^{-4}	3.5427×10^{-4}
	2	0.0018	4.2030×10^{-6}	4.9176×10^{-4}	2.9455×10^{-4}
	3	0.0018	4.1063×10^{-6}	4.5436×10^{-4}	2.8031×10^{-4}
	4	0.0018	5.4738×10^{-6}	4.4375×10^{-4}	2.8361×10^{-4}
8	1	0.0020	3.6311×10^{-7}	6.0901×10^{-4}	3.5427×10^{-4}
	2	0.0018	4.2030×10^{-6}	4.2856×10^{-4}	2.9585×10^{-4}
	3	0.0018	3.3571×10^{-6}	4.1259×10^{-4}	2.9617×10^{-4}
	4	0.0018	1.1027×10^{-6}	4.1099×10^{-4}	2.9763×10^{-4}
16	1	0.0020	3.6311×10^{-7}	6.0901×10^{-4}	3.5427×10^{-4}
	2	0.0018	2.1714×10^{-6}	4.0784×10^{-4}	3.0243×10^{-4}
	3	0.0019	1.8242×10^{-6}	4.0658×10^{-4}	3.0686×10^{-4}
	4	0.0018	2.8310×10^{-6}	4.0276×10^{-4}	3.0723×10^{-4}
20	1	0.0020	3.6311×10^{-7}	6.0901×10^{-4}	3.5427×10^{-4}
	2	0.0018	6.6695×10^{-7}	4.0415×10^{-4}	3.0435×10^{-4}
	3	0.0019	1.7187×10^{-6}	4.0651×10^{-4}	3.0917×10^{-4}
	4	0.0018	4.9368×10^{-7}	4.0146×10^{-4}	3.0913×10^{-4}

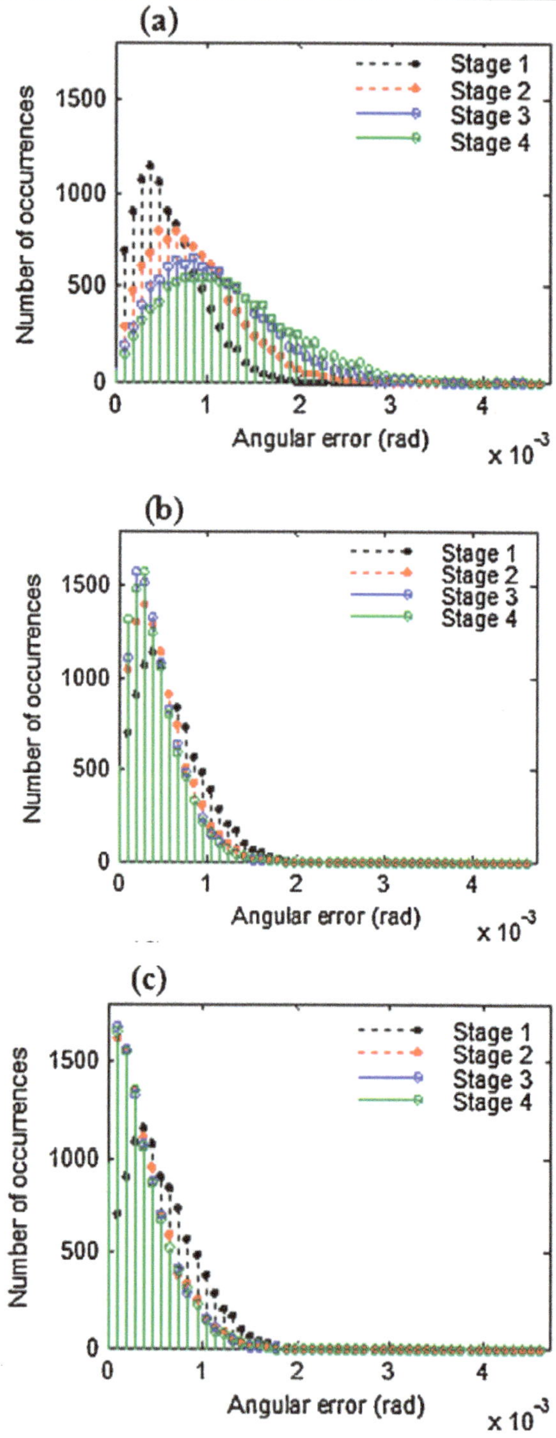

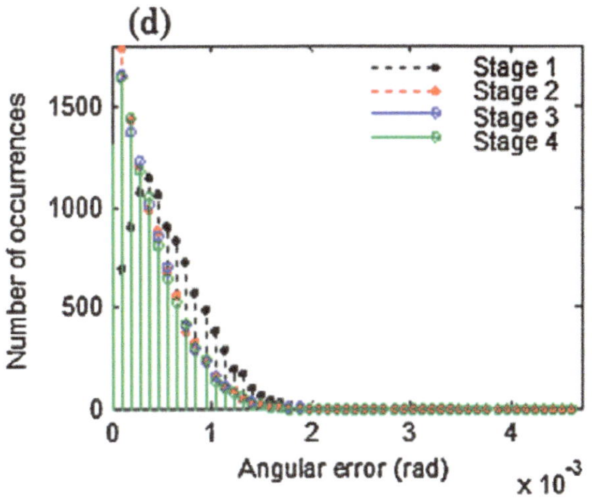

Figure 9: Histogram of angular error. (a) 0 orientation, (b) 4 orientations, (c) 8 orientations, (d) 16 orientations.

To compare the two assembly procedures, angular error for the straight-build assembly and radial variation for the parallelism-build assembly are also calculated, and are given in Section 4.2 below.

All of the results indicate that the error propagation varies significantly with the assembly procedures and the number of orientations.

Influence of orientations on variation propagation for straight-build assembly

From Fig. 6 and Table 1, the average radial variations tend to increase with the stage for the different number of orientations – there is a slight decrease between Stages 1 and 2, where the number of orientations are equal to or more than 4, as shown in Fig. 6(a). The average radial variations decrease with the number of orientations for the same stage, as shown in Fig. 6(b). The results illustrate that the variation propagation can be reduced by increasing the number of orientations. It can be seen that the variation propagation is significantly reduced from 0 to 4 orientations. However, as the number of orientations increases, the variation continues to decrease, but at a less significant rate. For example, the variation at the final stage is reduced by 48% from 0 to 4 orientations, by 10% from 4 to 8 orientations, and by 3% from 8 to 16 orientations. In Fig. 6(b), it is seen that the variation propagation decreases exponentially, rather than proportionally to the number of orientations. This indicates that there is limit to how much the variation can be reduced. It also means that the minimum variation can be achieved at reduced cost by properly selecting the number of orientations.

From Table 1, it is found that the standard deviation of radial variation increases with the stage as well. This indicates that the consistency of the straight-build assembly will decrease with the stage. It is also found that the standard deviation of radial variation decreases significantly from 0 to 4 orientations; while as the number of orientations increases, the standard deviation of radial variation is comparable to that with 4 orientations. This shows that the straight-build assembly has a good consistency with the number of orientations.

From Fig. 7(a)–(d), the procedure with 16 orientations produces the smallest levels of radial variation, with most values occurring in the range 0–0.1 mm. These results further demonstrate that the assembly quality and consistency increase with the number of orientations for the straight-build assembly.

Influence of orientations on variation propagation for parallelism-build assembly

From Fig. 8 and Table 2, the average angular error increases with the stage in the absence of orientations, while it decreases monotonically with stage in the presence of orientations. It also shows that the orientations can significantly reduce the angular error for the parallelism-build assembly, compared to the assembly without orientations. For example, the angular error at the final stage is reduced by 63% from 0 to 4 orientations. However, as the number of orientations increases, the angular error continues to decrease, but at a lower rate. For example, the angular error at the final stage is reduced by 7.4% from 4 to 8 orientations, and by 2% from 8 to 16 orientations. From Fig. 8(b), the angular error propagation appears to decrease exponentially, rather than proportionally to the number of orientations, as described earlier for straight-build assembly. This also indicates that there is a limitation on reducing the angular error, and a proper selection of the number of orientations can achieve the minimum angular error with reduced cost.

Table 2 indicates that the standard deviation of the angular error increases with stage without orientations, while it is smaller in the presence of orientations. These results confirm that the parallelism-build assembly can be improved by using the orientations.

From Fig. 9(a)–(d), the parallelism-build assembly with 16 orientations produces the smallest levels of angular error, with most values occurring in the range 0–0.001 rad. These results further demonstrate that a proper selection of the number of orientations can minimize the angular error at reduced cost for the parallelism-build assembly.

In summary, the average radial variation increases with stage for the straight-build assembly, while the average angular error decreases with stage for the parallelism-build assembly in the presence of orientations. The statistical results above indicate that the quality of the two assembly procedures can be improved by increasing the number of orientations.

Discussion

As mentioned earlier, to compare the two assembly procedures and study the effect of using different number of orientations on the error propagation, the radial variation for parallelism-build assembly and angular error for straight-build assembly are calculated. The results have a similar pattern for the different orientations and are only shown for components with 4 orientations. Comparisons of average radial variation and angular error are shown in Fig. 10(a) and (b), respectively, for the different assembly procedures. The mean and standard deviations of the variations and angular error are provided inTable 3 and Table 4, together with the maximum and minimum values, for parallelism-build assembly and straight-build assembly, respectively. The histograms of the radial variations and angular error for parallelism-build assembly and straight-build assembly are shown in Fig. 11(a) and (b), respectively.

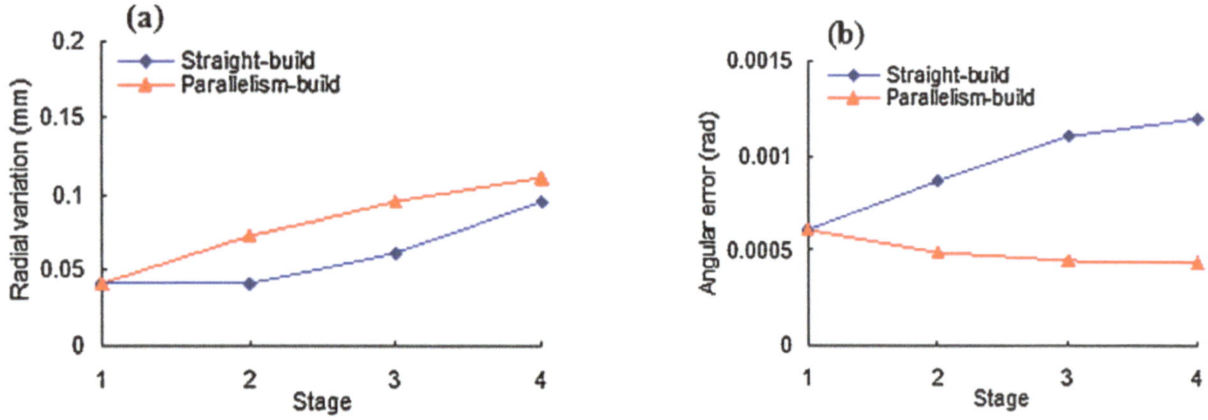

Figure 10: Radial variation and angular error for the different assembly procedures. (a) Radial variation, (b) angular error.

Table 3: Radial variation propagation from the table axis for the different assembly procedures

Procedure	Stage	Max variation (mm)	Min variation (mm)	Average variation (mm)	Standard deviation of variation (mm)
Straight-build assembly	1	0.0999	4.0078×10^{-4}	0.0412	0.0208
	2	0.1782	3.3417×10^{-4}	0.0405	0.0249
	3	0.3069	6.1712×10^{-4}	0.0605	0.0439
	4	0.5117	3.4981×10^{-4}	0.0946	0.0726
Parallelism-build assembly	1	0.0999	4.0078×10^{-4}	0.0412	0.0208
	2	0.2589	3.3417×10^{-4}	0.0721	0.0376
	3	0.3343	9.6801×10^{-4}	0.0949	0.0529
	4	0.4487	3.4981×10^{-4}	0.1109	0.0626

Table 4: Angular error for the different assembly procedures

Procedure	Stage	Max error (rad)	Min error (rad)	Average error (rad)	Standard deviation of error (rad)
Straight-build assembly	1	0.0020	3.6311×10^{-7}	6.0901×10^{-4}	3.5427×10^{-4}
	2	0.0030	4.2030×10^{-6}	8.7187×10^{-4}	4.7413×10^{-4}
	3	0.0039	1.8941×10^{-5}	0.0011	5.7701×10^{-4}
	4	0.0047	9.4271×10^{-6}	0.0012	6.6464×10^{-4}
Parallelism-build assembly	1	0.0020	3.6311×10^{-7}	6.0901×10^{-4}	3.5427×10^{-4}
	2	0.0018	4.2030×10^{-6}	4.9176×10^{-4}	2.9455×10^{-4}
	3	0.0018	4.1063×10^{-6}	4.5436×10^{-4}	2.8031×10^{-4}
	4	0.0018	5.4738×10^{-6}	4.4375×10^{-4}	2.8361×10^{-4}

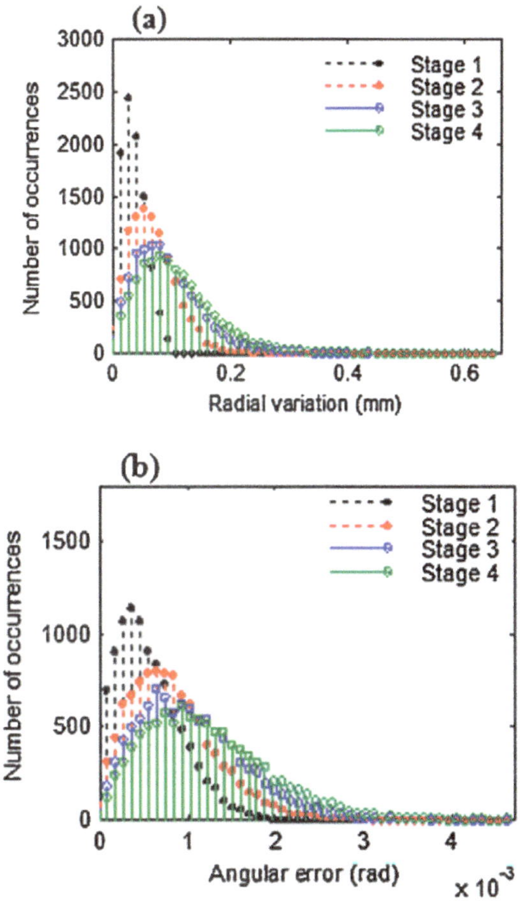

Figure 11: Histograms of radial variation for parallelism-build assembly and angular error for straight-build assembly. (a) Radial variation for parallelism-build assembly, (b) angular error for straight-build assembly.

From Fig. 10, it is seen that the straight-build assembly has smaller radial variation than the parallelism-build assembly from the table-axis, while the parallelism-build assembly provides reduced angular error, compared to the straight-build assembly. It is very interesting to notice that the radial variation increases rapidly after Stage 2 for the straight-build assembly, but the radial variation changes much more gently for the parallelism-build assembly, as shown in Fig. 10(a). This indicates that the minimal radial variation could be achieved by constraining the angular error in the straight-build assembly. Further studies to investigate this are under way, and will be reported at a later date.

From Table 3, the parallelism-build assembly has higher radial variation propagation; for example, the radial variation in the parallelism-build assembly is increased by 17% at Stage 4, compared to the straight-build assembly. Fig. 6 and Fig. 11 further demonstrate that the parallelism-build assembly increases radial variation propagation, where most values of the radial variation occurs in the range 0–0.1 mm for the straight-build assembly and in the range 0–0.15 mm for the parallelism-build assembly, respectively.

From Table 4, the parallelism-build assembly gives much lower angular error than the straight-build assembly. For instance, the angular error in the parallelism-build assembly is reduced by 63% at Stage 4, compared to the straight-build assembly. From Fig. 9 and Fig. 11, the parallelism-build assembly in the presence of orientations produces smaller levels of angular error, with most values occurring in the range 0–0.001 rad; while most values of angular error occur in the range 0–0.002 rad for the straight-build assembly. This indicates that the assembly procedure needs to be selected carefully in accordance with the particular industrial application.

From the average radial variations for the component with 4 orientations in Fig. 10(a), and with 0 orientation in Fig. 6(a), it is interesting to notice that not only the straight-build assembly but also the parallelism-build assembly can have reduced radial variation with orientations, the radial variation is reduced by 48% and 39% in the straight-build assembly and the parallelism-build assembly, respectively, compared to the procedure without orientations. This indicates that the orientations can improve the assembly quality.

CONCLUSIONS

Two assembly methods, straight-build assembly and parallelism-build assembly, have been analyzed for different number of orientations. The average radial variations and angular errors decrease with the number of available orientations for the same stage. The simulation results obtained indicate that the two assembly methods have the potential to significantly reduce manufacturing assembly errors. For the example considered, the first method reduces the radial variation by 48%, whist the second the angular error by 63%, where the component has 4 orientations, when compared to assemblies without orientations.

The straight-build assembly produces the minimum radial variation; and the parallelism-build assembly has the minimum angular error. For the first method, the average radial variations tend to increase with the stage for the different number of orientations; however, for the second method, the average angular error increases with the stage in the absence of orientations, while it decreases with the stage in the presence of orientations.

The two assembly methods have a good consistency, for instance, the maximum standard deviation is 0.0726 mm in the first method, and 3.5×10^{-4} rad in the second method in the presence of orientations.

The variation not only depends on the assembly procedures, but also on the number of orientations. The variation propagation decreases monotonously and exponentially with the number of orientations. Assembly quality could be improved by increasing number of orientations, however, the variation reduction is limited when the number of orientations becomes larger. Therefore, the best assembly and minimal manufacturing cost could be achieved by optimizing the selection of the number of orientations and assembly procedures.

The procedure of the two assembly methods is very simple, and can be easily used to analyze the variation propagation in manufacturing processes.

ACKNOWLEDGEMENTS

The authors gratefully acknowledge financial support from the Engineering and Physical Sciences Research Council through NIMRC project for this work. The authors also gratefully acknowledge help and advice from Mat Yates and Steve Slack at Rolls-Royce plc.

APPENDIX A. STATE TRANSITION MODELS

The state transition models are developed from robot manipulator [32], and consist of (i) modeling individual frame errors, (ii) total differential translation and rotation transformation, and (iii) calculation of Cartesian position and orientation errors in the world coordinate. The models are slightly different from those developed by Mantripragada and Whitney [1] and [33]. Since the variation in the world coordinates system is not only related to the differential rotation and translation in individual frame, but also dependent on nominal orientation and transition relative to the world coordinates system, the differential rotation and translation in individual frame must be converted into those in the world coordinates system through the transformation with respect to the world coordinates system, so that the variation is allowed to be the sum of those differential rotations and translations in the world coordinates system. However, Mantripragada and Whitney›s models regard relative transformation of two consecutive frames as that in the world coordinates system, and convert the differential position and orientation vectors in individual frame into the world coordinates system through the relative transformation. Their models of the variation propagation based on the wrong conversion cannot handle the misorientation and misposition of the components. Therefore, this paper presents a detailed derivation of the state transition models. The corrections of Mantripragada and Whitney›s models are also given in Appendix A.2below.

A.1. Modeling individual frame errors

Any manipulator can be considered to consist of a series of links connected together by joints. A coordinate frame can be embedded in each link of the manipulator, and then the relative position and orientation between these coordinate frames can be described by using homogeneous transformations [33]. The homogeneous transformation describing the relation between one link and the next has been called an A matrix [34]. An A matrix is simply a homogeneous transformation describing the relative translation and rotation between link coordinate systems. For example, A_1 describes the position and orientation of the first link. A_2 describes the position and orientation of the second link with respect to the first. Thus the position and orientation of the second in the base coordinates are given by the matrix product.

$$T_2 = A_1 A_2 \tag{A-1}$$

Similarly, Ai describes the ith link in terms of the (i – 1), the position and orientation of the ith link relative to the base coordinates are given by the matrix product as follows[32] and [35]:

$$T_i = A_1 A_2 \ldots A_i = \begin{bmatrix} R_i & p_i \\ 0 & 1 \end{bmatrix} \tag{A-2}$$

where i = 1,2, … , N, Ri is the rotational matrix, pi is the positional vector, $A_i = (A_i)_{Trans}(A_i)_{RotX}(A_i)_{RotY}(A_i)_{RotZ}$,

$$(A_i)_{Trans} = \begin{bmatrix} 1 & 0 & 0 & X_i \\ 0 & 1 & 0 & Y_i \\ 0 & 0 & 1 & Z_i \\ 0 & 0 & 0 & 1 \end{bmatrix},$$

$$(A_i)_{RotX} = \begin{bmatrix} 1 & 0 & 0 & 0 \\ 0 & \cos(x_i) & -\sin(x_i) & 0 \\ 0 & \sin(x_i) & \cos(x_i) & 0 \\ 0 & 0 & 0 & 1 \end{bmatrix}$$

$$(A_i)_{RotY} = \begin{bmatrix} \cos(y_i) & 0 & \sin(y_i) & 0 \\ 0 & 1 & 0 & 0 \\ -\sin(y_i) & 0 & \cos(y_i) & 0 \\ 0 & 0 & 0 & 1 \end{bmatrix},$$

$$(A_i)_{RotZ} = \begin{bmatrix} \cos(z_i) & -\sin(z_i) & 0 & 0 \\ \sin(z_i) & \cos(z_i) & 0 & 0 \\ 0 & 0 & 1 & 0 \\ 0 & 0 & 0 & 1 \end{bmatrix},$$

Xi, Yi and Zi are the displacements along the x-, y- and z-axes between frames i andi – 1. xi, yi and zi are the angles between frames i and i – 1 rotating about x-, y- and z-axes, respectively.

Thus Ti represents the position and orientation of frame i with respect to the world coordinates, if the base coordinates are assumed to be the world coordinates, as described in [35] and [36].

The homogeneous transformation matrix Ai can be rewritten as:

$$A_i = \begin{bmatrix} \cos(y_i)\cos(z_i) & -\cos(y_i)\sin(z_i) & \sin(y_i) & X_i \\ \sin(x_i)\sin(y_i)\cos(z_i)+\cos(x_i)\sin(z_i) & \cos(x_i)\cos(z_i)-\sin(x_i)\sin(y_i)\sin(z_i) & -\sin(x_i)\cos(y_i) & Y_i \\ \sin(x_i)\sin(z_i)-\cos(x_i)\sin(y_i)\cos(z_i) & \cos(x_i)\sin(y_i)\sin(z_i)+\sin(x_i)\cos(z_i) & \cos(x_i)\cos(y_i) & Z_i \\ 0 & 0 & 0 & 1 \end{bmatrix} \equiv \begin{bmatrix} R_i^- & p_i^- \\ 0T & 1 \end{bmatrix} \tag{A-3}$$

If there are errors in the dimensional relationships between two consecutive coordinate frames, there will be a differential change dAi between the two coordinate frames, and the correct relation between the two coordinates will be equal to

$$A_i^C = A_i + dA_i \tag{A-4}$$

where Ai is the relationship between the frame coordinates i – 1 and i, and dAi is the differential change in their relationship.

The differential change dAi is estimated by the following linear form.

$$dA_i = \frac{\partial A_i}{\partial X_i}\Delta X_i + \frac{\partial A_i}{\partial Y_i}\Delta Y_i + \frac{\partial A_i}{\partial Z_i}\Delta Z_i + \frac{\partial A_i}{\partial x_i}\Delta x_i + \frac{\partial A_i}{\partial y_i}\Delta y_i + \frac{\partial A_i}{\partial z_i}\Delta z_i \tag{A-5}$$

where ΔXi, ΔYi, ΔZi, Δxi, Δyi and Δzi are small errors in the kinematic parameters.

If δAi is defined as an error matrix transform with respect to frame i − 1, the following equations can be obtained.

$$dA_i = \delta A_i * A_i \tag{A-6}$$

$$\delta A_i = D_X \Delta X_i + D_Y \Delta Y_i + D_Z \Delta Z_i + D_x \Delta x_i + D_y \Delta y_i + D_z \Delta z_i \tag{A-7}$$

The D matrices are defined as follows:

$$D_X = \left(\frac{\partial A_i}{\partial X_i}\right) Ai - 1, \quad D_Y = \left(\frac{\partial A_i}{\partial Y_i}\right) A_i^{-1}, \quad D_Z = \left(\frac{\partial A_i}{\partial Z_i}\right) A_i^{-1},$$

$$D_x = \left(\frac{\partial A_i}{\partial x_i}\right) A_i^{-1}, \quad D_y = \left(\frac{\partial A_i}{\partial y_i}\right) A_i^{-1}, \quad D_y = \left(\frac{\partial A_i}{\partial z_i}\right) A_i^{-1}$$

Then

$$D_X = \begin{bmatrix} 0 & 0 & 0 & 1 \\ 0 & 0 & 0 & 0 \\ 0 & 0 & 0 & 0 \\ 0 & 0 & 0 & 0 \end{bmatrix}, \quad D_Y = \begin{bmatrix} 0 & 0 & 0 & 0 \\ 0 & 0 & 0 & 1 \\ 0 & 0 & 0 & 0 \\ 0 & 0 & 0 & 0 \end{bmatrix},$$

$$D_Z = \begin{bmatrix} 0 & 0 & 0 & 0 \\ 0 & 0 & 0 & 0 \\ 0 & 0 & 0 & 1 \\ 0 & 0 & 0 & 0 \end{bmatrix} \tag{A-8}$$

$$D_x = \begin{bmatrix} 0 & 0 & 0 & 0 \\ 0 & 0 & -1 & Z_i \\ 0 & 1 & 0 & -Y_i \\ 0 & 0 & 0 & 0 \end{bmatrix} \tag{A-9}$$

$$D_y = \begin{bmatrix} 0 & -\sin(x_i) & \cos(x_i) & Y_i \sin(x_i) - Z_i \cos(x_i) \\ \sin(x_i) & 0 & 0 & -X_i \sin(x_i) \\ -\cos(x_i) & 0 & 0 & X_i \cos(x_i) \\ 0 & 0 & 0 & 0 \end{bmatrix} \tag{A-10}$$

$$D_z = \begin{bmatrix} 0 & -\cos(x_i)\cos(y_i) & -\sin(x_i)\cos(y_i) & Y_i \cos(x_i)\cos(y_i) + Z_i \sin(x_i)\cos(y_i) \\ \cos(x_i)\cos(y_i) & 0 & -\sin(y_i) & -X_i \cos(x_i)\cos(y_i) + Z_i \sin(y_i) \\ \sin(x_i)\cos(y_i) & \sin(y_i) & 0 & -X_i \sin(x_i)\cos(y_i) - Y_i \sin(y_i) \\ 0 & 0 & 0 & 0 \end{bmatrix} \tag{A-11}$$

Eq. (A-7) can be rewritten as

$$\delta A_i = D_X \Delta X_i + D_Y \Delta Y_i + D_Z \Delta Z_i + D_x \Delta x_i + D_y \Delta y_i + D_z \Delta z_i$$

$$= \begin{bmatrix} \delta R_i & d_i \\ 0 & 0 \end{bmatrix} = \begin{bmatrix} 0 & -\delta z_i & \delta y_i & dx_i \\ \delta z_i & 0 & -\delta x_i & dy_i \\ -\delta y_i & \delta x_i & 0 & dz_i \\ 0 & 0 & 0 & 0 \end{bmatrix} \tag{A-12}$$

From Eqs. (A-8), (A-9), (A-10), (A-11) and (A-12), the following equations can be obtained:

$$d_i = \begin{bmatrix} dx_i \\ dy_i \\ dz_i \end{bmatrix} = \begin{bmatrix} 1 \\ 0 \\ 0 \end{bmatrix} \Delta X_i + \begin{bmatrix} 0 \\ 1 \\ 0 \end{bmatrix} \Delta Y_i + \begin{bmatrix} 0 \\ 0 \\ 1 \end{bmatrix} \Delta Z_i + \begin{bmatrix} 0 \\ Z_i \\ -Y_i \end{bmatrix}$$

$$\Delta x_i + \begin{bmatrix} Y_i \sin(x_i) - Z_i \cos(x_i) \\ -X_i \sin(x_i) \\ X_i \cos(x_i) \end{bmatrix}$$

$$\Delta y_i + \begin{bmatrix} [Y_i \cos(x_i) + Z_i \sin(x_i)]\cos(y_i) \\ -X_i \cos(x_i)\cos(y_i) + Z_i \sin(y_i) \\ -X_i \sin(x_i)\cos(y_i) - Y_i \sin(y_i) \end{bmatrix} \Delta z_i \tag{A-13}$$

$$\delta i = \begin{bmatrix} \delta x_i \\ \delta y_i \\ \delta z_i \end{bmatrix} = \begin{bmatrix} 1 \\ 0 \\ 0 \end{bmatrix} \Delta x_i + \begin{bmatrix} 0 \\ \cos(x_i) \\ \sin(x_i) \end{bmatrix}$$

$$\Delta y_i + \begin{bmatrix} \sin(y_i) \\ -\sin(x_i)\cos(y_i) \\ \cos(x_i)\cos(y_i) \end{bmatrix} \Delta z_i \tag{A-14}$$

Eqs. (A-13) and (A-14) can be rewritten as Eqs. (A-15) and (A-16), respectively.

$$d_i = m_{1i} \Delta X_i + m_{2i} \Delta Y_i + m_{3i} \Delta Z_i + m_{4i} \Delta x_i + m_{5i} \Delta y_i + m_{6i} \Delta x_i \tag{A-15}$$

$$\delta_i = m_{7i} \Delta x_i + m_{8i} \Delta y_i + m_{9i} \Delta z_i \tag{A-16}$$

where

$$m_{1i} = \begin{bmatrix} 1 \\ 0 \\ 0 \end{bmatrix}, \quad m_{2i} = \begin{bmatrix} 0 \\ 1 \\ 0 \end{bmatrix}, \quad m_{3i} = \begin{bmatrix} 0 \\ 0 \\ 1 \end{bmatrix}, \quad m_{4i} = \begin{bmatrix} 0 \\ Z_i \\ -Y_i \end{bmatrix},$$

$$m_{5i} = \begin{bmatrix} Y_i \sin(x_i) - Z_i \cos(x_i) \\ -X_i \sin(x_i) \\ X_i \cos(x_i) \end{bmatrix},$$

$$m_{6i} = \begin{bmatrix} [Y_i \cos(x_i) + Z_i \sin(x_i)]\cos(y_i) \\ -X_i \cos(x_i)\cos(y_i) + Z_i \sin(y_i) \\ -X_i \sin(x_i)\cos(y_i) - Y_i \sin(y_i) \end{bmatrix}, \quad m_{7i} = \begin{bmatrix} 1 \\ 0 \\ 0 \end{bmatrix},$$

$$m_{8i} = \begin{bmatrix} 0 \\ \cos(x_i) \\ \sin(x_i) \end{bmatrix} \text{ and } m_{9i} = \begin{bmatrix} \sin(y_i) \\ -\sin(x_i)\cos(y_i) \\ \cos(x_i)\cos(y_i) \end{bmatrix}$$

A.2. Total differential translation and rotation transformation

The transformation from the end to the base frames for N frames can be expressed as:

$$T_N + dT_N = \sum_{i=1}^{N} (A_i + dA_i) \tag{A-17}$$

where dTN represents the total differential change due to the kinematic errors.

By expanding Eq. (A-17), and ignoring the higher order differential changes, the following linear result is obtained.

equation(A-18)

$$T_N + dT_N = T_N + \sum_{i=1}^{N}(A_1 \cdots A_{i-1} dA_i A_{i+1} \cdots A_N) \tag{A-18}$$

Substituting Eq. (A-6) into Eq. (A-18) gives

$$dT_N = \sum_{i=1}^{N}(A_1 \cdots A_{i-1} \delta A_i A_i A_{i+1} \cdots A_N) \tag{A-19}$$

Using the definition of Ti_{-1} in Eq. (A-2), Eq. (A-19) can be expressed as

$$dT_N = \sum_{i=1}^{N}(T_{i-1})\delta A_i T_{i-1}^{-1} T_N) \tag{A-20}$$

By defining an error matrix transform δT_N^w with respect to the world coordinates

$$dT_N = \delta T_N^w T_N \tag{A-21}$$

and

$$\delta T_N^w = \begin{bmatrix} 0 & -\delta z_N^w & \delta y_N^w & dx_N^w \\ \delta z_N^w & 0 & -\delta x_N^w & dy_N^w \\ -\delta y_N^w & \delta x_N^w & 0 & dz_N^w \\ 0 & 0 & 0 & 0 \end{bmatrix} \tag{A-22}$$

Then Eq. (A-20) can be rewritten as:

$$\delta T_N^w = \sum_{i=1}^{N}(T_{i-1}\delta A_i T_{i-1}^{-1}) \tag{A-23}$$

From Eqs. (A-3), (A-22) and (A-23), the combined vectors d_N^w and δ_N^w associated with the total transformation of N frames can be calculated as:

$$d_N^w = \sum_{i=1}^{N}[R_{i-1}d_i + p_{i-1} \times (R_{i-1}\delta_i)]$$

$$\delta_N^w = \sum_{i=1}^{N} R_{i-1}\delta_i \tag{A-24}$$

From Eq. (A-24), it is easy to see that the variation in the world coordinates system is not only related to the differential rotation δi and translation di in individual frame, but also dependent on rotational matrix (Ri_{-1}) and transitional vector (pi_{-1}) relative to the base frame. The differential translational and rotational vectors in the individual frame are converted those in the world coordinate system by Eq. (A-24), where Ri_{-1} is the 3 × 3 rotational matrix and pi_{-1} is the 3 × 1 positional vector of Ti_{-1} defined by Eq. (A-2)above and the matrix Ti_{-1} is a transformation between the (i – 1)th and the base frames. However, in Mantripragada and Whitney›s work [1] and [37], the positional vector and rotational matrix are taken to be p_i and R_i instead, respectively, which are defined by relative transition and rotation from the transformation Ai_{-1} between the (i – 2)th and the (i – 1)th frames in Eq. (A-3), thus the differential translational and rotational vectors in the individual frame cannot be transformed into those in the world coordinate system. The further variation propagation control, which is based on this conversion, cannot handle the misorientation and misposition of the components.

Eq. (A-24) can be rewritten as:

$$d_N^w = \sum_{i=1}^{N}\{[R_{i-1}m_{1i}\Delta X_i + R_{i-1}m_{2i}\Delta Y_i + R_{i-1}m_{3i}\Delta Z_i] + [R_{i-1}m_{4i} + p_{i-1} \times (R_{i-1}m_{7i})]\Delta x_i + [R_{i-1}m_{5i} + p_{i-1} \times (R_{i-1}m_{8i})]\Delta y_i + [R_{i-1}m_{6i} + p_{i-1} \times (R_{i-1}m_{9i})]\Delta z_i\} \tag{A-25}$$

$$\delta_N^w = \sum_{i=1}^{N} R_{i-1}m_{7i}\Delta x_i + \sum_{i=1}^{N} R_{i-1}m_{8i}\Delta y_i + \sum_{i=1}^{N} R_{i-1}m_{9i}\Delta z_i \tag{A-26}$$

Eqs. (A-25) and (A-26) can also be rewritten as follows:

$$d_N^w = W_1\Delta X + W_2\Delta Y + W_3\Delta Z + W_4\Delta x + W_6\Delta y + W_8\Delta z \tag{A-27}$$

$$\delta_N^w = W_5\Delta x + W_7\Delta y + W_9\Delta z \tag{A-28}$$

where $\Delta X = \begin{bmatrix}\Delta X_1 & \cdots & \Delta X_N\end{bmatrix}^T$, $\Delta Y = \begin{bmatrix}\Delta Y_1 & \cdots & \Delta Y_N\end{bmatrix}^T$, $\Delta Z = \begin{bmatrix}\Delta Z_1 & \cdots & \Delta Z_N\end{bmatrix}^T$, $\Delta x = \begin{bmatrix}\Delta x_1 & \cdots & \Delta x_N\end{bmatrix}^T$, $\Delta y = \begin{bmatrix}\Delta y_1 & \cdots & \Delta y_N\end{bmatrix}^T$ and $\Delta z = \begin{bmatrix}\Delta z_1 & \cdots & \Delta z_N\end{bmatrix}^T$.

Combining Eqs. (A-27) and (A-28) gives:

$$\begin{bmatrix} d_N^w \\ \delta_N^w \end{bmatrix} = \begin{bmatrix} W_1 \\ 0 \end{bmatrix}\Delta X + \begin{bmatrix} W_2 \\ 0 \end{bmatrix}\Delta Y + \begin{bmatrix} W_3 \\ 0 \end{bmatrix}\Delta Z + \begin{bmatrix} W_4 \\ W_5 \end{bmatrix}\Delta x + \begin{bmatrix} W_6 \\ W_7 \end{bmatrix}\Delta y + \begin{bmatrix} W_8 \\ W_9 \end{bmatrix}\Delta z \tag{A-29}$$

Eqs. (A-29) can also be expressed as:

$$\begin{bmatrix} dx_N^w \\ dy_N^w \\ dz_N^w \\ \delta x_N^w \\ \delta y_N^w \\ \delta z_N^w \end{bmatrix} = \begin{bmatrix} W_{(1,1),1} & W_{(1,1),2} & \cdots & W_{(1,1),i} & \cdots & W_{(1,1),N} \\ W_{(1,2),1} & W_{(1,2),2} & \cdots & W_{(1,2),i} & \cdots & W_{(1,2),N} \\ W_{(1,3),1} & W_{(1,3),2} & \cdots & W_{(1,3),i} & \cdots & W_{(1,3),N} \\ 0 & 0 & 0 & 0 & 0 & 0 \\ 0 & 0 & 0 & 0 & 0 & 0 \\ 0 & 0 & 0 & 0 & 0 & 0 \end{bmatrix} \begin{bmatrix} \Delta X_1 \\ \Delta X_2 \\ \vdots \\ \Delta X_i \\ \vdots \\ \Delta X_N \end{bmatrix}$$

$$+ \begin{bmatrix} W_{(2,1),1} & W_{(2,1),2} & \cdots & W_{(2,1),i} & \cdots & W_{(2,1),N} \\ W_{(2,2),1} & W_{(2,2),2} & \cdots & W_{(2,2),i} & \cdots & W_{(2,2),N} \\ W_{(2,3),1} & W_{(2,3),2} & \cdots & W_{(2,3),i} & \cdots & W_{(2,3),N} \\ 0 & 0 & 0 & 0 & 0 & 0 \\ 0 & 0 & 0 & 0 & 0 & 0 \\ 0 & 0 & 0 & 0 & 0 & 0 \end{bmatrix} \begin{bmatrix} \Delta Y_1 \\ \Delta Y_2 \\ \vdots \\ \Delta Y_i \\ \vdots \\ \Delta Y_N \end{bmatrix}$$

$$
+\begin{bmatrix} W_{(3,1),1} & W_{(3,1),2} & \cdots & W_{(3,1),i} & \cdots & W_{(3,1),N} \\ W_{(3,2),1} & W_{(3,2),2} & \cdots & W_{(3,2),i} & \cdots & W_{(3,2),N} \\ W_{(3,3),1} & W_{(3,3),2} & \cdots & W_{(3,3),i} & \cdots & W_{(3,3),N} \\ 0 & 0 & 0 & 0 & 0 & 0 \\ 0 & 0 & 0 & 0 & 0 & 0 \\ 0 & 0 & 0 & 0 & 0 & 0 \end{bmatrix} \begin{bmatrix} \Delta Z_1 \\ \Delta Z_2 \\ \vdots \\ \Delta Z_i \\ \vdots \\ \Delta Z_N \end{bmatrix}
$$

$$
+\begin{bmatrix} W_{(4,1),1} & W_{(4,1),2} & \cdots & W_{(4,1),i} & \cdots & W_{(4,1),N} \\ W_{(4,2),1} & W_{(4,2),2} & \cdots & W_{(4,2),i} & \cdots & W_{(4,2),N} \\ W_{(4,3),1} & W_{(4,3),2} & \cdots & W_{(4,3),i} & \cdots & W_{(4,3),N} \\ W_{(5,1),1} & W_{(5,1),2} & \cdots & W_{(5,1),i} & \cdots & W_{(5,1),N} \\ W_{(5,2),1} & W_{(5,2),2} & \cdots & W_{(5,2),i} & \cdots & W_{(5,2),N} \\ W_{(5,3),1} & W_{(5,3),2} & \cdots & W_{(5,3),i} & \cdots & W_{(5,3),N} \end{bmatrix} \begin{bmatrix} \Delta\theta x_1 \\ \Delta\theta x_2 \\ \vdots \\ \Delta\theta x_i \\ \vdots \\ \Delta\theta x_N \end{bmatrix}
$$

$$
+\begin{bmatrix} W_{(6,1),1} & W_{(6,1),2} & \cdots & W_{(6,1),i} & \cdots & W_{(6,1),N} \\ W_{(6,2),1} & W_{(6,2),2} & \cdots & W_{(6,2),i} & \cdots & W_{(6,2),N} \\ W_{(6,3),1} & W_{(6,3),2} & \cdots & W_{(6,3),i} & \cdots & W_{(6,3),N} \\ W_{(7,1),1} & W_{(7,1),2} & \cdots & W_{(7,1),i} & \cdots & W_{(7,1),N} \\ W_{(7,2),1} & W_{(7,2),2} & \cdots & W_{(7,2),i} & \cdots & W_{(7,2),N} \\ W_{(7,3),1} & W_{(7,3),2} & \cdots & W_{(7,3),i} & \cdots & W_{(7,3),N} \end{bmatrix} \begin{bmatrix} \Delta\theta y_1 \\ \Delta\theta y_2 \\ \vdots \\ \Delta\theta y_i \\ \vdots \\ \Delta\theta y_N \end{bmatrix}
$$

$$
+\begin{bmatrix} W_{(8,1),1} & W_{(8,1),2} & \cdots & W_{(8,1),i} & \cdots & W_{(8,1),N} \\ W_{(8,2),1} & W_{(8,2),2} & \cdots & W_{(8,2),i} & \cdots & W_{(8,2),N} \\ W_{(8,3),1} & W_{(8,3),2} & \cdots & W_{(8,3),i} & \cdots & W_{(8,3),N} \\ W_{(9,1),1} & W_{(9,1),2} & \cdots & W_{(9,1),i} & \cdots & W_{(9,1),N} \\ W_{(9,2),1} & W_{(9,2),2} & \cdots & W_{(9,2),i} & \cdots & W_{(9,2),N} \\ W_{(9,3),1} & W_{(9,3),2} & \cdots & W_{(9,3),i} & \cdots & W_{(9,3),N} \end{bmatrix} \begin{bmatrix} \Delta\theta z_1 \\ \Delta\theta z_2 \\ \vdots \\ \Delta\theta z_i \\ \vdots \\ \Delta\theta z_N \end{bmatrix} \quad \text{(A-30)}
$$

where

$$
W_{li}(3 \times 1\,vector) = \begin{bmatrix} W_{(l,1),i} \\ W_{(l,2),i} \\ W_{(l,3),i} \end{bmatrix}, \quad l = 1, 2, \ldots, 9,
$$

$W_1i = Ri_{-1}m_1i$, $W_2i = Ri_{-1}m_2i$, $W_3i = Ri_{-1}m_3i$, $W_4i = Ri_{-1}m_4i + pi_{-1} \times (Ri_{-1}m_7i)$, $W_5i = Ri_{-1}m_7i$, $W_6i = Ri_{-1}m_5i + pi_{-1} \times (Ri_{-1}m_8i)$, $W_7i = Ri_{-1}m_8i$, $W_8i = Ri_{-1}m_6i + pi_{-1} \times (Ri_{-1}m_9i)$ and $W_9i = Ri_{-1}m_9i$.

Eq. (A-30) can be expressed as state transition model as follows:

$$
\begin{bmatrix} dx_k^w \\ dy_k^w \\ dz_k^w \\ \delta x_k^w \\ \delta y_k^w \\ \delta z_k^w \end{bmatrix} = \begin{bmatrix} dx_{k-1}^w \\ dy_{k-1}^w \\ dz_{k-1}^w \\ \delta x_{k-1}^w \\ \delta y_{k-1}^w \\ \delta z_{k-1}^w \end{bmatrix}
$$

$$
+\begin{bmatrix} W_{(1,1),k} & W_{(2,1),k} & W_{(3,1),k} & W_{(4,1),k} & W_{(6,1),k} & W_{(8,1),k} \\ W_{(1,2),k} & W_{(2,2),k} & W_{(3,1),k} & W_{(4,2),k} & W_{(6,1),k} & W_{(8,2),k} \\ W_{(1,3),k} & W_{(2,3),k} & W_{(3,1),k} & W_{(4,3),k} & W_{(6,1),k} & W_{(8,3),k} \\ 0 & 0 & 0 & W_{(5,1),k} & W_{(7,1),k} & W_{(9,1),k} \\ 0 & 0 & 0 & W_{(5,2),k} & W_{(7,2),k} & W_{(9,2),k} \\ 0 & 0 & 0 & W_{(5,3),k} & W_{(7,3),k} & W_{(9,3),k} \end{bmatrix}
$$

$$
\times \begin{bmatrix} \Delta X_k \\ \Delta Y_k \\ \Delta Z_k \\ \Delta\theta x_k \\ \Delta\theta y_k \\ \Delta\theta z_k \end{bmatrix} \quad \text{(A-31)}
$$

Eq. (A-31) relates the total differential rotation vector and differential translation vector.

A.3. Cartesian position and orientation errors in the world coordinate

The correct position is:

$$
T_N^c = T_N + dT_N = (I + \delta T_N^w)T_N \quad \text{(A-32)}
$$

To separate the rotational components and translational components of the correct position, as described in [35], TN can be written in the following form

$$
T_N = \begin{bmatrix} R_N & p_N \\ 0 & 1 \end{bmatrix} \quad \text{(A-33)}
$$

and δT_N^w can be written as

$$
\delta T_N^w = \begin{bmatrix} \delta R_N^w & d_N^w \\ 0 & 0 \end{bmatrix} \quad \text{(A-34)}
$$

From Eqs. (A-32), (A-33) and (A-34), the resulting rotational components and translational components can be written as

$$
R_N^c = (I + \delta R_N^w)R_N \quad \text{(A-35)}
$$

and

$$
p_N^c = p_N + (\delta_N^w \times p_N) + d_N^w \quad \text{(A-36)}
$$

If the correct Cartesian position is expressed as

$$
p_N^c = p_N + dp \quad \text{(A-37)}
$$

Then from Eqs. (A-36) and (A-37) the first-order differential error in the Cartesian position of the end-effector as follows:

$$
dp = (\delta_N^w \times p_N) + d_N^w \quad \text{(A-38)}
$$

Thus the error of final matrix transform in the world coordinate system is:

$$
\delta T = \begin{bmatrix} 0 & -\delta z_N^w & \delta y_N^w & dp_N^X \\ \delta z_N^w & 0 & -\delta x_N^w & dp_N^Y \\ -\delta y_N^w & \delta x_N^w & 0 & dp_N^Z \\ 0 & 0 & 0 & 0 \end{bmatrix} \quad \text{(A-39)}
$$

and final matrix transform can be obtained by Eqs. (A-30), (A-38) and (A-39).

Thus Cartesian position error in the world coordinate system at Stage k is:

$$dp_k = \begin{bmatrix} dp_k^X \\ dp_k^Y \\ dp_k^Z \end{bmatrix} = \delta_k^w \times p_k + d_k^w \quad \text{(A-40)}$$

Cartesian orientation error in the world coordinate system at Stage k is:

$$\delta_k^w = \begin{bmatrix} \delta x_k^w \\ \delta y_k^w \\ \delta z_k^w \end{bmatrix} \quad \text{(A-41)}$$

REFERENCES

1. Mantripragada R, Whitney DE. Modeling and controlling variation propagation in mechanical assemblies using state transition models. IEEE Transactions on Robotics and Automation 1999;15:124–40.
2. Carnelio JA, Yim H. Identification of dimensional variation patterns on compliant assemblies. Journal of Manufacturing Systems 2006;25:65–76.
3. Huang W, Kong Z. Simulation and integration of geometric and rigid body kinematics errors for assembly variation analysis.Journal ofManufacturing Systems 2008;27:36–44.
4. Camelio JA, Jack Hu S, Ceglarek D. Impact of fixture design on sheet metal assembly variation. Journal of Manufacturing Systems 2004;23:182–93.
5. Cai W. A new tolerance modeling and analysis methodology through a twostep linearization with applications in automotive body assembly. Journal of Manufacturing Systems 2008;27:26–35.
6. Lin CY, Huang WH, Jeng MC, Doong JL. Study of an assembly tolerance allocation model based on Monte Carlo simulation. Journal of Materials Processing Technology 1997;70:9–16.
7. Thimm G, Britton G, Cheong FS. Controlling tolerance stacks for efficient manufacturing. The International Journal of Advanced Manufacturing Technology 2001;18:44–8.
8. Forouraghi B. Worst-case tolerance design and quality assurance via genetic algorithms. Journal of Optimization Theory and Applications 2002;13:251–68.
9. Dantan JY, Qureshi AJ. Worst-case and statistical tolerance analysis based on quantified constraint satisfaction problems and Monte Carlo simulation. Computer-Aided Design 2009;41:1–12.
10. Spolts MF. Design of machine elements. Englewood Cliffs, New Jersey: PrenticeHall; 1971.
11. Arron DD, Walter JM, Charles EW. Machine design theory and practices. New York: MacMillan; 1975.
12. He JR. Estimating the distributions of manufactured dimensions with the beta probability density function. International Journal of Machine Tools and Manufacture 1991;31:383–96.
13. Nigam SD, Turner JU. Review of statistical approaches to tolerance analysis. Computer-Aided Design 1995;27:6–15.
14. Mease D, Nair VN, Sudjianto A. Selective assembly in manufacturing. Technometrics 2004;46:165–75.
15. Rachuri S, Han YH, Foufou S, Feng SC, Roy U, Wang F, et al. A model for capturing product assembly information. Journal of Computing and Information Science in Engineering 2006;3:11–21.
16. Jin J, Shi J. State space modeling of sheet metal assembly for dimensional control. Journal of Manufacturing Science and Engineering 1999;121:756–62.
17. Shi J. Stream of variation modeling and analysis for multistage manufacturing processes. CRC Press, Taylor & Francis Group; 2006.
18. ShiJ, ZhouS. Quality control and improvementformultistage systems: a survey. IIE Transactions on Quality and Reliability Engineering 2009;41:744–53.
19. Liu J, Jin J, Shi J. State space modeling for 3-dimensional variation propagation in rigid-body multistage assembly processes. IEEE Transactions on Automation Science and Engineering 2010;7:274–90.
20. Ding Y, Ceglarek D, Shi J. Modeling and diagnosis of multistage manufacturing process. Part I. State space model. In: Japan/USA symposium on flexible automation. 2000.
21. Ragu K, Mohanram PV. Tolerance design of multistage radial flow submersible pumps. Mechanika 2007;63:64–70.
22. Gunston B. Rolls-Royce aero engines. England: Patrick Stephens Limited; 1989.
23. Klocke F, Veselovac D, Auerbach T, Seidner R. Intelligent assembly for aero engine components. In: Intelligent robotics and applications. Berlin/Heidelberg: Springer; 2008.
24. Mears ML. Geometry estimation and adaptive actuation for centering preprocessing and precision measurement. Dissertation, Mechanical Engineering, Georgia Institute of Technology; 2006.
25. Roy U, Li B. Representation and interpretation of geometric tolerances for polyhedral objects. I. Form tolerances. Computer-Aided Design 1998;30:151–61.
26. Roy U, Li B. Representation and interpretation of geometric tolerances for polyhedral objects. II. Size, orientation and position tolerances. Computer-Aided Design 1999;31:273–85.
27. Cai M, Yang JX, Wu ZT. Mathematical model of cylindrical form tolerance. Journal of Zhejing University Science 2004;5:890–5.
28. Whitney DE. Mechanical assemblies. Oxford: Oxford University Press; 2004.
29. Chen KZ, Feng XA, Lu QS. Intelligent location-dimensioning of cylindrical surfaces in mechanical parts. Computer-Aided Design 2002;34:185–94.
30. Whitney DE, Gilbert OL. Representation of geometric variations using matrix transforms for statistical

tolerance analysis in assemblies. In: Proceedings of the IEEE international conference on robotics and automation. 1993.

31. Whitney DE, Gilbert OL, Jastrzebski M. Representation of geometric variations using matrix transforms for statistical tolerance analysis in assemblies. Research in Engineering Design 1996;6:191–210.
32. Paul RP. Robot manipulators: mathematics, programming, and control. Cambridge: The MIT Press; 1981.
33. Pieper DL. The kinematics of manipulators under computer control. Stanford Artificial Intelligence Laboratory, Stanford University, AIM 72; 1968.
34. Denavit J, Hartenberg RS. A kinematic notation for lower-pair mechanisms based on matrices. Journal of Applied Mechanics 1955;6:215–21.
35. Veitschegger WK, Wu CH. Robot accuracy analysis based on kinematics. IEEE Journal of Robotics and Automation 1986;2:171–9.
36. Wu CH. A CAD tool for the kinematic design of robot manipulator. In: International conference on advanced automation. 1983.
37. Mantripragada R. Assembly oriented design: concepts, algorithms and computational tools. Ph.D. thesis, Mechanical Engineering, Massachusetts Institute of Technology; 1998.

CHAPTER 5 A PRODUCT-DRIVEN SYSTEM APPROACH FOR MULTILEVEL DECISIONS IN MANUFACTURING PLANNING AND CONTROL

Carlos Herrera[a], André Thomas[b] and Victor Parada[c]

[a] Departamento de Ingeniería Industrial, Universidad de Concepción, Edmundo Larenas, 215, Concepción, Chile

[b] CRAN, Université de Lorraine, Nancy, France

[c] Departamento de Informática, Universidad de Santiago de Chile, Santiago de Chile, Chile

ABSTRACT

Decision-making in Manufacturing Planning and Control Systems uses processes that consider several levels of product aggregation and different time horizons. Decisions rendered on each level do not always have similar goals. The problem is that building Intelligent Manufacturing Systems (IMS) involves coordinate decisions on different levels to achieve a common objective. There is no current research on IMS regarding coordination among decision levels in Product-driven Control Systems, hence simulations of the planning and control processes should be conducted to analyze the behavior of multilevel objectives. Accordingly, in this paper a simulation model gives account of different coordination issues between tactical and operational levels. At the tactical level, production plans are generated by an Advanced Planning and Scheduling system, whereas at the operational level, a distributed decision rule is used. The whole simulation makes decentralized decisions that are managed by production lots represented by holons. Results for our simulation indicate that coordination among active lots is capable of making effective multilevel distributed decisions when compared with conventional approaches.

INTRODUCTION

Holonic Manufacturing Systems (HMS) show promising results toward improving features such as flexibility and adaptability due to the complexity and dynamism of current manufacturing systems (Valckenaers, Brussel, Verstraete, Germain, & Hadeli, 2007; Valckenaers et al., 2006). In HMS, entities (i.e. machines, robots, AGVs, or workers) are modeled as holons, which consist of a physical component and an informationprocessing component (Pannequin, Morel, & Thomas, 2007, 2009). The majority of the manufacturing applications use different approaches to model HMS, focusing only on operational and control levels, specifically to model and implement manufacturing execution systems (MES). For instance, ExPlanTech is an agentbased technology for planning and production control (Marik, Pechoucek, Stepankova, & Lazansky, 2000; Pechoucek, Rehak, Charvat, Vlcek, & Kolar, 2007), which was developed as a multiagent system for project-based production systems. ExplanTech uses a community of autonomous agents that represent entities or information production entities in which no centralized decision mechanism is used.

Citation: Carlos Herrera, André Thomas & Victor Parada, A product-driven system approach for multilevel decisions in manufacturing planning and control, http://dx.doi.org/10.1080/21693277.2014.949895.

On the other hand, hierarchical architectures such as PABADIS Promise allows for production control based on multiple levels of automation (Wunsch & Bratukhin, 2007), so that decision centralization is avoided by locating decision levels closer to work-flow levels. Decision levels correspond to the ERP (Enterprise Resource Planning) at tactical level, MES at operational level, and control levels, and communication between ERP and MES is based on 'web services' using Agent Communication Language. While these approaches consider planning and control levels, they do not include products as central entities in the decision-making process. On the contrary, Product-driven Control Systems (PDCS) (McFarlane, Sarma, Chirn, Wong, & Ashton, 2002; Morel, Panetto, Zaremba, & Mayer, 2003; Morel, Valckenaers, Faure, Pereira, & Diedrich, 2007) transforms products into active agents in the decision-making process, in which products can also be represented as holons. Decisions made at the planning level must include medium-term horizons to prevent 'myopia,' so resources required for production (i.e. personnel, labor, raw materials, and machinery maintenance) should be planned in advance. Conversely, operational-level decisions are concerned with short-term horizons, which are inherently 'myopic,' and must respond quickly and efficiently to disturbances (i.e. production blocking, machine breakdowns, and demand changes). Thus, planning and control systems should be robust, flexible, and reactive with respect to short-, medium-, and long-term decisions.

In order to deal with these issues, architecture to model a holarchy from products and sets of products at each level has been proposed (Herrera 2011; Herrera, Belmokhtar, & Thomas, 2012). This approach allows coordination among multiple decision levels and their associated decision horizons while focusing on the work flow (products). Like other related manufacturing approaches (Tang, Gu, Wang, & Zheng, 2011), this architecture allows for recursion, that is each composition level of the holarchy organization exhibits the same structure and organization on each of its levels. In addition, recursion enables the replication of the same functions at each level by doing slight modifications to the objectives and decision methods. Accordingly, this work aims to analyze the coordination between centralized planning and decentralized control decisions in PDCS. To achieve this objective, decentralized decisions are assumed to be performed by numerous holons, which detect disturbances in planning and trigger local changes that affect central planning. To deal with this, an agent-based simulation model is proposed. At the planning level, the goal is to preserve the stability of the plans. At the operational level, the goal is to minimize makespan (maximum completion time or C_{max}) degradation by satisfying buffer-stock constraints. Our claim is that a product-driven system may become more effective than conventional production planning and control approaches to coordinate multiple objectives and tactical decision levels.

This paper is organized as follows: Section 2 describes the main components and methods of our HMS-based simulation approach, Section 3 discusses the results of the simulation experiment, Section 4 analyzes the main results, and Section 5 highlights the conclusions and further work.

MATERIALS AND METHODS

Centralized and distributed decisions

At different levels of Manufacturing Planning and Control Systems (MPCS), decisions are made by considering a rolling horizon, and the levels of the architecture are associated with different degrees of aggregation for products (i.e. product families, production orders, lots, finished products, etc.). A major challenge for these systems is to preserve the coherence of decisions among the levels. However, whenever disturbances occur, the objectives for each level are not easily achieved, and disturbances may cause major planning changes. Note that frequent changes can be the source of considerable instability. In addition, these effects often cause reduced efficiency and poor productivity. Short-term changes are more frequent and can significantly reduce system performance. Thus, MPCS should provide sufficient flexibility at the operational levels and ensure consistency with objectives at the upper levels.

Accordingly, our approach considers two decision levels: tactical and operational. At the tactical level, a decision concerns the production quantities for every item within a product family and for each period on a planning time horizon. This problem is generally associated with the Master Production Schedule and is usually represented using a lot-sizing model (Pochet & Wolsey, 2006). This model aims to minimize production costs by defining a set of parameters such as marginal costs and system capacity. During the first period of this planning horizon and once each quantity has been obtained, these quantities must be divided and sequenced to be incorporated into the manufacturing system. This problem has been called the lot-streaming problem (Sarin & Jaiprakash, 2007) whose objectives are to reduce the total production time (C_{max}). This model is usually applied to manufacturing systems that contain parallel manufacturing processes. The decision at this level is comprised of a sequence of sublots that correspond to the weekly planning, which considers the production start time and the quantity of each product to be manufactured. Since lot-streaming assumes constant production rates, various disturbances (i.e. machine blocking, machine breakdowns, and accidents) may affect the rates. Thus, changes of the parameters of the model may affect planning efficiency by reducing production capacity and increasing the gap between the planning and the launched production (system nervousness).

In our approach, products or sublots are modeled as holons which can modify their environment. Holons are assumed capable of making a single distributed decision that is stopping its production at a certain stage and heuristically reassigning the remaining quantity. Reassignment consists of assigning the quantity that has not been manufactured to another scheduled lot (one or many) which modifies the planning. Until the new assignment is completed, a part of the sublot remains in an intermediate stock (buffer). Splitting decision is then dependent on the remaining production and stock capacities and the existence of similar types of sublots that were previously planned. Once divisions have been made, a holon sublot evaluates the variation in planning by using a replanning linear-programming model. The model seeks to replace the partitioned lots and assess different choices that will minimize the increase of C_{max}, which correspond to different subsets of the same reference that will increase their size, to be placed in the queue submodule.

Study case

In order to perform our simulation, a company was selected as a case study. It manufactures turbochargers for the automotive industry and produces a maximum of 10,000 products per day with hundreds of references. The plant is divided into production cells, which encompass all stages of production that are required to produce a finished product. Some production cells are dedicated to a specific customer.

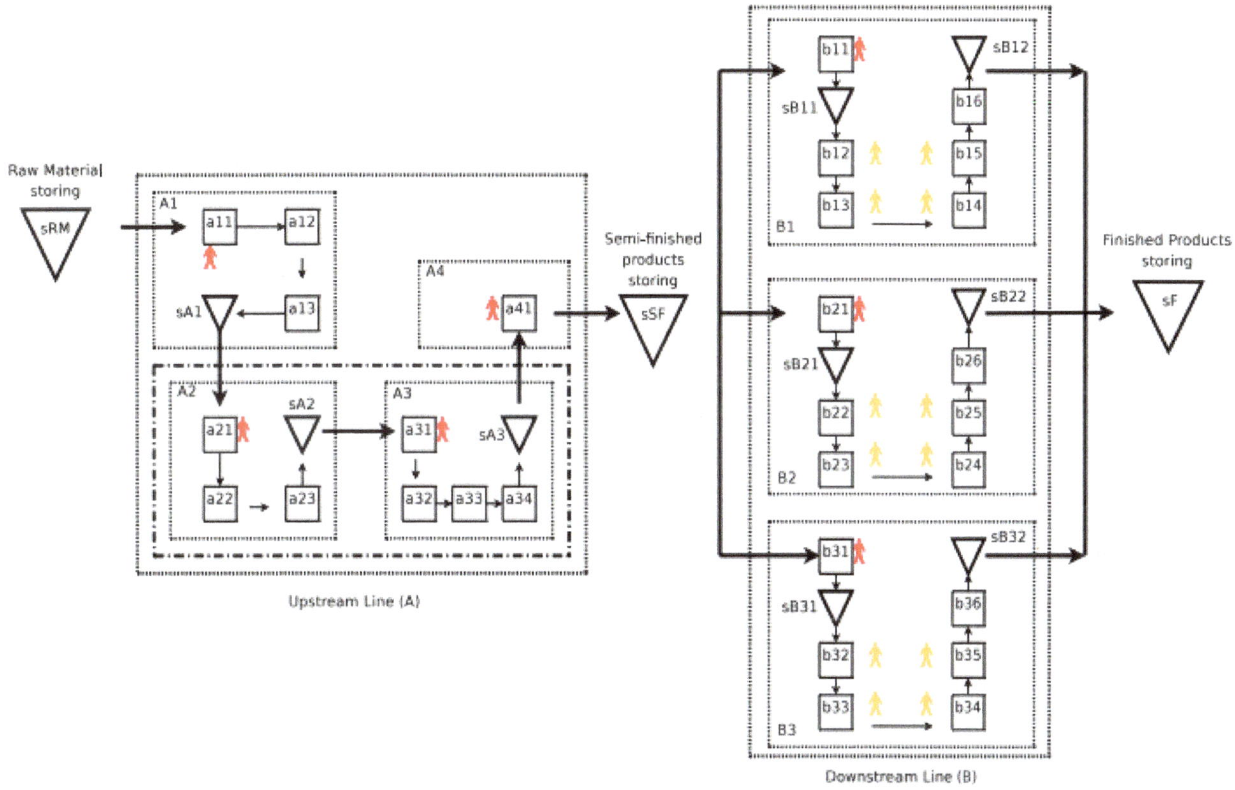

Figure 1: Case study diagram.

The production cell includes storage of raw materials, semifinished (buffer), and finished products. In addition, the cell-based production process is divided into two stages. An initial set of operations are performed in the first line (module A), generating semi- finished products. These products are then assembled into three independent assembly submodules (module B) (see Figure 1).

Indexes

$l = 1, 2, \ldots, L$	lots
$i \in \Omega_l$	sublots in lot l
$j = 1, 2, \ldots, J$	sequence positions
$k = 1, 2, \ldots, K$	cells at B

Variables

C_{max}	makespan
xb_{ijk}	replanned sublot quantity of item l in sequence position j assigned to module k of stage B
STA_j	start time at stage A of sublot in position j
STB_{jk}	start time at module k of sublot in position j

Distributed decision rule

The decision rule is based on a Linear Programming (LP) model that allows to evaluate the global effects on C_{max} with respect to the changes in the size of the sublots. Since that at each time, the holon sublot detects a change in processing times; the rule is used to approximate the overall effect of the change in the planned quantities. Rules using LP are specified as follows:

Parameters

x_{cut}	quantity to be replanned
q_i	minimum sublot size of item i
TPA_i	marginal production time at A of item i
TPB_i	marginal production time at B of item i
SA_i	setup time at A of item i
SB_i	setup time at B of item i
$I = \sum_{i=1}^{L} \lvert\Omega_l\rvert$	number of sublots
L	number of items
K	number of submodules
$n_l = \lfloor Q_l / q_l \rfloor$	maximum number of sublots in lot l
$\Omega_l = \{1, 2, \ldots, n_l\}$	set of sublots in lot l
xf_{ijk}	fixed sublot quantities
yf_{ijk}	fixed sequence
t_{new}	new start time at A for the first sublot in the planning after disturbance detection

$$(P_0)\ \min C_{max} \tag{1}$$

$$\sum_{i \in \Omega_w} \sum_{j=1}^{J} \sum_{k=1}^{K} xb_{ijk} = x_{cut} \tag{2}$$

$$xb_{ijk} = 0, i\epsilon\Omega_l, \quad \forall l : l \neq w, \forall j, \quad \forall k \tag{3}$$

$$\mathrm{STA}_1 = t_{\mathrm{new}} \tag{4}$$

$$\mathrm{STA}_j = \mathrm{STA}_{j-1} + \sum_{i\epsilon\Omega_w} \sum_{j=1}^{J} \sum_{k=1}^{K} \mathrm{TPA}_l \cdot xf_{i(j-1)k} + \mathrm{SA}_l \cdot yf_{i(j-1)k}, \quad \forall j : j > 1 \tag{5}$$

$$\mathrm{STB}_{jk} \geq \mathrm{STA}_j + \sum_{l=1}^{L} \sum_{i\epsilon\Omega_l} \mathrm{TPA}_l \cdot yf_{ijk}, \quad \forall j, \quad \forall k \tag{6}$$

$$\mathrm{STB}_{jk} \geq \mathrm{STB}_{(j-1)k} + \sum_{l=1}^{L} \sum_{i\epsilon\Omega_l} \mathrm{TPB}_l \cdot (xf_{i(j-1)k} + xb_{i(j-1)k}) + \mathrm{SB}_l \cdot yf_{i(j-1)k}, \quad \forall \ j : j > 1, \ \forall k \tag{7}$$

$$C_{\max} \geq \mathrm{STB}_{Jk} + \sum^{L} \sum \mathrm{TPB}_l \cdot (xf_{iJk} + xb_{iJk}) + \mathrm{S} \tag{8}$$

$$xb_{ijk} \in \mathbb{Z}^+; \mathrm{STA}, \mathrm{STB}, C_{\max} \geq 0 \tag{9}$$

In our decision model, the objective function (1) minimizes C_{max} represented by the end date of the last piece in the sequence. Constraint (2) ensures that the sum of the reassignments (xb_{ijk}) will be equal to the remaining quantity in the intermediate stock (xcut). Constraint (3) establishes that the reassignment can only be performed for the planned sublots that belong to the same lot that was previously divided. The start time of the sublot in position j is set to t_{new} by constraint (4). Constant t_{new} represents the new start time of the first sublot after disturbance detection. This sublot corresponds to the first sublot in the planned sequence (not yet in production). The recursive relationship in constraint (5) expresses that the start time of module A for the sublot in the jth position must be equal to the start time of the previous sublot (sublot in position j 1) plus its setup and production time, which is determined by considering a fixed sequence (xf_{ijk} and yf_{ijk}). Constraint (6) ensures that the start time of module B will always be greater than the start time of module A plus the production time for module A corresponding product. Constraint (7) considers that the production time for module B must be increased proportionally by the reassigned quantities. The makespan is defined by constraint (8).

Simulation settings

In order to simulate the distributed decision model, two decision levels have been defined for our system:

(1) Tactical level: It uses an integer programming model that defines quantities as produced by item and period in a rolling horizon (Herrera & Thomas, 2009). Quantities are divided into sublots during the first period, and the sequence to be used in the manufacturing process must be defined using an integer programming model that solves the lot-streaming problem. Unlike P_0 models, the quantities and sequences are variable, which increases the execution time. However, this is performed only at the beginning of the operation period.

(2) Operational level: During the production period (week), variations in the production times are simulated for different modules (i.e. disturbances are simulated as blocking and breakdowns). In order to react to perturbations, P_0 is solved to determine if a certain quantity of items is placed into stock so as to determine whether this decision improves the planning with respect to the initial situation. This distributed decision process depends on the variation between the planned waiting time and the real waiting time of a product in the queue of module B. Table 1 shows the main parameters considered in the simulation.

RESULTS

The simulation is performed considering a horizon of one year by obtaining weekly operational results. Some results consider the distributed decision and others disregard the distributed decision. Stability is achieved at the tactical level using a nervousness measure (Herrera & Thomas, 2009) which quantifies the variation in the planned quantities on a weekly basis. At the operational level, the obtained C_{max} and Work-In-Process (WIP) are compared.

Table 1: Simulation settings

Parameters	Values
Items	12
Planning horizon	8
Simulation horizon	52
Demand	Normal (1200, σ)
Production	Uniform ([5,15])
Stock	Uniform ([10,25])
Backlog	Uniform ([20,40])
Setup	Uniform ([1900,2200])

Nervousness

The results of the decision processes are shown in Figure 2, considering nervousness and comparing centralized and hybrid decision approaches. The cases represent situations in which the product is active (hybrid) and situations in which the product is inactive (centralized). The complete experiment is discussed in Herrera (2011). The centralized case considers a model that reduces the nervousness of the plan; thus, its shape in Figure 2 represents a 'stable plan.' These results show the difference between the launched production and the weekly planned production for a one-year operational horizon (60 periods are covered according to a transient period of 8 weeks).

In order to capture major differences between the two decision systems, Figure 3 shows the same results by applying a Savitzky–Golay filter, which preserves the features of the initial distribution and the width of the peaks. Table 2 displays the results of a statistical hypothesis test that was employed to verify if differences exist between the series. The H_0 hypothesis was described as 'significant differences exist between both cases with respect to the nervousness results,' and the H_1 hypothesis was described as 'significant differences do not exist between both cases respect to the nervousness results.' The results reveal no changes in stability for the plans in which the products are active.

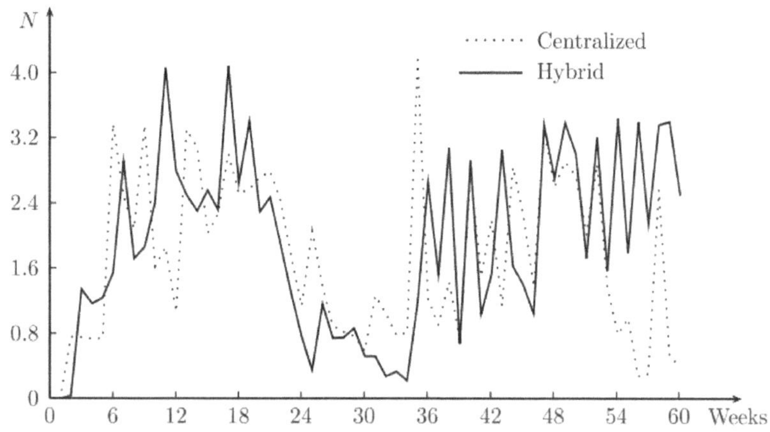

Figure 2: Nervousness comparison between centralized and distributed approaches.

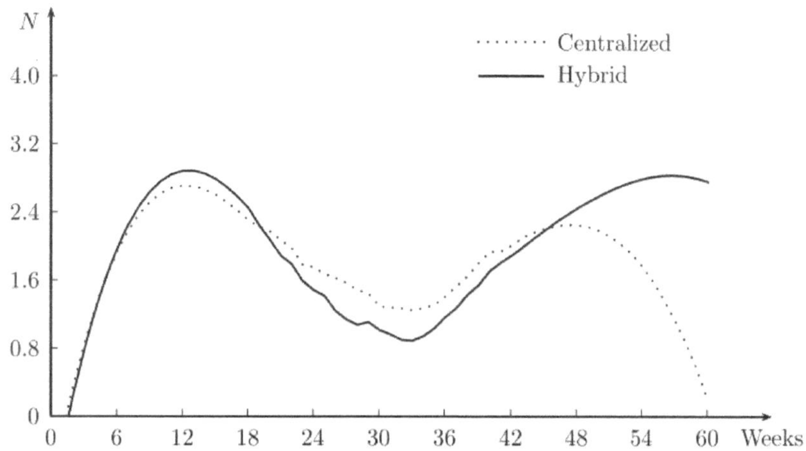

Figure 3: Nervousness comparison between centralized and distributed approaches filtered.

Table 2: ANOVA table for differences between centralized and distributed approaches considering nervousness

Source	SS	DF	MS	F	Prob. > F
Columns	1.007	1	1.0069	18.79	0.3382
Error	128.511	118	1.08907		
Total	129.518	119			

C_{max}

Figure 4 displays the results comparing C_{max} by using centralized and hybrid approaches. Figure 5 shows the same result applying a Savitzky–Golay filter. Table 3 displays the results of a statistical hypothesis test that was employed to verify if differences exist between the series. The H_0 hypothesis was described as 'significant differences exist between both cases with respect to the C_{max} results,' and the H_1 hypothesis was described as 'significant differences do not exist between both cases respect to C_{max} results.' The results reveal a statically significant difference between centralized and the proposed approach reducing C_{max} deterioration.

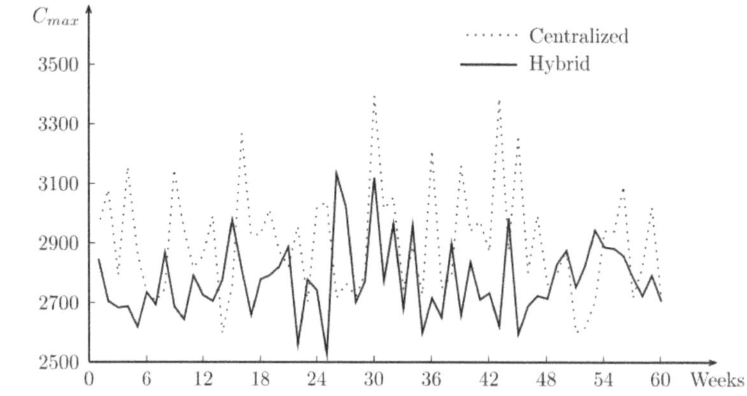

Figure 4: Makespan comparison between centralized and distributed approaches.

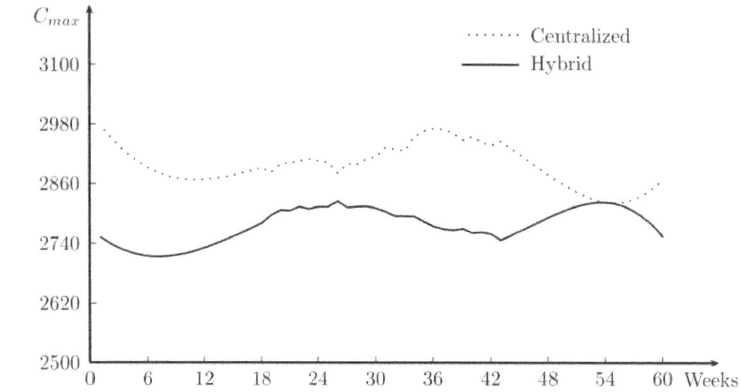

Figure 5: Makespan comparison between centralized and distributed approaches filtered.

Table 3: ANOVA table for differences between centralized and distributed approaches considering makespan

Source	SS	DF	MS	*F*	Prob. > *F*
Columns	471729.2	1	471729.2	18.79	3.09027e-5
Error	29622734.4	118	25107.9		
Total	3434463.6	119			

WIP

Figure 6 displays the results showing the WIP for both approaches. This measure represents the average stock of all references at the end of the week. Figure 7 shows the same result as that of Figure 6, after applying a filter. The H_0 hypothesis was described as 'significant differences exist between both cases with respect to the WIP results,' and the H_1 hypothesis was described as 'significant differences do not exist between both cases respect to WIP results.' These results show that the intermediate stock level is used more frequently when the products are active (see Table 4).

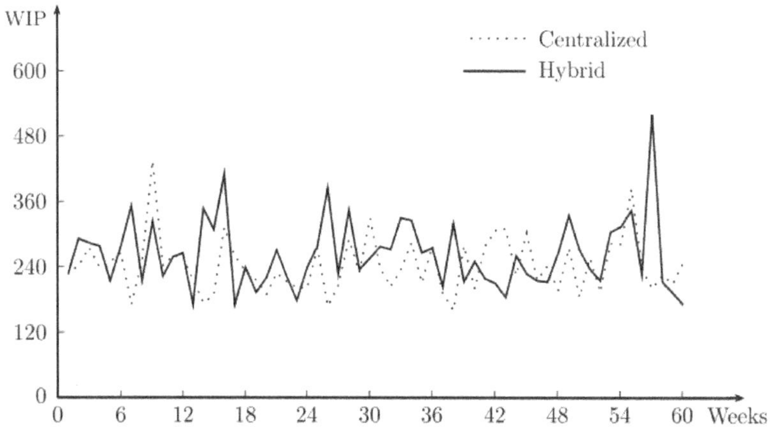

Figure 6: WIP comparison between centralized and distributed approaches.

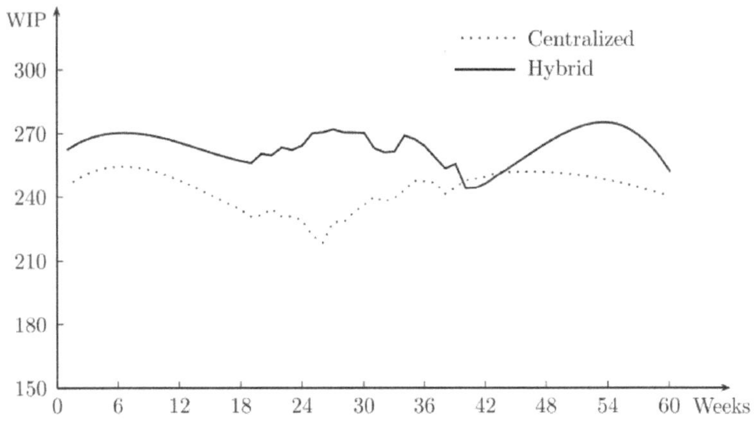

Figure 7: WIP comparison between centralized and distributed approaches filtered.

Table 4: ANOVA table for differences between centralized and distributed approaches considering WIP

Source	SS	DF	MS	*F*	Prob. > *F*
Columns	11787.8	1	11787.8	3.44	0.066
Error	403804.4	118	3422.1		
Total	415592.1	119			

CONCLUSION AND FUTURE WORK

In this paper, a multiple decision-level simulation MPCS approach is proposed which is aimed at coordinating decisions at different levels using centralized and distributed methods. Here, local decisions represent decisions made by 'holons sublots' in the context of PDCS.

Experiments show our coordination between central and local decisions for a PDCS based approach is efficient. Stable planning at the tactical level in the middle term is assessed indicating a significant performance in reactivity at the operation level in the short term (operational level). The results of C_{max} are particularly interesting as they reveal a net gain with no drop in stability, and show that 'robustness' can be achieved in different types of decisions. On the other hand, results show that for stock costs are paid for gain in nervousness and C_{max}. In addition, C_{max} efficiency is directly related to the increase and even saturation of the intermediate stock. Thus, the proposed model shows promise to enable efficient use of stock. Also, mathematical programming-based models and methods are proved to enable acceptable approximations to our optimization problem when compared with collaborative strategies.

As further work, our system should be assessed in large-production environments or in situations where centralized decisions are not feasible at all. For this, new formulations are needed to capture the global effects of local decisions in terms of computational efficiency. Furthermore, the adaptation of the proposed system should be investigated in order to be applied to Chilean industrial areas.

REFERENCES

1. Herrera, C. (2011). Cadre générique de planification logistique dans un contexte de décisions centralisées et distribuées [A generic modelling and optimisation framework for decentralised logistic decisions in product-driven manufacturing systems] (PhD thesis). Université Henri Poincaré, Nancy I.
2. Herrera, C., Belmokhtar, S., & Thomas, A. (2012). Viable system model approach for holonic product-driven manufacturing systems. In T. Borangiu, A. Thomas, & D. Trentesaux (Eds.), Service orientation in holonic and multi-agent manufacturing control, Studies in Computational Intelligence (SCI) (Vol. 402, pp. 169–181). Berlin: Springer.
3. Herrera, C., & Thomas, A. (2009). Simulation of less Master Production Schedule nervousness model. Proceedings of the 13th IFAC Symposium on Information Control Problems in Manufacturing, Moscow, Russia, pp. 1585–1590.
4. Marik, V., Pechoucek, M., Stepankova, O., & Lazansky, J. (2000). Proplant: Multiagent system for production planning. Applied Artificial Intelligence, 14, 727–762.
5. McFarlane, D., Sarma, S., Chirn, J., Wong, C., & Ashton, K. (2002). The intelligent product in manufacturing control. Journal of EAIA, 5464, 1–21.
6. Morel, G., Panetto, H., Zaremba, M., & Mayer, F. (2003). Manufacturing enterprise control and management system engineering: Paradigms and open issues. Annual Reviews in Control, 27, 199–209.
7. Morel, G., Valckenaers, P., Faure, J. M., Pereira, C. E., & Diedrich, C. (2007). Manufacturing plant control challenges and issues. Control Engineering Practice, 15, 1321–1331.
8. Pannequin, R., Morel, G., & Thomas, A. (2007). Benchmarking issues for product-driven decision-making. Proceedings of the 9th International Conference on the Modern Information Technology in the Innovation

Processes of the Industrial Enterprise MITIP'2007, Florence, Italy.

9. Pannequin, R., Morel, G., & Thomas, A. (2009). The performance of product-driven manufacturing control: An emulation-based benchmarking study. Computers in Industry, 60, 195–203.

10. Pechoucek, M., Rehak, M., Charvat, P., Vlcek, T., & Kolar, M. (2007). Agent-based approach to mass-oriented production planning: Case study. IEEE Transactions on Systems, Man, and Cybernetics, Part C, 37, 386–395.

11. Pochet, Y., & Wolsey, L. (2006). Production planning by mixed integer programming. New York, NY: Springer.

12. Sarin, S., & Jaiprakash, P. (2007). Flow shop lot streaming. New York, NY: Springer.

13. Tang, D., Gu, W., Wang, L., & Zheng, K. (2011). A neuroendocrine-inspired approach for adaptive manufacturing system control. International Journal of Production Research, 49, 1255–1268. doi:10.1080/00207543.2010.518734.

14. Valckenaers, P., Brussel, H. V., Verstraete, P., Germain, B. S., & Hadeli. (2007). Schedule execution in autonomic manufacturing execution systems. Journal of Manufacturing Systems, 26, 75–84.

15. Valckenaers, P., Cavalieri, S., Germain, B., Verstraete, P., Hadeli, Bandinelli, R., ... Brussel, H. (2006). A benchmarking service for the manufacturing control research community. Journal of Intelligent Manufacturing, 17, 667–679.

16. Wunsch, D., & Bratukhin, A. (2007). Multilevel order decomposition in distributed production. In IEEE Conference on Emerging Technologies and Factory Automation (ETFA) (pp. 872–879). IEEE, Patras, Greece.

CHAPTER 6 END MILL TOOLS INTEGRATION IN CNC MACHINING FOR RAPID MANUFACTURING PROCESSES: SIMULATION STUDIES

Muhammed Nafis Osman Zahid[a], Keith Case[a] & Darren Watts[a]

[a]Mechanical and Manufacturing Engineering, Loughborough University, Loughborough, Leicestershire LE11 3TU, UK

ABSTRACT

Computer numerical controlled (CNC) machining has been recognized as a manufacturing process that is capable of producing metal parts with high precision and reliable quality, whereas many additive manufacturing methods are less capable in these respects. The introduction of a new layer-removal methodology that utilizes an indexing device to clamp the workpiece can be used to extend CNC applications into the realm of rapid manufacturing (CNC-RM) processes. This study aims to improve the implementation of CNC machining for RM by formulating a distinct approach to integrate end mill tools during finishing processes. A main objective is to enhance process efficiency by minimizing the staircasing effect of layer removal so as to improve the quality of machined parts. In order to achieve this, different types of end mill tools are introduced to cater for specific part surfaces during finishing operations. Virtual machining simulations are executed to verify the method and the implications. The findings indicate the advantages of the approach in terms of cutting time and excess volume left on the parts. It is shown that using different tools for finishing operations will improve the capabilities of CNC machining for rapid manufacturing applications.

INTRODUCTION

In machining processes, it is crucial to ensure good surface finish quality as this will influence the performance of machined parts and production costs (Davim, 2001). Improving surface finish has become a major concern particularly in rapid manufacturing processes where the objective is to achieve final geometry directly. A major consideration in rapid manufacturing processes is that planning time needs to be substantially reduced so that overall time from design to finished part is also reduced. In additive processes (AM), relatively long fabrication times are compensated by the limited planning time required. In traditional computer numerical controlled (CNC) the opposite situation exists, and whilst machining can be very rapid, the planning process may be complex and time-consuming. In trying to use CNC for rapid applications, the main effort is in maintaining a simple approach to planning whilst possibly sacrificing some of the available machining efficiency.

An approach that has some of the advantages of additive manufacturing involves the use of an indexable fourth axis device to hold and clamp the workpiece onto the table of a three-axis CNC milling machine (Figure 1). This approach allows a

Citation: Muhammed Nafis Osman Zahid, Keith Case & Darren Watts, End mill tools integration in CNC machining for rapid manufacturing processes: simulation studies, DOI:10.1080/21693277.2015.1094756.

layer-removal process to take place at various rotations about one axis and complex shapes and features can be created by layer-based removal of material (Frank, Wysk, & Joshi, 2006). The indexable device permits rotation about one axis while also clamping the workpiece. Layers are removed from several orientations to reveal all surfaces of the part without refixturing (Frank et al., 2006). Determination of the necessary orientations is central to the process and this is achieved by visibility algorithms (a minimum set of orientations are sought where the surfaces are 'visible' from the direction of the tool).

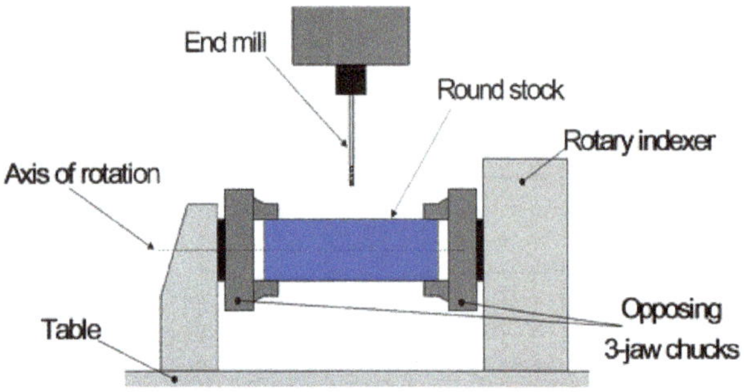

Figure 1: Setup for CNC-RP (Frank, Wysk, & Joshi, 2004).

Within each orientation, roughing and finishing operations are performed sequentially using different tool sizes and machining parameters. Machining starts with the first roughing operation that cuts until the furthest possible surface is reached or the workpiece is fully cut (Frank, 2007). Then, finishing operations are executed no further than the central axis of the cylindrical workpiece. In order to reduce effort in the planning tasks, feature recognition on machined parts is avoided. Therefore, the selection of a cutting tool becomes straightforward and is based on the smallest diameter and a flat end mill tool is most likely to be selected.

Despite the effectiveness of reducing complexities in the planning phase, the use of a single end mill tool in finishing processes leads to several drawbacks. First, due to the geometry of flat end mill cutters, a staircasing effect is visible on part surfaces particularly on contours and non-flat regions. This effect still can be seen even if very small cutting layer depths are adopted (Frank, Joshi, & Wysk, 2002). In addition, using small cutting depths will obviously increase cutting times considerably. Another limitation can be seen in the assessment of tool accessibility of part features. Analysis that focuses on machinability has resulted in some uncut material being left on the part due to inaccessibility of the cutting tool (Li & Frank, 2006) and inevitably this will affect dimensional accuracy. Figure 2 illustrates potential non-machined regions generated on a toy jack part. To date, no specific method has been developed to cater for the integration of different end mills in rapid machining processes. Therefore, this study attempts to show that using different end mill tools in finishing operations can improve surface finish by reducing the excess volume left on the parts. The next section of the paper will review the previous approach in cutting tools selection for CNC-RM processes. A methodology to integrate end mill tools is then described considering the implications for process planning and cutting operations. A framework for the simulation studies is established that is later used to generate the results to validate the proposed method. The findings are described further to highlight the achievements and limitations of the approach before drawing conclusions.

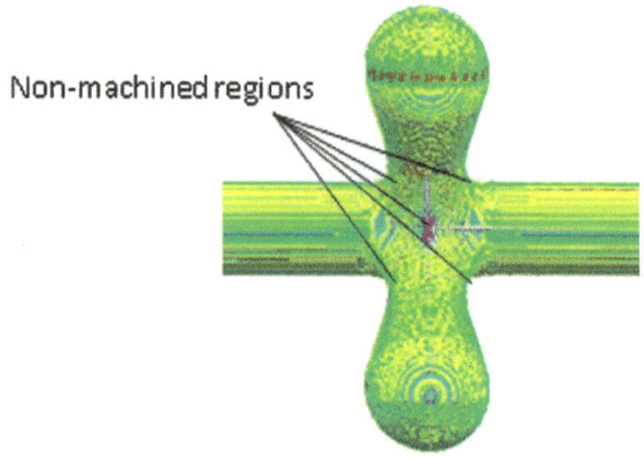

Figure 2: Possible non-machined regions using a flat end mill tool (Li & Frank, 2006).

CUTTING TOOLS AND MACHINING OPERATIONS

Initially, due to the objective of simplifying process planning, only a single cutting tool with the smallest possible diameter is used in CNC machining for rapid prototyping processes (Frank, 2003). Since the visibility analysis is based on 2D cross-sectional layers of part geometries, a flat end mill tool is usually selected to finish cut the part. Consequently, the machined parts exhibit the same staircase effect as presented in most of the AM processes. Figure 3 shows the staircase effect formed on contoured surfaces of a part. However, CNC machines are capable of reducing the layer thickness down to 0.0127 mm. Therefore, the step appearance is minimized within an acceptable range for making prototypes. From a planning perspective, the use of a single cutting tool has generalized the cutting areas to cover all surfaces in each orientation. This avoids any feature recognition tasks during machining process. Only one cutting operation is performed in each orientation that considerably reduces the number of operations and planning complexities. However, a small cutting tool requires more passes, which will result in longer machining times. Besides, the removal of thin layers is inefficient especially when the bulk of the material needs to be removed during early phases of machining. These considerations tend to restrict the capabilities of CNC machining even if some level of automation is embedded in the process planning.

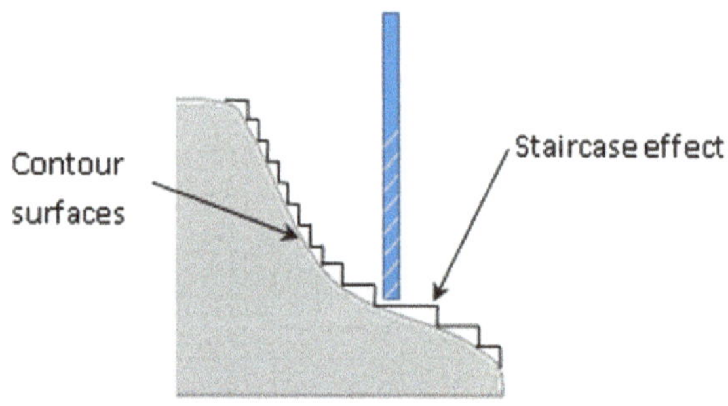

Figure 3: Staircase effect presented on contoured surfaces.

Due to the inefficiency caused by a single cutting tool, later developments suggest the use of different tool sizes to handle roughing and finishing operations (Frank, 2007). As a result, roughing operations utilize larger tool diameters selected based on the size of the workpiece. Hence, high volumes of material can be removed during roughing operations whilst leaving sufficient material for finishing processes. Meanwhile, the approach of selecting the smallest tool diameter remains the same for executing finishing operations. This allows the cutting tool to reach all features presented on the part. Both roughing and finishing operations employ the same cutting areas. Thus, within one cutting orientation, two different toolpaths are developed and this directly increases the number of machining operations. Additionally, tool changes are required in each orientation and can be carried out through automated systems built into the CNC machine. This combination of different tool sizes based on roughing and finishing operations succeeds in enhancing the application of CNC machines in producing parts rapidly.

The introduction of different cutting tool sizes has resulted in a further study that attempts to assist in the selection of a proper combination of end mill tool sizes for roughing and finishing operations. Different tool sizes possess different abilities in removing the material and shaping the part. Large tools are capable of removing more material in shorter time, but are restricted in accessing small cutting regions. To the contrary, small tools have much improved accessibility but very inferior rate of material removal. Therefore, an appropriate combination of tool sizes is necessary to machine the workpiece effectively. The region that is accessible to different tool sizes can be calculated through the Tool Access Volume (TAV) method. Then, the analysis is extended to identify appropriate tool selection and sequencing by using a method known as Relative Delta Volume Clearance (RDVC) (Lim, Corney, Ritchie, & Clark, 2001). Based on information gained from these analyses, tool sizes and a set of cutting orientations can be identified by applying a genetic algorithm technique (Renner,2008). Accordingly, the machining times are substantially reduced by using carefully selected combinations of tool sizes.

A recent study has proposed another approach to further improve the machining operations in CNC-RM (Osman Zahid, Case, & Watts, 2014). It was successfully carried out by manipulating the orientations for roughing operations. Instead of performing roughing and finishing operations using the same set of orientations, rough cuts were extracted and performed through another set of orientations. This set of orientations was based on a combination of four cutting orientations, 0–90–190–270°. A customized programme was developed to simulate the cutting process and propose efficient cutting directions. This method reduced the required cutting tool depth during roughing operations for avoidance of tool breakage. Since roughing and finishing operations are based on different sets of orientations, tool changes occur only once throughout the process. The time spent on roughing operations frequently increases, but it was found that this resulted in a reduction of total machining time.

Over the years, much research has been aimed at reducing machining time by manipulating cutting parameters as a part of process planning (Bouzid, 2005; Lavernhe, Tournier, & Lartigue, 2008; Palanisamy, Rajendran, & Shanmugasundaram, 2007). A good surface finish with low machining time can be obtained through the proper selection of machining parameters (Bharathi Raja & Baskar, 2012). The prominent parameters to be controlled are cutting speed, feed rate and depth of cut. Based on a multiple regression model, a previous study indicated that the depth of cut was the most influential parameter in determining the level of surface roughness achieved (Agarwal, 2012). Therefore, it is undeniable that cutting parameters are among the major factors that influence surface quality and machining time. However, in CNC-RM applications, the aim is to rapidly generate machining operations that reflect the planning and execution stages. Optimizing cutting parameters would require the handling of a large number of variables, and this tends to complicate the planning tasks and limits the level of automation that can be adopted in the planning stage. Therefore, the study conducted here integrates different end mill tools and observes the implications for machining time and surface quality. However, efficient cutting parameters suggested from previous studies could still be incorporated during the planning stage. Based on specific materials and tool sizes, a set of efficient parameters could be incorporated using decision-making tools in the development of machining plans.

Generally, the developments in selection of cutting tools and machining operations have enhanced the ability of CNC machining for rapid processes. The time required to machine parts has been substantially reduced with the introduction of roughing and finishing operations, appropriate tool size combinations and different rough cut sets of orientations. All these have enhanced the application of end mill tools in CNC machining, particularly for rapid manufacturing. However, far too little attention has been paid to the geometries of the end mill tool while developing the machining operations. Hence, there are no clear guidelines on how to combine the different tool geometries in the processing steps of CNC-RM. Nevertheless, several studies have strongly addressed the significance of cutting tool geometries on the level of quality achieved in the machining process (Patel, 2010; Reddy & Rao, 2005). Even if cutting can proceed at a very small depths of cut, neglecting the tool geometries tends to constrain the abilities of CNC machining in producing high quality parts. Therefore, considering the rapid manufacturing application, integrating different types of end mill tools is thought to be an approach to fulfil the quality requirements of end-use products.

METHODOLOGY

In attempts to implement CNC machining for rapid manufacturing processes, finishing tool selection is strictly defined based on the smallest diameter single end mill tool. This approach minimizes the tasks required in the planning phase and realizes the rapid generation of machining operations. Nevertheless, there is an implication for part quality as a single end mill tool is not capable of catering for a wide range of part surfaces. Therefore, integrating different end mill tools is a possible way of resolving the problem, but the guidelines must be clearly defined so that process planning can be handled effectively. Several important tasks are identified including separation of cutting regions and the types of end mill tool selected.

Cutting tools and surface classification

In the CNC-RM approach, finishing operations are executed through predetermined orientations and aim to shape the material completely. Hence, this process cuts the material with very small depths of cut to obtain a fine surface finish. Previous developments have implemented different cutting tools to finish cut parts built by welding process through a layer deposition method (Akula & Karunakaran, 2006). However, this approach utilized various type of tools and several surface categories, and seems impractical as it tends to complicate the planning tasks. However, some modifications to minimize and eliminate irrelevant setups will help to adopt this method in CNC-RM processes.

From one particular cutting direction, surfaces of the part can be classified into flat and non-flat surfaces. A flat surface is defined as a surface that is perpendicular to the cutting tool direction whereas the rest of the surfaces are categorized as non-flat surfaces. These classifications allow specific cutting tools to be assigned to machine the part effectively. Therefore, one orientation will perform up to two cutting operations if both types of surface are present. The first areas that will be cut are the flat surfaces using a flat end mill cutter. The process continues by machining non-flat surfaces using a ball nose end mill. Based on Figure 4, assuming cutting starts at 0° cutting orientation, the tool is engaged with the workpiece from ZC direction. In this particular direction, the dark areas represent flat surfaces that are perpendicular to the cutting tool direction. The first finishing operation is performed on this surface using a flat end mill tool. Then, the CNC machine replaces the current tool with a similar-sized ball nose end mill. This tool is used to machine the other surfaces shown as light grey areas on the part.

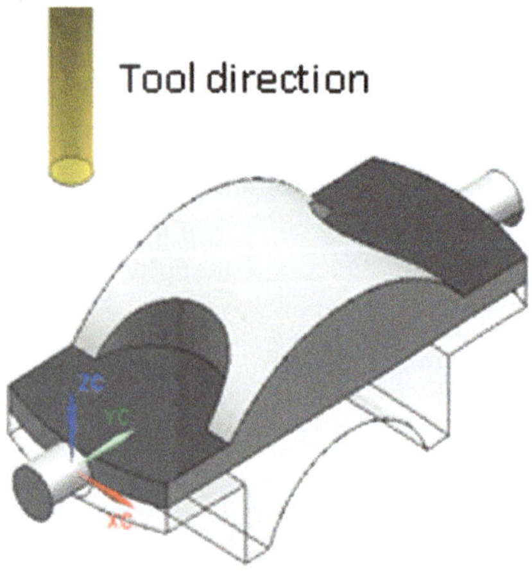

Figure 4: Classification of flat and non-flat surfaces in one cutting direction.

Generally, there are three common end mill tools (Figure 5) that differ based on the shapes of the tool tip. These are flat, bull nose and ball nose end mills. In this study, only two tools are selected to be utilized in finishing operations. This selection is expected to greatly reduce the complications for process planning as only two variables are present instead of three. A flat end mill is capable of machining plane surfaces with a minimum scallop size compared to other tools (Elber, 1995; Ryu, Choi, & Chu, 2006). Therefore, this is exactly the right tool to cut the flat surfaces classified on parts. The second selection utilizes a ball nose tool to cater for non-flat surfaces. Based on previous studies, the capabilities of this tool in handling sculptured surfaces are proven and high-quality surface finish can be produced (Chen, Huang, & Chen, 2005; Chen & Shi, 2008; Vijayaraghavan, Hoover, Hartnett, & Dornfeld, 2008). Moreover, this tool can be easily guided to engage with part surfaces and produces simpler NC code for machining (Chen & Shi, 2008). Meanwhile, a bull nose end mill is capable of removing high volumes of material with efficient machining times (Patel, 2010). But, due to the tool tip geometry, there is limited access to certain part features and this can result in high volumes of excess material. This limitation is shown in Figure 6. Therefore, in this study, only flat and ball nose end mills are selected for cutting tool integration in the CNC-RM processes.

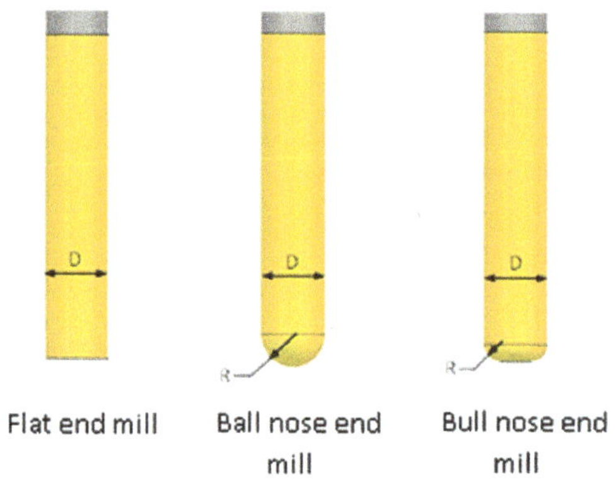

Figure 5: Three common shapes of end mill tool (Engin & Altintas, 2001).

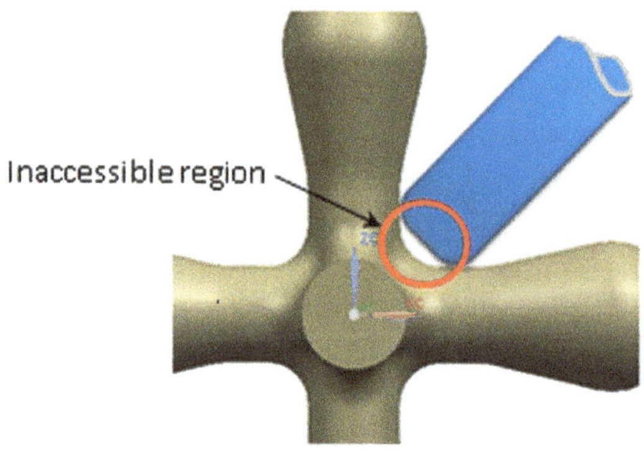

Figure 6: Restricted accessible regions for bull nose end mill tool.

Machining and verification processes

The introduction of different end mill tools requires a revision of the machining parameters, especially the cutting levels. Similar to roughing operations, finishing operations continue until the central axis of the cylindrical workpiece is reached. However, this is only applicable if a flat end mill tool is used in the operation. In the case where both tools are used, the ball nose cutter requires extended cutting levels to completely remove the remaining material. This is critically important for a part that contains closed regions that only permit tool access in limited directions. Figure 7 illustrates the situation where there is excess material due to there being insufficient cutting levels. Therefore, the cutting levels for a ball nose tool are extended by at least half the cutting tool diameter from the centre of the workpiece. Other parameters such as cutting feed and spindle speed are dependent on the diameter of the end mill tool. The finishing operations in this study employ the same tool diameter regardless of the type of tool used. Hence, both share the same value of spindle speed and feed rate. In this study, spindle speed and feed rates were determined based on the size of the cutting tools. The parameters are proposed by the CAD/CAM software itself and are not necessarily optimum cutting parameters. Depending on the models, depths of cut ranged between 1 and 1.5 mm for rough machining and 0.1 and 0.3 mm for finishing operations.

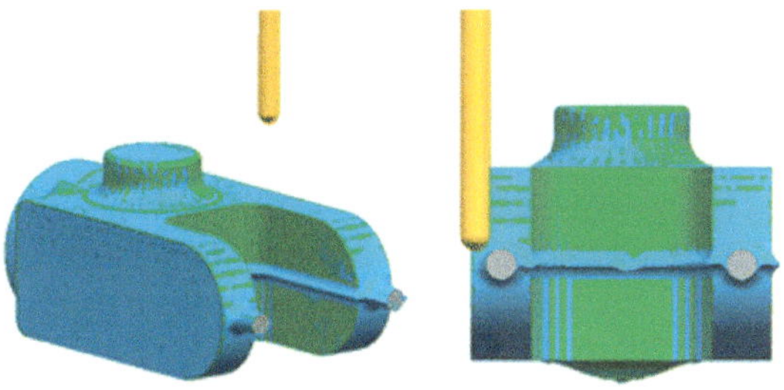

Figure 7: Remaining uncut materials left by ball nose end mill tool.

Table 1: Results for drive shaft and knob model

Approach / Criteria		Drive shaft		Knob	
		Approach 1	Approach 2	Approach 1	Approach2
Machining time	h:min:s	04:13:20	04:02:35	03:50:09	03:52:40
Finishing time		02:38:30	02:28:39	02:44:49	02:27:16
Non-cutting time		00:16:58	00:16:08	00:09:11	00:10:20
Roughing time		01:17:52	01:17:48	01:15:04	01:15:04
Cutting orientations		90–180–280–0–32–180–0°	90–180–280–0–32–180–0°	90–180–280–0–180–45–315°	90–180–280–0–180–45–315°
Machined volume	mm^3	150,273.23	150,432.83	112,946.97	113,203.60
Current part volume		53,244.05	53,084.45	21,660.63	21,403.91
Excess volume		514.82	355.22	527.99	271.27

The simulation analysis was carried out with NX 7.5 software using the manufacturing application. A series of machining operations were created and the software used to estimate machining time to fabricate the parts. In order to calculate the excess volume left, the analysis was extended by using CGTech VERICUT ® 7.2.3 (CGTech, 2012). Recently, this software has been interfaced to NX which provides convenient access to run the simulation once operations have been created. There are several pre-processing steps while calculating the excess volume. First, the machining operations are translated into a cutter location source file (.clsf). Then, the file is run on VERICUT® software to simulate the cutting operations. Once completed, the 'X-caliper' tool is activated to obtain volumetric information. This includes stock volume, machined volume and current part volume. Prior to that, the machined part volume is calculated by a 'measure bodies' tool available in the NX system. All this information is gathered and the excess volume is calculated by subtracting the current part volume from the machined part volume. Finally, the results obtained are used to evaluate the effectiveness of integrating end mill tools in the machining processes.

RESULTS AND DISCUSSION

The simulation studies were conducted using different models including drive shaft, knob, salt bottle and toy jack models. Figure 8 illustrates the models adopted for the simulation studies. To visualize the implications of the suggested approach, similar cutting operations relying totally on the use of a flat end mill tool were developed. Therefore, Approach 1 was based on a process that utilized a single flat end mill whereas Approach 2 represented the method developed in this study that integrates different cutting tools. In this study, machining parameters were determined based on the size of the cutting tools. Since there are two types of operations (roughing and finishing), two sets of cutting parameters were involved. Meanwhile in the finishing operations, two different tools are introduced (flat and ball nose end mill). However, the machining parameters are similar since only one size of cutting tool is used. It is important to note that all the machining parameters are constant between Approach 1 and 2. Thus, the improvements gained from this study totally relied on the implication of different end mills during finishing operations. The study is not meant to validate the optimum cutting parameters, but the values could be adopted in the planning stage.

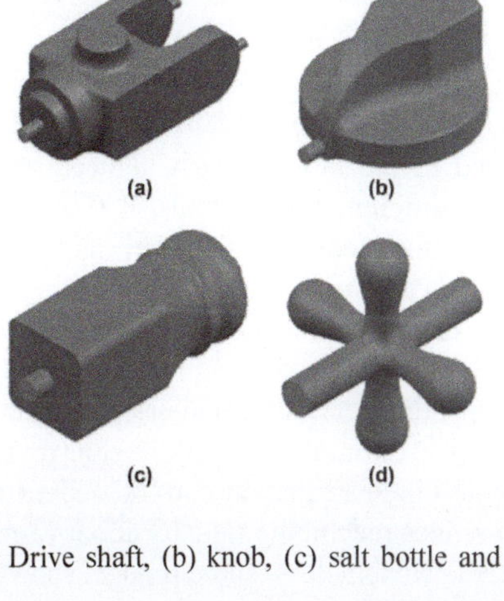

Figure 8: (a) Drive shaft, (b) knob, (c) salt bottle and (d) toy jack model.

There are two main evaluation criteria that are based on machining time and excess volume. Tables 1 and 2 summarize the results produced from simulation studies. Machining time is recorded in (h:min:s) format and volume information is in mm^3. The amount of material removed from the workpiece is presented as machined volume in the table. Part and stock volumes are calculated earlier based on the virtual models. It is apparent from these tables that there are considerable differences between Approaches 1 and 2. Out of the four models, three indicate a reduction of cutting time by implementing the approach developed in this study. Interestingly, all models achieved low excess volumes once different cutting tools were used in finishing operations. Assigning specific tools to particular surfaces, the ranges of excess volume reduce from 0.9–3.3% down to 0.7–1.6% of the total part volume.

Table 2: Results for salt bottle and toy jack model

Approach Criteria		Salt bottle		Toy jack	
		Approach 1	Approach 2	Approach 1	Approach 2
Machining time	h:min:s	04:32:02	04:23:26	03:25:22	03:21:42
Finishing time		02:38:44	02:29:33	02:16:04	02:12:44
Non-cutting time		00:11:57	00:12:28	00:14:06	00:13:44
Roughing time		01:41:21	01:41:26	00:55:12	00:55:12
Cutting orientations		0–90–190–270–45–180–270°	0–90–190–270–45–180–270°	50–140–240–320–49–140–228–320°	50–140–240–320–49–140–228–320°
Machined volume	mm^3	76,676.80	76,845.85	109,662.15	109,792.30
Current part volume		34,499.39	34,330.33	7769.86	7639.71
Excess volume		417.56	248.5	252.74	122.59

Table 3: Excess materials distribution based on studied models

Drive shaft		Knob	
Approach 1	Approach 2	Approach 1	Approach 2
514.82 mm^3	355.22 mm^3	527.99 mm^3	271.27 mm^3
	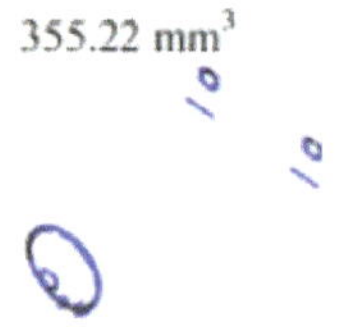		
Salt bottle Approach 1 417.56 mm^3	Approach 2 248.5 mm^3	Toy jack Approach 1 252.74 mm^3	Approach 2 122.59 mm^3
	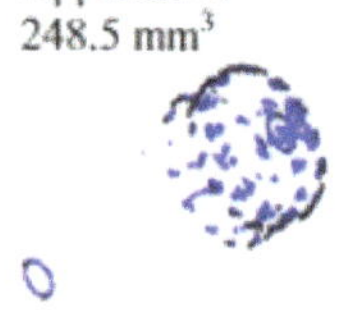	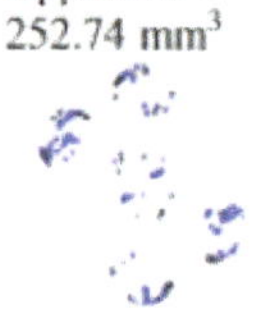	

Implications for machining time

Data gathered in the tables clearly indicate the benefits of implementing different cutting tools in CNC-RM processes. The majority of the models studied show a decreasing trend of machining time compared to a single tool approach. Finishing cutting times contribute most of the savings. This signals the effectiveness of cutting tools removing material on assigned surfaces. Depending on part geometries, cutting times differ from 2 min up to 11 min. For example, results for the drive shaft model in Table 1 indicate that the current method in Approach 2 manages to reduce machining time by about 11 min compared to Approach 1. Nonetheless, some models show only a small reduction of machining time such as the toy jack model. In this model, only ball nose tools were used as only non-flat surfaces were present on the part. For a medium production run, this reduction might still be important in minimizing cost and time. Oppositely, the knob model indicated a slightly longer machining time compared to Approach 1. The variety of flat and non-flat surfaces present on the part are possibly the main factors influencing cutting time.

The simulation studies adopted different sets of orientations for roughing and finishing operations as proposed from previous research (Osman Zahid et al., 2014). Therefore, integrating different end mill tools did not add to the number of orientations, but it increased the number of operations and tool changes. The presence of flat and non-flat surfaces in one cutting orientation will require both end mill tools and one tool change operation. Taking the salt bottle as example, there are two orientations consisting of flat and non-flat surfaces. Hence, four finishing operations are required within the two cutting directions. Meanwhile, in a single tool approach (Approach 1), only one finishing operation is needed in each orientation and this is quite straightforward compared to Approach 2. However, considering the effectiveness of material removal, it is worthwhile integrating the cutting tools.

Volume of excess materials

Previous studies have validated the capabilities of ball nose cutters on sculptured surfaces. Hence, assigning the tools based on flat and non-flat surfaces manages to minimize the excess volume on parts. Surprisingly, simulation results consistently indicate that excess volume was reduced for all models studied. Some models show a considerable reduction. For instance, excess volume for the knob model decreases dramatically from 527 mm^3 down to 271 mm^3. In CNC machining, uncut materials can be reduced by modifying cutting parameters in finishing operations including depth of cut, number of passes, feed rates and speed. However, in the CNC-RM approach, all these parameters remain constant except for the type of cutting tools used. Therefore, the achievements of this study show that the cutting tool combination is the main factor that influences the results.

Furthermore, the implications can be visualized by analysing the material distribution diagram as shown in Table 3. Comparing Approach 1 and 2, it can be seen that most of the excess material on non-flat surfaces of the part have been reduced. This indicates that the ball nose cutter effectively minimizes the staircasing effect. Keeping the flat end mill cutter to machine flat surfaces is useful in overcoming the weakness of ball nose tools.

LIMITATIONS AND FUTURE WORK

Using multiple end mill tools in finishing operations manages to greatly reduce the excess volume left on the machined parts. The results of simulation studies indirectly portray the benefits of the approach in terms of surface quality. But, observation of the cutting simulations has revealed a limitation in shaping the sacrificial support of the part. If the support structure is connected to a flat vertical surface, a ball nose end mill faces a difficulty in completely shaping the structure. An excess volume (fillet shape) is created on the edge of the support and part. This problem is shown in Figure 9. On the other hand, a flat end mill is capable of forming the sharp edge, but is limited to areas that are perpendicular to the cutter direction. Excess volume will still be present in the rest of the connection area. Currently, the only way to minimize the problem is by using the smallest possible end mill tool during finishing processes. Larger tools will increase the excess volume and cause more work at the post-processing stage.

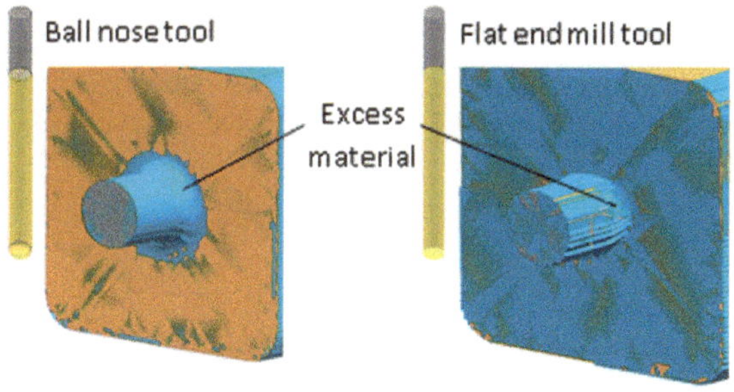

Figure 9: Excess material formed at the edge of sacrificial support.

In real applications, several variations can be expected on the method proposed in this study. First, the rate of material removal is highly dependent on the size of cutting tools adopted in roughing and finishing processes. The implications can be seen based on accessibility and machining time. Selection of the smallest cutting tool in finishing operations will guarantee the accessibility, but machining time will probably be extended. Using optimization tools like TAV and RDVC that have been discussed earlier could be helpful in reducing the variations. Another factor that will cause variation between simulation and real machining is process interventions. In the simulation study, the estimated cutting time represents the complete operation including cutting and non-cutting time, but times for tool changes and workpiece rotations were not considered. If both are fully automated, it would have less impact on the cutting time. However, if the operations are carried out manually, which requires the machine to stop, then the estimated time is likely to be extended. For this reason, it is highly recommended that CNC-RM operations are executed in an automated manner.

The integration of cutting tools has benefited the production stage of the CNC-RM, but, it affects the planning stage as more variables need to be handled. Previously, process planning in CNC machining restricted implementation in RM as it involves many manual tasks and is highly dependent on human skill (Anderberg, Beno, & Pejryd, 2009; Relvas & Simoes, 2004). However, a recent development in CAD/CAM has introduced several features that can be used to effectively handle the process planning tasks (Agrawal, Soni, & Dwivedi, 2013). Therefore, an assisting tool can be developed to integrate the cutters while constructing the machining operation in the system and make it adaptable to any kind of part features. On the other hand, the identification of flat and non-flat surfaces within the cutting orientations needs to be performed correctly to ensure the right end mill tools are selected. Flat surfaces are not only the horizontal and vertical surfaces of the CAD model, but, depending on cutting direction, inclined surfaces can also be considered as flat surfaces if they are perpendicular to the direction of tool movement. Considering these issues, the medium of communication between the user and the system must be preserved to assist decision-making during process planning. To realize this, customized programming tools and computerized systems can be implemented in the CNC-RM process plan.

CONCLUSION

This simulation study has revealed the advantages of using different cutting tools in finishing operations particularly for CNC-RM processes. The present study was designed to improve part quality by reducing uncut material left after completion of machining. Generally, using different tools to cater for flat and non-flat surfaces is an effective way to reduce excess volume. Furthermore, machining times also can be reduced depending on part geometries. In order to visualize the implications, the results were compared to a previous method that adopted a single flat end mill tool throughout the cutting operations. Distinctly, the comparisons show the method developed in this study is capable of enhancing part quality and appearance. Further studies could focus on integrating this approach into CNC-RM process planning. The surface classification task needs to be handled carefully so that it will not complicate the planning task. Also, further verification is still required through machining experiments so that real parts can be fabricated and analysed completely. In the simulation studies, the excess volume was a key indicator to determine the surface quality of the part. However, in real machining operations, the surface quality needs to be examined through roughness analysis.

REFERENCES

1. Agarwal, N. (2012). Surface roughness modelling with machining parameters (speed, feed and depth of cut) in CNC milling.*MIT International Journal of Mechanical Engineering, 2*, 55–61.
2. Agrawal, A., Soni, R. K., & Dwivedi, N. (2013). Development of integrated CNC-RP system through CAD/CAM environment.*International Journal of Mechanical and Production Engineering Research and Development, 3*, 1–10.
3. Akula, S., & Karunakaran, K. (2006). Hybrid adaptive layer manufacturing: An Intelligent art of direct metal rapid tooling process. *Robotics and Computer-Integrated Manufacturing, 22*, 113–123.10.1016/j.rcim.2005.02.006 ,
4. Anderberg, S., Beno, T., & Pejryd, L. (2009). CNC machining process planning productivity – A qualitative survey. *Proceedings of the International 3rd Swedish Production Symposium*, Göteborg, Sweden, 228–235.
5. Bharathi Raja, S., & Baskar, N. (2012). Application of particle swarm optimization technique for achieving desired milled surface roughness in minimum machining time. *Expert Systems with Applications, 39*, 5982–5989.10.1016/j.eswa.2011.11.110,
6. Bouzid, W. (2005). Cutting parameter optimization to minimize production time in high speed turning. *Journal of Materials Processing Technology, 161*, 388–395.10.1016/j.jmatprotec.2004.07.062 ,
7. CGTech. (2012). *VERICUT 7.2: Release notes.*
8. Chen, J., Huang, Y., & Chen, M. (2005). A study of the surface scallop generating mechanism in the ball-end milling process.*International Journal of Machine Tools and Manufacture, 45*, 1077–1084.10.1016/j.ijmachtools.2004.11.019 ,
9. Chen,T.,&Shi,Z.(2008).Atoolpathgenerationstrategyfor three-axis ball-end milling of free-form surfaces. *Journal of Materials Processing Technology, 208*, 259–263.10.1016/j.jmatprotec.2007.12.142 ,
10. Davim, J. P. (2001). A note on the determination of optimal cutting conditions for surface finish obtained in turning using design of experiments. *Journal of Materials Processing Technology, 116*, 305–308.10.1016/S0924-0136(01)01063-9 ,
11. Elber, G. (1995). Freeform surface region optimization for 3-axis and 5-axis milling. *Computer-Aided Design, 27*, 465–470.10.1016/0010-4485(95)00019-N ,
12. Engin, S., & Altintas, Y. (2001). Mechanics and dynamics of general milling cutters: Part I: Helical end mills. *International Journal of Machine Tools and Manufacture, 41*, 2195–2212.10.1016/S0890-6955(01)00045-1 ,
13. Frank, M. C. (2003). *The development of a rapid prototyping process using computer numerical controlled machining* (PhD thesis). The Pennsylvania State University.
14. Frank, M. C. (2007). Implementing rapid prototyping using CNC machining (CNC-RP) through a CAD/CAM interface.*Proceedings of the Solid Freeform Fabrication Symposium*, Autin, TX, 112–123.
15. Frank, M., Joshi, S. B., & Wysk, R. A. (2002). CNC-RP: A technique for using CNC machining as a rapid prototyping tool in product/process development. *Proceedings of the 11th Annual Industrial Engineering Research Conference*, Orlando, FL, 19–22.
16. Frank, M. C., Wysk, R. A., & Joshi, S. B. (2004). Rapid planning for CNC milling - A new approach for rapid prototyping.*Journal of Manufacturing Systems, 23*, 242–255.10.1115/1.2039100 ,
17. Frank, M. C., Wysk, R. A., & Joshi, S. B. (2006). Determining setup orientations from the visibility of slice geometry for rapid computer numerically controlled machining. *Journal of Manufacturing Science and Engineering, 128*, 228–238.10.1115/1.2039100,
18. Lavernhe, S., Tournier, C., & Lartigue, C. (2008). Optimization of 5-axis high-speed machining using a surface based approach. *Computer-Aided Design, 40*, 1015–1023.10.1016/j.cad.2008.08.006 ,
19. Li, Y., & Frank, M. C. (2006). Machinability analysis for 3-axis flat end milling. *Journal of Manufacturing Science and Engineering, 128*, 454–464.10.1115/1.2137748 ,
20. Lim, T., Corney, J., Ritchie, J., & Clark, D. (2001). Optimizing tool selection. *International Journal of Production Research, 39*, 1239–1256.10.1080/00207540010020063
21. Osman Zahid, M. N., Case, K., & Watts, D. (2014). Optimization of roughing operations in CNC machining for rapid manufacturing processes. *Production & Manufacturing Research, 2*, 519–529. [Taylor & Francis Online]
22. Palanisamy, P., Rajendran, I., & Shanmugasundaram, S. (2007). Optimization of machining parameters using genetic algorithm and experimental validation

for end-milling operations. *The International Journal of Advanced Manufacturing Technology, 32*, 644–655.10.1007/s00170-005-0384-3 ,

23. Patel, K. (2010). *Web based automatic tool path planning strategy for complex sculptured surfaces* (Master thesis). University of Waterloo.

24. Reddy, N. S. K., & Rao, P. V. (2005). Selection of optimum tool geometry and cutting conditions using a surface roughness prediction model for end milling. *The International Journal of Advanced Manufacturing Technology, 26*, 1202–1210.10.1007/s00170-004-2110-y ,

25. Relvas, C., & Simoes, J. (2004). Optimization of computer numerical control set-up parameters to manufacture rapid prototypes. *Proceedings of the Institution of Mechanical Engineers, Part B: Journal of Engineering Manufacture, 218*, 867–874.10.1243/0954405041486055 ,

26. Renner, A. (2008). *Computer aided process planning for rapid prototyping using a genetic algorithm* (Master thesis). Iowa State University.

27. Ryu, S. H., Choi, D. K., & Chu, C. N. (2006). Roughness and texture generation on end milled surfaces. *International Journal of Machine Tools and Manufacture, 46*, 404–412.10.1016/j.ijmachtools.2005.05.010 ,

28. Vijayaraghavan, A., Hoover, A., Hartnett, J., & Dornfeld, D. (2008). Improving end milling surface finish by workpiece rotation and adaptive toolpath spacing. *International Journal of Machine Tools & Manufacture, 49*, 89–98. ,

CHAPTER 7 INVENTORY CLASSIFICATION BASED ON DECOUPLING POINTS

Joakim Wikner[a] & Eva Johansson[a]

[a] Department of Industrial Engineering and Management, School of Engineering, Jönköping University, Jönköping, Sweden

ABSTRACT

The ideal state of continuous one-piece flow may never be achieved. Still the logistics manager can improve the flow by carefully positioning inventory to buffer against variations. Strategies such as lean, postponement, mass customization, and outsourcing all rely on strategic positioning of decoupling points to separate forecast-driven from customer-order-driven flows. Planning and scheduling of the flow are also based on classification of decoupling points as master scheduled or not. A comprehensive classification scheme for these types of decoupling points is introduced. The approach rests on identification of flows as being either demand based or supply based. The demand or supply is then combined with exogenous factors, classified as independent, or endogenous factors, classified as dependent. As a result, eight types of strategic as well as tactical decoupling points are identified resulting in a process-based framework for inventory classification that can be used for flow design.

INTRODUCTION

Increasing customer requirements for customization, cost issues, and ever shortening product life cycles result in a dynamic behaviour and structural complexity of the logistics system that are very demanding. The decisions are not only a complex matter due to the intricacy of each individual decision but also the sheer volume of decisions adds weight to the burden. In addition, the reduction of inventories in the logistics systems leads to stronger dependencies across the material flows, which increases the risk for knock-on effects of disturbances as pointed out by, among others, Goldratt and Cox (1984). Considering these characteristics in combination, it is obvious that logistics managers need tools to simplify and automate decision-making. Important concepts in this context are abstraction, in order to retain only information that is relevant for a particular decision problem, and aggregation, to focus on groups of entities where the same decision can be implemented.

From a decision-maker's perspective, the implications of applying abstraction (selecting relevant information) and aggregation (grouping of information) are significant. Abstraction is usually applied through hierarchical decision-making. At the lowest level, detailed decisions are usually made with a local and short-term focus. At higher levels, issues of more strategic character are handled. Typical examples at the strategic level are to focus on constraining resources and materials critical to customer service. Aggregation can

Citation: Joakim Wikner & Eva Johansson, Inventory classification based on decoupling points, DOI:10.1080/21693277.2015.1067845.

be applied by aggregating decision problems and making a decision based on a key property of the aggregate, such as a product group, and then more or less automatically applying the decision to all products of that product group. The terminology used in this paper refers to this type of aggregation as classification. By classifying an item as a specific type due to some of its properties, a group of items sharing this property is created.

These two concepts, abstraction and aggregation, are the theme of this paper. Item and inventory classifications of different types are very common in logistics management (see, e.g. Broeckelmann, 1999), and our ambition is to further emphasize the importance of a process perspective in this context. Previously, much focus has been on classifications using item properties as a point of departure, e.g. ABC classification, but with an additional process perspective, further support can be provided to logistics management. The purpose of the research presented in this paper is to develop a process-based framework for inventory classification. The application of the framework in planning and control is discussed in order to show how the results can be used.

The paper is structured as follows. Next, a general overview of the inventory management context is provided in terms of decoupling points. The concept of supply system is introduced and some key properties of decoupling points related to replenishment and withdrawal are defined. Thereafter, a generic structural model for logistics systems is introduced to identify four key delivery strategies and their respective positions of the two critical decoupling points of strategic importance. Finally, a process-based approach to inventory classification is derived from the discussion on strategic and tactical decoupling points, followed by managerial implications for planning and scheduling, and some ideas for further research.

DECOUPLING POINTS AND MATERIAL FLOWS

The context of material flows is here referred to as a supply system that can be described as a set of resources transforming raw materials into finished goods. Throughout the material flow, resources interact in the transformation process to create value in terms of form, time, and place. In an ideal state, the material flow is continuous and one piece (Ohno, 1988, p. 96) meaning that there is a steady and even flow where all the resources are synchronized and performing with a minimum of balancing loss in the system. This ideal state is, however, rarely possible to realize in the complete supply system. Usually, this type of material flow is limited to a set of resources performing a number of steps continuously but in a context of only a manufacturing cell or a line. In the context of place transformation, the only case where continuous flow can be achieved, except for in pipeline transportation, is through aggregation of several material flows to, for instance, a truck load, thus providing one-piece flow for each separate material flow. Using common lean terminology, this kind of well-defined section of one-piece flow is referred to as a process (Rother & Shook, 2003, p. 18). Each process then represents transformation and replenishes a point of inventory downstream and withdraws materials from one or more other points of inventory upstream. At a 'micro' level, the material flows can therefore be modelled as a set of stock points, i.e. inventories, each connected with processes that withdraw materials and processes that replenish materials. Each inventory can hence be seen as a point that decouples two processes (replenishment and withdrawal) and enables them to act somewhat independently. This kind of decoupling point is referred to as a DP (this definition is in line with the general perception of the concept, see, e.g. Blackstone (2008). The supply system can hence be modelled as a set of DPs that interact through replenishment and withdrawal processes. Note that a process that is a replenishment process for one DP can be a withdrawal process for another DP. The only exceptions are the replenishment processes related to supply external to the supply system (e.g. external suppliers) and withdrawal processes that are external to the supply process (e.g. external customers).

Decoupling points categorization

The concept of DP is a core component in the management of material flows. Considering different combinations of unexpected and expected events, related to replenishment and withdrawal, for each and every DP results in a very demanding challenge for management since the number of DPs in an enterprise or supply chain is staggering. It is very difficult in all realistic settings, at least with today's technology, to control each and every DP to any greater extent. To handle this complexity, the DPs can be differentiated into categories. In particular three categories play a key role in materials management; they are here referred to as operational, tactical, and strategic:

- *Operational DPs* connect transformation points in the physical material flows and are, for example, known as queues in front of resources.
- *Tactical DPs* are related to materials given a unique identifier, usually referred to as item number or part number. Introducing this identity code enables item-level materials management.
- *Strategic DPs* are a subset of all tactical DPs that play a role of critical importance to the interface of the supply system and its context.

Decoupling points and planning and scheduling

Strategic DPs relate to items that are exposed to external influence on the withdrawal side, such as customer orders, or on the replenishment side, such as purchase orders. External demand is not possible to derive from demand for other items and is hence usually unpredictable in terms of both quantity and timing. Classifying items as exposed to external influence or only as internally affected corresponds to the classification in terms of dependent and independent demand, which is the fundamental principle behind materials planning (Vollmann, Berry, Whybark, & Jacobs, 2005, p. 134). Demand for an item that can be derived from demand for other items is in this context referred to as dependent demand (DD). In its most straightforward application, demand for a component is calculated as the demand for the parent item multiplied by the explosion quantity. Demand that cannot be calculated in this way but is directly affected by external demand is thus classified as independent demand (ID). This classification of demand has a long history and is particularly associated with material requirements planning, MRP (Orlicky, 1975), of which backward scheduling is a key concept. As this scheduling approach is performed, the demand for the next level of items is calculated in the bill-of-material (BoM) explosion process. Many items with DD may also have a certain proportion of ID, if they, for example, are sold separately as spare parts besides being used in the production of another parent item. In summary, DD items are exposed to internal demand and hence of a less strategic nature. ID items, on the other hand, are exposed to external demand, e.g. customer orders, and due to their strategic nature correspond to strategic DPs. The similarities between independent and strategic DPs have so far been covered from a demand-based perspective but, as will be shown below, the corresponding similarities can also be identified from a supply-based perspective. Integrating the demand-based and the supply-based perspectives results in a more comprehensive approach to the classification of DPs.

THE LOGISTICS SYSTEM

The basic supply system outlined above corresponds to an inventory subsystem related to one DP connected with a replenishment process and a withdrawal process. In isolation, this is a relatively simple entity where item-based inventory management can be applied. An inventory subsystem is, however, usually a part of a larger logistics system consisting of a set of connected inventory subsystems. As the complete logistics system is considered, a range of dependencies is introduced, which increases the complexity of the decision problem. Instead of dealing with single-item inventory management, the decision-maker is facing a scenario with a network of inventory subsystems that interact with each other, and to some extent also interact with the context of the complete logistics system. This type of system also corresponds to the logical entity introduced by Wikner (2014) as a supplement to the traditional actor-based approach to flow design. Such a complex decision problem is very difficult to handle with a monolithic model and therefore requires some kind of hierarchical approach to system analysis providing the system analyst with an appropriate abstraction level for each decision. Above, the lower abstraction level in terms of one individual DP was introduced. Next, the higher abstraction level is introduced involving the key concepts of strategic DPs and strategic supply systems. In particular, we are interested in DPs exposed to exogenous influence. Exogenous influence refers to external influence, of which the customer order is the most important. Internal influence, on the other hand, is dependent on other internal events and is thus endogenous.

Demand and supply categories

To discriminate between different DPs with the objective of identifying the vital few, it is necessary to identify the DPs exposed to exogenous influence. Dependent and independent demands (DD and ID) were described earlier and are classified in Table1. To the authors' knowledge, there is no corresponding general conceptual framework for bottom-up approaches where the supply is the driver in the process. More specific approaches do, however, exist, such as the forward flow scheduling approach of process flow scheduling (Taylor & Bolander, 1994). Taking supply as the point of reference would result in employing a forward scheduling approach to generate supply for the next higher level in the BoM. Note that a common denotation of levels in a BoM is that the top level is 0 and for each subsequent lower level the level number increases. In analogy with the demand-based concept, this is referred to as dependent supply (DS). Using the concept of forward scheduling, the supply of material on level i in the BoM would be dependent on supply on the next lower level, level $i + 1$, during the scheduling phase. In a backward scheduling scenario, the demand for items on level i would be dependent on the demand for items on the next higher level, level $i - 1$. The key property of DD and DS is that they both can be calculated based on the available endogenous information of the logistics system, such as lead times and available inventory.

Table 1: Summary of demand and supply categories

	Supply	Demand
Independent	Supply that must be estimated (purchase orders)	Demand that must be estimated (forecast and customer orders)
Dependent	Supply that can be calculated (forward scheduling)	Demand that can be calculated (backward scheduling)

The description above shows the clear analogy between DS and DD. The next step is to identify independent supply (IS) as an analogy to ID. This analogy is, however, not as clear-cut, but we define supply from a source outside the logistics system as IS since it is outside the strict control of the logistics system. Control can cover a complete manufacturing network or just a single site. The important aspect is that within the controlled supply system there is no need for traditional purchasing orders; instead internal orders (sometimes referred to as distribution orders) can be used covering both the demand process from one DP and the supply process of another DP. In summary, four types of drivers can be identified in the material flows as in Table 1, i.e. the intersections between supply/demand and independent/dependent.

Based on these two types of independent drivers (IS and ID), two types of corresponding DPs can be identified; they are discussed in the following two sections.

Demand-based decoupling point

Delivery strategies such as make-to-stock (MTS) and make-to-order (MTO) play a key role in providing an appropriate balance between efficiency and responsiveness in meeting market demand (see, e.g. Chopra & Meindl, 2007). The baseline for designing the appropriate strategy in this perspective is the relation between the delivery lead time (D), requested by the market, and the total supply lead time (S) within which the system needs to produce the output. The ratio between these two lead times is critical; it was first identified by Shingo (1981) as the D:P relation. Mather (1984) changed the relation to the inverse for practical reasons and since then it is known as the P:D ratio or P:D relation. Shingo used P as he was referring to product lead time. In a similar vein, the more general supply lead time is used and thus referred to as S (see, e.g. Wikner,2014). S more explicitly involves, for example, also purchasing and distribution lead times. If S is shorter than D, all supply activities can be performed while the customer is waiting. But when S is longer than D some activities must be performed in advance to create semi-processed parts to be used when the customer order arrives. The supply system is thus split into two parts corresponding to two subsystems. The subsystem supplying semi-processed parts is performed before the customer order arrives and is therefore forecast-driven. The second subsystem uses the semi-processed parts in stock to deliver on the customer order. This is sometimes referred to as customer-order-driven (see, e.g. Hoekstra & Romme, 1992) but also as demand-driven (see, e.g. Christopher, 1998). The semi-processed parts are kept in stock at a decoupling point that is known as the customer order decoupling point, CODP, (see, e.g. Giesberts & van der Tang, 1992) but also as order penetration point, OPP (Sharman, 1984) or simply as decoupling point (Hoekstra & Romme, 1992). The CODP has been defined in a number of different ways but, as summarized by Wikner (2014, p. 195), '[t]he customer order decoupling point (CODP) separates decisions about initiating flow based on speculation on future customer orders from commitment against actual customer orders'. Using these three building blocks (forecast-driven supply system, CODP, and customer-order-driven supply system), the supply system can be depicted as in Figure 1, where a consumption system has been introduced as an origin of demand and to represent the customers of the supply system. The diamond is used to represent the CODP in line with the guidelines suggested by Wikner (2014).

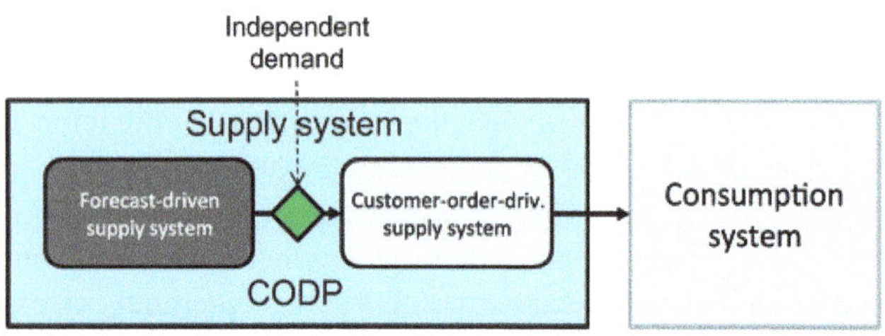

Figure 1: The CODP perspective of the supply system.

Considering a customer order arriving in the supply system of Figure 1, the customer must accept to wait during the time it takes to perform the customer-order-driven supply system (during the lead time D). This system lead time could involve activities outside manufacturing (see, e.g. Rudberg & Wikner, 2004). To replenish the CODP, some kind of forecast is necessary since customer orders are assumed to arrive in time to perform only the customer-order-driven supply system in Figure 1. The forecasting can be performed at the end-item level but if any parts are shared between different end items it is easier, due to aggregation effects reducing the coefficient of variation in demand, to forecast at the CODP on semi-processed parts. This corresponds to defining the CODP parts as subject to ID as shown in Figure 1. Activities downstream of the CODP may be performed under certainty of demand since a customer order is driving the activities. The parts at the CODP must, however, be produced to forecast since they are expected to be available at the moment the customer order arrives.

Supply-based decoupling point

The CODP emphasizes the relation between the delivery lead time D and the supply lead time S, and hence investigates the demand-related properties of S but does not provide any further information on what actually constitutes supply. Taking a supply perspective, an important concept is to what extent the complete supply system is integrated. In this case, the focus is on inventory and materials management and therefore also the level of integration is defined based on the level of control that can be exercised on the supply system. In an integrated supply system, the complete set of supply activities would be controlled as one entity. If, however, the supply system is not integrated, the supply system could be separated into two subsystems, here referred to as one controlled subsystem and one uncontrolled subsystem, see Figure 2. A hexagon is used to illustrate the purchase order decoupling point (PODP) in line with Wikner (2014). The uncontrolled subsystem is out of control reach for the controlled subsystem.

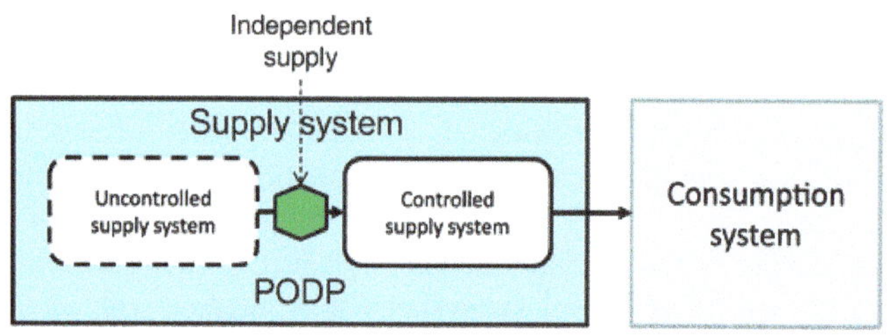

Figure 2: The PODP perspective of the supply system.

An example is a component supplier of a company that supplies parts per order and where there is no further integration. A purchase order is sent to replenish the inventory when required. In this case the supplier would be modelled as an uncontrolled system including not only the supplier but also all the supplier's suppliers, etc. Using this approach, a recursive systems definition is obtained where the controlled supply system and how it interacts with the consumption system is the focus. Remember that from the uncontrolled supply system perspective, the controlled supply system would represent the consumption system as depicted in Figure 3. This is because all upstream activities not covered by the controlled supply system are included in the uncontrolled supply system.

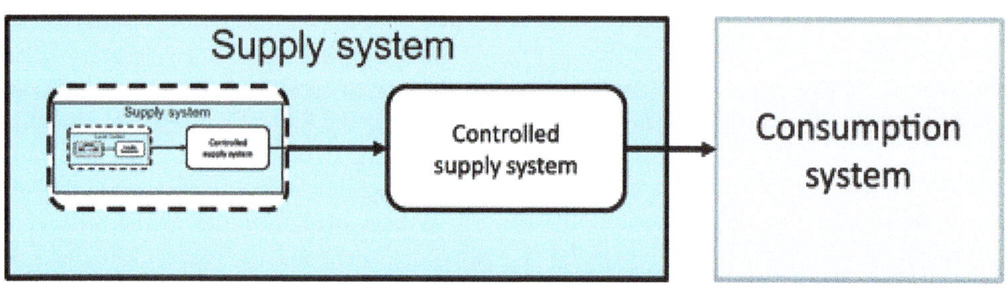

Figure 3: Recursive model of a supply system.

Within the controlled supply system, there is full transparency in terms of planning and execution which enables a cost-efficient allocation of decoupling points, i.e. buffers, from a system's perspective. The uncontrolled supply system does, however, not provide this level of transparency by definition. Since control cannot be exercised on this system by the controlled supply system, the only way to manage unexpected events occurring in this system is to introduce a buffer explicitly targeting these properties and decoupling the controlled system to protect it. As shown by Lee (2002), this relation must be carefully designed and alternative strategies for minimizing unexpected events should be considered, but if uncertainty prevails it must be handled with some type of buffer, otherwise customer service and/or cost efficiency will suffer. Since the relation with the uncontrolled supply system usually is through some kind of purchase order, we introduce the term PODP for this kind of buffer, as in Figure 2. Activities within the controlled supply system are usually coordinated by production orders and distribution orders but when it comes to coordinating with an external system, the relation is usually handled through a purchase order. This model is also related to the lead time based analysis performed by Wallin, Rungtusanatham, and Rabinovich (2006). They related the delivery lead time D to production time P in the spirit of Shingo (1981), but then made a distinction between the part of P covered by the supplier and the part covered by the focal company. Using this approach, different strategies for inventory management of purchased material were identified, but here the emphasis is on the complete supply system structure and not only purchased materials at the PODP. In Wallin et al. (2006), the focal company is the unit of analysis. Here, however, the unit of analysis is a supply system, and to emphasize what specific supply system is considered it may be referred to as the focal supply system. In Figure 2 the focal supply system would correspond to the controlled supply system. A similar approach to dividing the supply system into subsystems based on exogenous influence has also been investigated by Wikner (2005), but then from the perspective of efficiency and responsiveness.

The PODP can hence be defined as follows: '[t]he PODP separates decisions about delimiting flow based on what is external to the logical entity from what is internal and hence controllable' (Wikner, 2014, p. 199). This indicates that the uncontrolled supply system provides input to the controlled supply system. This input contains more uncertainty compared to what is perceived from the controlled supply system and thus requires some kind of estimation (forecasting), which also is the key property of IS, see Table 1.

Modelling the logistics system based on delivery strategies

The fundamental building blocks of the logistics system have been defined above and are summarized in Figures 1 and 2. Basically, the properties of the supply systems highlighted are type of driver, related to CODP, and level of control, related to PODP. Each particular logistics system has its own unique characteristics based on its history, market conditions, available resources, etc. There are, however, some generic configurations that can provide a baseline for illustrating how the constructs introduced above can be employed to represent an enterprise from a logistics perspective. The two fundamental configurations used here are the completely customer-order-driven system (MTO) and the completely forecast-driven system (MTS). The third system (assemble-to-order, ATO) represents a compromise system with elements of both MTO and MTS (Wemmerlöv, 1984). The fourth system, finally, involves also the possibility of outsourcing not just forecast-driven activities to uncontrolled supply systems but also customer-order-driven activities; this is referred to as purchase-and-make-to-order (PMTO) (Hoekstra & Romme, 1992).

Make-to-stock (MTS)

When the MTS strategy is used, all activities are forecast-driven (with the possible exception of some picking and delivery activities). In this case, the CODP is positioned as far downstream in the material flow as possible. No activities are customer-order-driven and usually the master schedule (see, e.g. MRPII) or pacemaker (see, e.g. rate-based master scheduling as in lean) is positioned at the end-item level, where ID is present. This allows for a materials planning approach based on DD (backward scheduling), which is the most common approach in this setting, for all items down to the PODP. The PODP hence represents the extension of the integrated planning and control approach as shown in Figure 4. In Figures 4–7, the dashed vertical arrows represent exogenous input to the supply system. The dashed horizontal lines represent endogenous propagation of demand and supply respectively, in the direction indicated by the arrow.

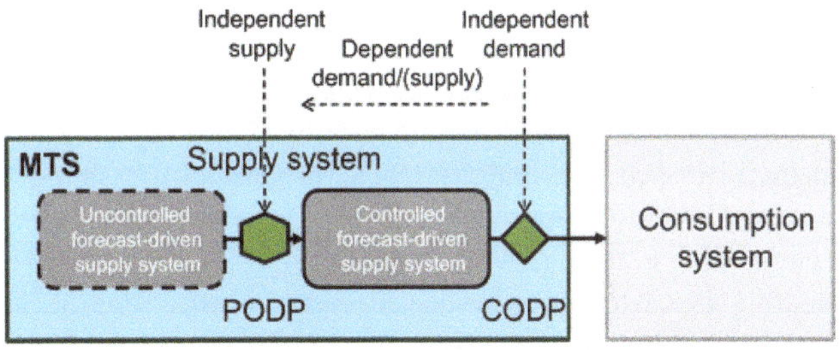

Figure 4: Modelling make-to-stock (MTS) supply systems.

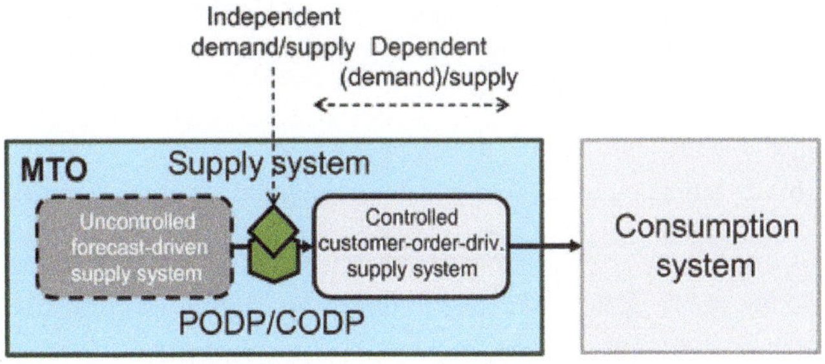

Figure 5: Modelling make-to-order (MTO) supply systems.

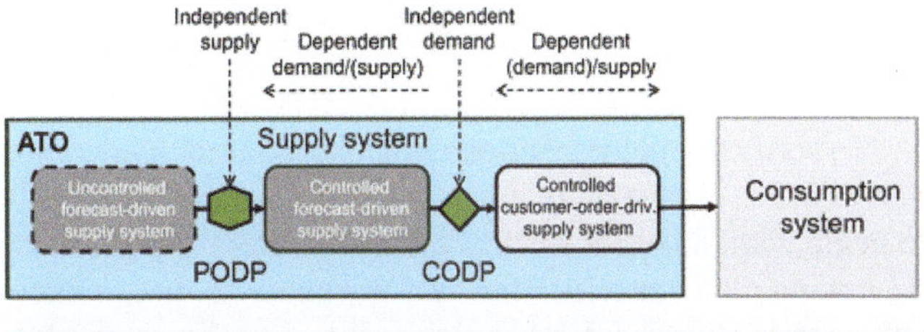

Figure 6: Modelling assemble-to-order (ATO) supply systems.

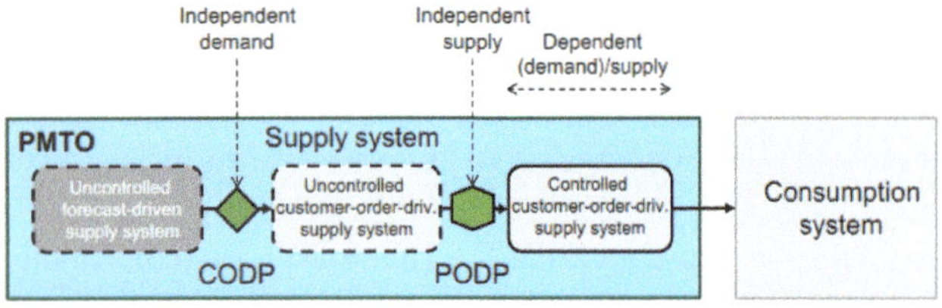

Figure 7: Modelling purchase-and-make-to-order (PMTO) supply systems.

Make-to-order (MTO)

The second fundamental configuration corresponds to the MTO strategy, where all activities of the controlled system are customer-order-driven. When the customer order is received, the customer waits until the product is finished. The planning approach can be based on DD (as in MRP), where the end item is master scheduled and DD is calculated for all other items down to the CODP. A preferred scenario would, however, in some cases be to start with the CODP items availability and then to forward schedule all items (DS) up to the end item to provide, for example, information about delivery capability. This process would then preferably also involve finite loading considering any finite capacity to provide reliable delivery dates. Since the CODP here is positioned at the beginning of the controlled supply system, it coincides with the PODP as shown in Figure 5.

Assemble-to-order (ATO)

A preferred strategy in many cases is to establish a level in the BoM that allows for modularization. Different modules can then be used in a multitude of configurations enabling a mass customization approach. Modules are then produced to forecast using a standard MTS 'sub-strategy', whereas the final assembly is customer-order-driven and similar to the MTO approach. Setting the two models of MTS (Figure 4) and MTO (Figure 5) in tandem results in an ATO scenario as in Figure 6. The CODP is positioned between the forecast-driven and the customer-order-driven supply systems, and the PODP is also in this case positioned at the interface between the uncontrolled and the controlled supply systems.

Purchase-and-make-to-order (PMTO)

If only the CODP is considered, the three strategies identified above, MTS, MTO, and ATO, are key scenarios. Each strategy consists of a unique combination of forecast-driven and customer-order-driven supply systems. Taking a broader perspective and also considering that the delivery lead time D may extend beyond the lead time of the controlled supply system provide leeway for a fourth scenario. This scenario involves an external supplier (here referred to as an uncontrolled supply system) that is customer-order-driven, as illustrated in Figure 7. There are now two customer-order-driven supply systems separated by the PODP, and the CODP is positioned upstream of these two supply systems. This is a possible, if not always desirable, strategy as some activities performed while the customer, i.e. the consumption system, is waiting are performed beyond the control of the supply system responsible for delivery to the customer. A possible approach in this case would be to seek to integrate, i.e. 'in-source', the uncontrolled customer-order-driven supply system and, if this is not possible, to establish a safety lead time at the PODP, unless all uncertainties between the CODP and the PODP can be eliminated.

Summary of the four logistics system configurations

To conclude the discussion on logistics systems configured based on the concepts of customer-order/forecast-driven and uncontrolled/controlled, the configurations are summarized in Table 2. The four concepts provide four possible configurations of supply systems as represented by the columns. In addition, it is also required, by definition, that certain precedence constraints are fulfilled. A forecast-driven supply system should not be positioned downstream of a customer-order-driven supply system. This is because the supply systems are positioned along a time axis and if it is possible to be customer-order-driven, this is assumed to always be the preferred option. Why forecast-driven if actual demand is available? The second precedence constraint is that uncontrolled supply systems are positioned upstream from controlled supply systems. One exception to this logic is when certain operations within a routing are outsourced. This precedence constraint is nevertheless assumed here.

Table 2: Summary of the four possible configurations of the logistics supply system

	Uncontrolled forecast-driven	Controlled forecast-driven	Uncontrolled customer-order-driven	Controlled customer-order-driven
Config. MTS	X	X	–	–
Config. ATO	X	X	–	X
Config. MTO	X	–	–	X
Config. PMTO	X	–	X	X

To verify that there are no other fundamental configurations regarding these constraints, a simple analysis can be performed. To begin with, it is assumed that there is always the possibility of having uncontrolled forecast-driven activities which corresponds to purchasing materials to forecast. This means that the configurations of the leftmost column of Table 2 are valid (uncontrolled forecast-driven). Configurations MTS and ATO in Table 2 both involve a forecast-driven supply system that also is controlled. This means that there can be no uncontrolled customer-order-driven activities without invalidating any of the precedence rules. The only option available is thus to include or exclude controlled customer-order-driven activities, which results in MTS and ATO. If there are no controlled forecast-driven activities, as in configurations MTO and PMTO, that is only possible with two different configurations. Either all customer-order-driven activities are controlled (MTO) or parts of them are uncontrolled (PMTO). If all activities were uncontrolled, the investigated system would be empty and here considered as void.

These four configurations of logistics supply systems highlight the structural importance of the two types of decoupling points: CODP and PODP. Based on this insight, we return to the complete set of DPs to introduce a classification that takes this insight as a point of departure.

PROCESS-BASED INVENTORY CLASSIFICATION USING DECOUPLING POINTS

Inventory classification is usually based on some properties of the items in inventory such as value, weight, or perishability. In the case of classification based on, e.g. withdrawal frequency or supplier characteristics, more emphasis is placed on a process perspective but one of the most important properties of processes is still not considered. As processes represent all the value-adding and non-value-adding activities performed in an enterprise, a key factor is how and why these processes are initiated, i.e. what triggers a process. In the analysis of the logistics system above, two important triggers were identified. Basically, the triggers can be characterized as exogenous or endogenous. An exogenous trigger refers to external influence, of which the customer order is the most important, and an endogenous trigger refers to internal influence, corresponding to DD and DS. Finally, a process is also vulnerable when it crosses organizational boundaries, and the relation to external suppliers is of particular interest since it is a consequence of a make/buy decision and represents the reach of what is within process control of the business. In all, there are four process-based concepts that can be used in classifying the DPs and hence inventory. They are related to dependent or independent, which can be combined with a supply or demand perspective in four different ways. These four concepts (dependent, independent, demand, and supply) are combined below and then classified as strategic (related to the choice of delivery strategy and level of outsourcing) or tactical (related to other item-based DPs of importance when synchronizing the material flows).

Strategic process-based decoupling points

The customer order triggers the processes in the customer-order-driven supply system. From an inventory perspective, the DP of outstanding importance is the CODP which is subject to ID and crucial to the service level that can be achieved. The CODP can be seen as the incarnation of the delivery strategy since the position of the CODP is based on the delivery lead time. The positioning of the CODP has been identified as a key strategic decision, for example, in terms of the CODP itself (Olhager, 2003), as a consequence of balancing efficiency and responsiveness (Chopra & Meindl, 2007), as the point where lean meets agile (leagility) (Mason-Jones, Naylor, & Towill, 2000), or where the physical and innovative parts of the supply chain meet (Olhager, Selldin, & Wikner, 2006).

The positioning of the PODP is also strategic in the sense that it represents the selected level of vertical integration, i.e. the level of outsourcing. It is also crucial for buffering external disturbances and hence for providing the internal flows with the best possible conditions. From an inventory perspective, a second important property of the PODP is the amount of control that can be exercised on a process, and it is assumed that uncontrolled subsystems constitute a risk and may have a negative impact on the controlled supply system, which is downstream from the PODP.

The inventory at the CODP enables the forecast-driven supply system to focus on efficiency since the inventory at the CODP buffers against seasonal variations as well as short-term fluctuations related to demand. The dimensioning of the CODP inventory is usually quantity based, as it is the impact of the demand variations on the service level that is in focus. At the PODP, on the other hand, the situation is more concerned with the reliability of the supply lead times and hence one is more inclined to use safety lead time at this DP. In addition, the PODP also handles the lot-size inventory due to the delivery frequency from the external suppliers.

In summary, the strategic DPs are either driven by downstream exogenous factors (related to external customers through customer orders) or driven by upstream exogenous factors (related to external suppliers through purchase orders). The strategic DPs are categorized in Table 3 based on how they are related to these two types of exogenous triggers. If the trigger is only related to ID with no independent supply (NIS), the decoupling point is of the type NIS–ID, better known as CODP (see Figure 1). The reverse, IS but no independent demand (NID), is indicated as IS–NID and defined above as PODP (see Figure 2). These two types of strategic DPs may be separate but still present in the same flow as shown in Figures 4, 6, and 7. However, in some particular cases, a strategic DP can be exposed to both independent supply and independent demand (IS–ID), which represents the case where a DP is both a CODP and a PODP (see Figure 5). The fourth case finally, with NIS and NID, is not included in terms of strategic DPs but will be further analysed next at the tactical level.

Table 3: Strategic process-based decoupling points

	No independent demand (NID)	Independent demand (ID)
No independent supply (NIS)	–	NIS–ID
Independent supply (IS)	IS–NID	IS–ID

Tactical process-based decoupling points

Tactical decisions should follow as a consequence of the decided strategy. In the same way the positioning of DD and DS is a direct consequence of ID and/or IS. The strategic DPs create a structural framework based on the delivery strategy and the outsourcing strategy. The objective of the positioning of DD and DS DPs is to support this strategy. These DPs are endogenously triggered to support the ID of all related items using the bill-of-material/distribution as the road map. In all, three DP classes can be identified at the tactical level as shown in Table 4. DPs with no dependent supply (NDS) but with DD, i.e. type NDS–DD, are backward scheduled based on the demand downstream, and DPs with DS but no dependent demand (NDD), i.e. type DS–NDD, are forward scheduled based on the supply upstream. Since backward and forward scheduling in practice are difficult to combine at the same DP, the combination of dependent supply and dependent demand (DS–DD) is rarely established. An example could however be when one item is part of two different products one of which is made to customer order and the other is sold from finished goods, i.e. a hybrid of MTS (see Figure 4) and MTO (see Figure 5) for that particular item. The combination involving no dependent supply and no dependent demand (NDS–NDD) is not included at the tactical level since a tactical driver is not present. DPs classified as tactical are generally used during the planning phase to establish the appropriate level of inventory over time at these decoupling points.

Table 4: Tactical process-based decoupling points

	No dependent demand (NDD)	Dependent demand (DD)
No dependent supply (NDS)	–	NDS–DD
Dependent supply (DS)	DS–NDD	(DS–DD)

Summary of process-based inventory classifications

Three types of DPs were identified above as strategic and related to ID (CODP-based) and IS (PODP-based). If NID and NIS were present (upper left cell of Table 3), the DPs would be classified as tactical and could still be differentiated as exposed to DD or DS as in Table 4. In total, these two tables highlight six different types of DPs. By combining these two tables, two of the strategic DPs, IS–NID and NIS–ID, may be further differentiated. Compared to Table 3, NID is differentiated into no demand (ND) and DD. If there is NID there could still be DD. NIS is differentiated into no supply (NS) and DS in a corresponding way as in Table 5. This results in five different types of strategic DPs, as highlighted in Table 5. In combination with the three types of DPs in Table 4, there are in total eight types of DPs (inventory) that have been identified using the process-based approach suggested here, with a foundation in planning and control. Five of them are of a strategic character involving either IS or ID. Three classes of DPs are of a tactical character involving NIS and NID as summarized in Table 5. Table 5 is a combination of Table 3 (mainly represented by the white area) and Table 4 (represented by the shaded area). 'No Demand (ND)' in Table 5represents that the use of inventory in the DP is not directly driven by demand. Correspondingly, 'No Supply (NS)' refers to the fact that the replenishment of the DP is not driven by supply pushing items into the DP. A DP with a combination of ND and NS would be completely disconnected from the system and is therefore not classified in Table 5.

Table 5: Process-based inventory classifications

	No demand (ND)	Dependent demand (DD) (backward scheduled)	Independent demand (ID) (CODP based)
No supply (NS)	–	NS–DD	NS–ID
Dependent supply (DS) (forward scheduled)	DS–ND	(DS–DD)	DS–ID
Independent supply (IS) (PODP based)	IS–ND	IS–DD	IS–ID

In essence, Table 5 represents eight (nine) classes of decoupling points, i.e. inventory points. Hence, the table constitutes a classification of inventory based on the intersection between the demand perspective (the columns) and the supply perspective (the rows). Note that these DPs are tactical or strategic meaning that all DPs are related to items. From this perspective, Table 5 can be seen as providing eight different boxes where all items in an enterprise can be allocated to at least one box. In some cases, one individual item can be related to two boxes if it, for example, is used in two different end products where one is MTS and the other is MTO. It is important to note that Table 5 does not provide any information on if an item is forecast-driven or customer-order-driven, except for the CODP items which are forecast-driven since they are expected to be in stock when the customer order arrives. The table does not provide a set of possible systems but rather the building blocks that can be used for designing any type of system. The NS–ND category should be empty since all items are either driven by demand or supply.

MANAGERIAL IMPLICATIONS

The eight different classes of DPs in Table 5 represent an abstraction in the sense that these are the DPs that should be handled at a strategic and at a tactical level. They also represent an aggregation in the sense that each cell in Table 5represents a type, or group, of DPs that have certain prerequisites concerning, for example, planning and control. Using the terminology of the ABC classification (see, e.g. Vollmann et al., 2005), the process-based classification in Table 5 can be summarized as generating three classes of inventory (DPs) to focus on. Class A is considered as the most important class and is fundamental for the order fulfilment process. Class B is the second in rank and relates to exogenous uncertainty that is difficult to control, and Class C, finally, is of least importance since it is within the controlled system:

- Class A: CODP inventory (column 3 in Table 5), since it is key to the delivery strategy chosen and a foundation for the delivery reliability to the customer.
- Class B: PODP inventory (row 3 in Table 5), since it is key to the sourcing strategy chosen and based on the delivery reliability of the suppliers.
- Class C: Tactical DPs (grey cells in Table 5), as they represent the link between PODPs, CODPs, and the customers.

Interpreting the different types of DPs in terms of planning and control methods provides some further insights into the significance of the eight types of DPs. The details of this mapping are not the subject of this paper but some guidelines have already been outlined above and some further insights are summarized below.

The framework suggested above and summarized in Table 5 provides a baseline for materials management but does not in itself provide any methods for planning and control. It does, however, highlight some key properties of when a particular planning and control method is appropriate. A key aspect of the framework is process triggering; in materials management this aspect has received considerable interest and in general a process is assumed to be triggered in basically two ways (Hopp & Spearman, 2000, p. 340):

- A customer to the process demands that the process should be performed (frequently referred to as pull). The customer can either be an external customer or an internal customer represented by another process or a point of inventory.
- A schedule demands that the process should be performed (frequently referred to as push). In some instances it is assumed that the schedule in push is based on forecasted demand (see, e.g. Marchwinski, Shook, & Schroeder,2008), but this is a confusing and misleading assumption in this context. How should, for example, customer-order-driven manufacturing using scheduling software be classified under this assumption? The key issue here is rather that the mode is proactive and the schedule can be based on a wide range of information, whereas the pull approach operates in a reactive mode.

The terminology used in these two cases is somewhat different but still from a functional view they are perfectly in line with the framework introduced here, to which process triggering is central. To provide further support for this conclusion, three examples are presented of what kind of scenarios three different approaches are designed to handle. First, the standard MRPII approach is discussed, followed by takt-based planning and control, and finally an example from the process industry is presented.

The DPs NS–DD and NS–ID correspond to Orlicky's (1975) framework for MRP, where NS–ID is referred to as ID and should be master scheduled, whereas NS–DD is referred to as DD and should be handled by the MRP logic. In case there is NDD (only single-level bill-of-materials), the NS–ID could be managed using traditional inventory control methods such as an order point system. Another example is the classical ATO scenario where the order promising process considers stock availability of modules (NS–ID) at the CODP and the order promising logic checks availability downstream using forward scheduling (DS–ND) to determine possible delivery dates.

In the lean environment, the interpretation of Table 5 requires a distinction between how the system is dimensioned during planning (takt time analysis, line balancing, estimation of number of kanbans, and size of FIFO buffers, etc. (see, e.g. Duggan,2013) and how it actually works within those boundaries during execution. The pacemaker is frequently positioned at the CODP (see, e.g. Rother & Shook, 2003). Downstream from the pacemaker, the DPs would then be considered as DS–ND and upstream as NS–DD during the planning phase. During the execution phase, all DPs would act as NS–ID responding to pull signals from downstream processes considered basically as ID.

In some process industries, the incoming raw material is not controllable but rather a consequence of long-term contracts, such as in farming. The livestock may be supplied to the abattoir at a certain age and must be processed without delay and therefore represents IS–ND and requires a planning and control approach based on divergent product profiles. Referring to process flow scheduling (Taylor & Bolander, 1994), the DPs covered by reverse flow scheduling (referred to as backward scheduled here) would be NS–DD except for the final stage, which would be NS–ID. Forward flow scheduling (forward scheduled) would be DS–ND except for the first stage, which would be IS–ND or NS–ID depending on whether demand or supply drives the flow. Mixed flow scheduling finally would then be a combination with some central point in the flow being NS–ID and then upstream NS–DD and downstream DS–ND.

These examples highlight how inventory can be classified and related to different scheduling and control techniques. As can be seen from these examples, the approach for the operational DPs here follows from the approach selected for the corresponding item DP (which is not generally necessary if control can be differentiated on an operations level). This means that if the item DPs are of the type NS–DD, then also all the operations to produce that item are assumed to be DD (backward scheduled).

As shown above, the classification framework introduced can be used in a contextual way for a wide range of environments to classify different types of DPs. The approach is hence generic and not only applicable in certain environments. The approach rests on a set of fundamental system properties and can therefore be applied in most environments.

CONCLUSIONS AND FURTHER RESEARCH

Materials management is a comprehensive topic usually involving several thousands of items and a multitude of aspects to analyse such as risk, investment, durability, etc. Using abstraction and aggregation as outlined in this paper, the management of inventory can be more explicitly targeted at focusing on the inventory of strategic importance, i.e. the strategic DPs. Process-based classification enables evaluation of inventory management across different delivery strategies (different positions of the CODP), which are of particular interest to companies with hybrid strategies where, for example, both MTO and MTS products are present. From a customer service perspective the management of the CODP inventory is critical to all kinds of delivery strategies. In combination with evaluation of how the inventory at the PODP is managed, a comprehensive view of management of strategic inventory can be obtained. The CODP is also the key starting point for most planning and control methods for manufacturing and logistics. The framework does not in itself provide a new approach to materials management but it provides a structured way to better understand the role of different DPs, i.e. inventory points, and their role in materials management.

With this classification as a point of reference, there are many possible lines of research to pursue. One of the most important is probably that this classification may guide researchers on inventory and materials management in terms of which DPs to focus on. Also, in case of evaluation of manufacturing strategies across different delivery strategies, the strategic DPs may serve as a point of reference for delineation between lean and agility (see, e.g. Mason-Jones et al., 2000). Giesberts and van der Tang (1992) pointed out that in many cases companies have 'hybrid production situations' and they showed how information systems may take this into account. The impact of this classification scheme on hybrid strategies remains an area of interesting opportunities for further research that could also take the framework suggested by Wikner (2014) into consideration. In the same spirit, the practitioners can use this classification to focus their investment on inventory and materials management where it gives the best leverage. The classification may also work as a tool for categorizing different concepts such as planning and control methods to provide guidelines on how to implement these methods based on strategic as well as tactical decoupling points.

ACKNOWLEDGEMENTS

This research has been performed in collaboration with six companies in the projects KOPeration and KOPtimera.

REFERENCES

1. Blackstone, J. H. (Ed.). (2008). *APICS dictionary*. Chicago, IL: APICS The Association for Operations Management.
2. Broeckelmann, R. G. (1999). *Inventory classification innovation: Paving the way for electronic commerce and vendor managed inventory*. Boca Raton, FL: St. Lucie Press.
3. Chopra, S., & Meindl, P. (2007). *Supply chain management: Strategy, planning, and operation*. Upper Saddle River, NJ: Pearson Prentice Hall.
4. Christopher, M. (1998). *Logistics and supply chain management: Strategies for reducing cost and improving service* (2nd ed.). London: Financial Times/Prentice Hall.
5. Duggan, K. J. (2013). *Creating mixed model value streams: Practical lean techniques for building to demand* (2nd ed.). Boca Raton, FL: CRC Press.
6. Giesberts, P. M. J., & van der Tang, L. (1992). Dynamics of the customer order decoupling point: Impact on information systems for production control. *Production Planning & Control, 3*, 300–313.
7. Goldratt, E. M., & Cox, J. (1984). *The goal: A process of ongoing improvement*. Croton-on-Hudson, NY: North River Press.
8. Hoekstra, S., & Romme, J. (1992). *Integrated logistics structures: Developing customer oriented goods flow*. London: McGraw-Hill.
9. Hopp, W. J., & Spearman, M. L. (2000). *Factory physics* (2nd ed.). New York, NY: McGraw-Hill.
10. Lee, H. L. (2002). Aligning supply chain strategies with product uncertainties. *California Management Review, 44*, 105–119.10.2307/41166135
11. Marchwinski, C., Shook, J., & Schroeder, A. (Eds.). (2008). *Lean lexicon – A graphical glossary for lean thinkers* (4th ed.). Cambridge, MA: Lean Enterprise Institute.
12. Mason-Jones, R., Naylor, B., & Towill, D. R. (2000). Engineering the leagile supply chain. *International Journal of Agile Management Systems, 2*, 54–61.10.1108/14654650010312606
13. Mather, H. (1984). *Attack your P:D ratio*. APICS Conference Proceedings, Falls Church, VA, 84–88.
14. Ohno, T. (1988). *Toyota production system: Beyond large-scale production*. Cambridge, MA: Productivity Press.
15. Olhager, J. (2003). Strategic positioning of the order penetration point. *International Journal of Production Economics, 85*, 319–329.10.1016/S0925-5273(03)00119-1 ,
16. Olhager, J., Selldin, E., & Wikner, J. (2006). Decoupling the value chain. *International Journal of Value Chain Management, 1*, 19–32.10.1504/IJVCM.2006.009021
17. Orlicky, J. (1975). *Material requirements planning: The new way of life in production and inventory management*. New York, NY: McGraw Hill.
18. Rother, M., & Shook, J. (2003). *Learning to see: Value stream mapping to create value and eliminate muda*. Cambridge, MA: Lean Enterprise Institute.
19. Rudberg, M., & Wikner, J. (2004). Mass customization in terms of the customer order decoupling point. *Production Planning & Control, 15*, 445–458. ,
20. Sharman, G. (1984). The rediscovery of logistics. *Harvard Business Review, 62*, 71–79.
21. Shingo, S. (1981). *Study of 'Toyota' production system from industrial engineering viewpoint*. Tokyo: Japan Management Association.
22. Taylor, S. G., & Bolander, S. F. (1994). *Process flow scheduling: A scheduling framework for flow manufacturing*. Falls Church, VA: APICS.
23. Vollmann, T. E., Berry, W. L., Whybark, D. C., & Jacobs, F. R. (2005). *Manufacturing planning and control systems for supply chain management*. New York, NY: McGraw-Hill.
24. Wallin, C., Rungtusanatham, M. J., & Rabinovich, E. (2006). What is the 'right' inventory management approach for a purchased item? *International Journal of Operations & Production Management, 26*, 50–68. ,

25. Wemmerlöv, U. (1984). Assemble-to-order manufacturing: Implications for materials management. *Journal of Operations Management, 4*, 347–368.10.1016/0272-6963(84)90021-4

26. Wikner, J. (2005). *Hantering av osäkerhet i försörjningskedjor med utgångspunkt i frikopplingspunkter* [Management of uncertainty in supply chains from a decoupling point perspective]. Proceedings of Produktionslogistik 2005 – Plans forsknings-och tillämpningskonferens, Borås, 221–236.

27. Wikner, J. (2014). On decoupling points and decoupling zones. *Production & Manufacturing Research, 2*, 167–215.

28. Wikner, J., Johansson, E., & Persson, T. (2009). *Process based inventory classification.* The 21st Annual Nofoma Conference,Jönköping Sweden.

CHAPTER 8

PRODUCTION SCHEDULING FOR CONTINUOUS MANUFACTURING SYSTEMS WITH QUALITY CONSTRAINTS

Jalel Ben Hmida[a], Jim Lee[a], Xinchun Wang[a] & Fathi Boukadi[b]

[a] Systems Engineering, University of Louisiana at Lafayette, Rougeou, Hall Room # 244, Lafayette, LA 70504, USA

[b] Department of Petroleum Engineering, University of Louisiana at Lafayette, Madison Hall, Room # 134, Lafayette, LA 70504, USA

ABSTRACT

This research is motivated by a real world production scheduling problem in a continuous manufacturing system involving multiple objectives, multiple products and multiple processing lines with various inventory, production and quality constraints. Because of the conflicting objectives, a global optimization approach is considered as not feasible by the plant management. Given a customer demand forecast, two practical heuristic or sequential optimization algorithms are developed to generate daily production schedules for two primary objectives: minimize shipment delays (pull-backward procedure) and minimize average inventory levels (push-forward procedure). A third heuristic algorithm (reduce switch-over procedure) which is based on the current management practice is also developed to serve as a benchmark. A factorial experiment was performed to evaluate the performance of the heuristic procedures and to identify factors that might affect the performance differences among heuristics. Since each heuristic is designed to give priority to one of the three conflicting objectives, none of them is absolutely superior to the other algorithms in all aspects. However, the first two heuristic procedures performed better than the current management practice in shipment delays and average daily inventory. The production schedules generated by the two procedures also satisfy the quality constraints. The experimental results also showed that the performance of the algorithms is significantly affected by product mix, inventory levels, and demand pattern.

INTRODUCTION

Scheduling is concerned with efforts to see that the resources of a manufacturing system are well utilized so that the products are produced within reasonable accord with the customer demand. Numerous planning and scheduling methods have been developed for discrete part manufacturing systems, and published research has shown that the approach for scheduling production depends on the type of manufacturing system involved. Reviews of the characteristics of continuous manufacturing systems in Vollmann, Berry, and Whybark (1992), Askin and Standridge (1993) suggest that the continuous manufacturing systems require different modeling approaches.

This research is motivated by an actual production scheduling problem in a continuous manufacturing system of a

Citation: Jalel Ben Hmida, Jim Lee, Xinchun Wang & Fathi Boukadi, Production scheduling for continuous manufacturing systems with quality constraints, DOI:10.1080/21693277.2014.892846.

large producer of chemical products. On average, the chemical plant receives more than 200 customer orders in a month, totaling about four million pounds of chemical products. The production system is capable of producing a large number of different types of products through multiple processing lines. A large number of process constraints exist due to varied capabilities of the production lines and the processing requirements. Most of the products can be produced on more than one line and some of the processes require the sharing of special tools. Some products have precedence constraints, and some products have similar production conditions that should be scheduled for production consecutively.

Currently, scheduling daily production in the system is subjectively based on the management's experience. With an increasing emphasis on the multiple objectives of on-time shipment, low inventory, and production quality; the management of the plant needs a scheduling tool to improve the production scheduling for better system performance. The three objectives identified by the plant management are described as below:

- *On-time shipments*. The current overdue shipment rate (OSR) is unacceptable by the top management. The lack of on-time delivery may prevent the plant from being considered as a major supply candidate by the downstream industries. A measure of on-time shipment is commonly considered as the most important performance indicator of a continuous manufacturing system.
- *Low inventory*. To reduce overdue shipments and if processing lines are available, finished products are normally produced before the due dates. This results in high level of inventory. While "zero inventory" may not be practically feasible, efforts must be made to reduce the overdue shipments and the inventory level at the same time.
- *Product quality*. Production run length is defined as the number of days a processing line is scheduled to produce the same product type. Longer runs can produce finished products with consistent quality. Frequent product switchovers in the processing line can result in quality problems. However, a run length larger than necessary can increase the inventory level.

The management of continuous manufacturing systems of this type faces the dilemma of meeting customer delivery dates while operating the system efficiently. This involves conflicting objectives. The conflict arises because improvement in one objective can be made to the detriment of one or more of the other objectives. Moreover, various production and quality constraints need to be satisfied. All of these concerns are addressed in this research. Currently the approach used by the plant management gives priority to product quality where the product switchover is minimized. A scheduling tool, which can generate several daily production schedules based on algorithms with priorities given to different objectives, would be helpful to the plant management.

LITERATURE REVIEW

Production scheduling, perhaps because of its considerable quantifiable structure and richness of virtually endless variations, continues to attract many researchers for scholarly work. Patterson, Slowinski, Talbit, and Weglarz (1989) stated that most of the realistic scheduling problems could not be solved in a computationally efficient manner. Linear programming (LP) is commonly used to solve production planning and scheduling problems optimally, but the problem formulation is normally over-simplified and the solution is used only at the strategic level. Heuristic rules have been developed to deal with practical scheduling problems on the shop floor.

Deepark and Grossmann (1990) proposed a multi-period linear programming model for the simultaneous production planning and scheduling of multi-product batch manufacturing systems that may consist of several non-identical parallel lines. The model involves a simplistic representation of the scheduling problem where many possible constraints are not considered. Gupta, Bector, and Gupta (1990) provided an approach to determine the optimal sequence of products with due date in a single machine system. Linear programming models were developed for three different due date assignment methods. Minton, Johnston, Phillips, and Laird (1992) described a simple heuristic approach to solve large-scale constraint satisfaction and scheduling problems. Anwar and Nagi (1997) formulated an integrated scheduling and lot-sizing problem. An efficient heuristic is developed that schedules operations by exploiting the critical path of a network and iteratively groups in order to determine the lot-size that minimizes the make span as well as set-up and holding costs. Certain aspects of scheduling problems such as operational pre-conditions (e.g. product sequencing and cleaning requirements) are important to avoid compromising final product quality, safety and feasibility of operations. (Amaro & Barbosa-Póvoa, 1999) developed a Mixed Integer Linear Programming (MILP) formulation so as to account for operational conditions. The formulation is solved using a Branch and Bound (B&B) procedure. In some production systems, early shipments are forbidden. Scudder, Hoffmann, and Rohleder (1993) explored various policies on shipments using a simulation model and also came up with best policies.

Many scheduling problems are multi-objective that can be efficiently solved by a heuristic approach. Daniels (1994) described a multi-objective scheduling situation where information concerning the relative importance of the criteria is available through interaction with a given decision maker. Yim and Lee (1996) used Petri nets and heuristic search for multiple objectives scheduling in a flexible manufacturing systems (FMSs). Tung, Lin, and Nagi (1999) presented a hierarchical approach to scheduling in a FMS that pursues multiple performance objectives. Viana and de Sousa (2000) used heuristics to solve multi-objective resource constrained project scheduling problem. Lou, Liu, Zhou, Wang, and Sun (2012) present a method for developing a pro-active scheduling model that anticipates known uncertainties and a reactive model that accounts for and handles unknown uncertainties.

One approach to avoid frequent product switchover in production is to assign a long setup time between product switchovers. He and Kusiak (1992) considered a mixed-integer formulation of the single-machine scheduling program with sequence-dependent setup times and precedence constraints. Seki and Kogure (1996) developed a stochastic-demand model to deal with lot scheduling problem for multi-item continuous demand utilizing a single machine with setup times. A recent study by Wagner and Smits (2004) used local search heuristic for stochastic economic lot scheduling problem that considered multiple product in a single facility with limited capacity and significant setup times and costs.

Recent research in scheduling has also incorporated various management concepts. Modarress, Ansari, and Willis (2000) presented a technique using Just-In-Time (JIT) and Statistical Process Control (SPC) methods to minimize the short run production in an environment with demand uncertainties. Bose and Pekny (2000) presented Model Predictive Control (MPC) concept for planning and scheduling problem. A forecasting model was developed to calculate target inventory. The importance of Supply Chain (SC) production planning was studied in Caramanis, Anli, and Paschalidis (2003). Their model produces weekly production schedule that minimizes inventory and backlog costs subjected to non-linear constraints on production imposed by weekly varying dynamic lead times and inventory hedging policies.

Giannelos and Georgiadis (2003) investigated scheduling in a continuous manufacturing process. They developed a model for scheduling fast consumer good manufacturing process using MILP techniques combined with continuous representation of time so as to alleviate some of the computational problems incurred by discrete time-index formulations. Kogan, Leu, and Perkins (2002) suggested polynomial-time algorithm for the continuous-time scheduling problem in a manufacturing system where multiple product types are produced on a set of parallel machines. The main objective of the study was to schedule the production so that inventory, production, and planning horizon costs are minimized.

Review of scheduling literature indicates that most of the realistic scheduling problems are solved by efficient heuristic methods. These methods are developed primarily for discrete manufacturing systems and do not consider the effects of scheduling on product quality. In this paper, mathematical models which formulate the problem in continuous manufacturing systems with multiple objectives are presented first. Several heuristic algorithms are then proposed to solve the scheduling problem with priority given to different objectives. The management can determine an acceptable production schedule based on the proposed decision support procedures. The performance of the heuristic procedures is evaluated based on a simulation experiment.

PROBLEM DESCRIPTION

The scheduling problem addressed in this research can be stated as follows: *A* continuous manufacturing system can produce*M* types of products. There are *N* processing lines, each being capable of producing a subset of *M* products. If finished products are produced earlier than the due date, they are stocked in the warehouse where each product has a limited storage area. The production rate varies by products on a line, and only one product can be produced by one line at any time.

Table 1 shows a sample of the data of maximum inventory level allowed and production rate of each processing line by product. For the purpose of modeling, the setup time is included in our estimation of production rate by processing line. For example, if line *A* is used to produce product 1SB and the minimum run length is 4 days. The total production in this run of 4 days is estimated to be 100 (thousand pounds), and the daily production is considered to be 100/4=25 (thousand pounds). The real production time of the last day in a run is normally shorter than the other days, as the line must be cleaned and adjusted (i.e. setup time) for another run with a different product. As a result, the total estimation is normally slight lower than the actual production to cover the loss of setup time. When the run length is long, it is possible that the production run can be completed 20–30 min early. The effect of the difference does not significantly affect the estimation on the production rate.

Table 1: Sample data for inventory levels and processing line capabilities

		Production rate by line (×1000 lb/day)				
Product	Maximum inventory (×1000 lb)	*A*	*B*	*C*	*D*	*E*
1SB	150	25	25	0	0	0
1SC	262	25	25	0	0	0
2SB	80	24	0	24	0	0
2SC	126	24	0	24	0	0
3SB	150	23	0	0	23	0
3SC	189	23	0	0	23	0
4B	150	19	0	0	0	19
4C	112	19	0	0	0	19
5	180	34	0	0	0	0
6B	120	0	29	0	0	0
6C	180	0	29	0	0	0

For example, if line *A* is used to process product 1SB for one day, then 25 thousand pounds of 1SB can be produced. While most of the products can be produced on more than one line, some of the products can be produced only on one line (e.g. product 5 can be produced on Line *A* only). A sample of customer orders by days is shown in Table 2. Based on the customer orders, the scheduling problem is to determine the daily production schedule for each processing line with production, inventory and quality constraints.

Table 2: Sample monthly customer order (in thousand units)

	Product type												
Due day	1SB	1SC	2SB	2SC	3SB	3SC	4B	4C	5	6B	6C	7	8B
1	0	15	20	0	0	0	0	0	35	0	0	0	0
2	0	0	0	0	0	0	0	10	0	0	35	0	0
3	20	10	0	20	0	30	0	0	0	20	0	0	0
4	0	30	0	0	0	0	0	0	0	0	0	0	0
5	5	0	0	0	0	0	0	0	0	0	0	0	0
6	10	0	15	0	0	0	0	0	0	0	0	0	0
7	0	0	0	0	0	0	0	0	5	0	0	0	0
8	0	30	0	0	0	0	0	0	0	0	0	0	0
9	20	0	0	0	0	0	0	0	0	10	35	0	40
10	0	0	30	0	0	0	0	0	0	0	0	0	0

To formulate the problem we require the following notation:

T	planning horizon
t	day in planning horizon ($t=1,\ldots,T$)
M	number of product types
N	number of processing lines
i,j	product index ($i,j=1,\ldots,M$)
k	processing line index ($k=1,\ldots,N$)
L_i	waiting period for product i in days. Waiting period is the number of days that a product needs to stay in warehouse before shipment for quality assurance
U_i	actual due date of product i
E_i	adjusted due date of product i based on waiting period
HL_i	minimum run length (in days) of product i
IH_i	maximum inventory allowed for product i
I_{it}	inventory of product i in day t
D_{it}	demand of product i in day t
Q_{ikt}	production quantity of product i on line k in day t
QO_i	quantity of overdue product i
W_{kt}	times of product switchovers on line k in day t
C_{ik}	daily production rate for product i on line k
P	set of jobs $[i,j]$, where product i precedes product j
G_g	set of product group which should be produced sequentially
g	represents the type of groups
R_r	set of products that require the same special tool (resource) r for processing
r	represents the type of resource
RA_r	number of units of resources available for products in set R_r
H_{ik}	production run length (in days) of product i on line k
B_{ik}	production starting day of product i on line k
O_{it}	demand of product i that cannot be produced on time in day t
X_{ikt}	a zero-one variable. X_{ikt} is 1 if product i is scheduled on line k in day t

The three objectives to be considered in the production scheduling are: minimizing overdue shipment, minimizing inventory, and minimizing product switchovers. They are discussed below:

(1) OSR measures the amount of customer demands that cannot be met by a production schedule. In this research, it is expressed as ratio of overdue shipment amount to the total demands in the planning horizon. The overdue shipment of production day t (O_{it}) is demand (D_{it}) minus the inventory $I_{i,t1}$ and minus the production of product i on all lines (Q_{ikt}).

$$\mathrm{OSR} = \sum_i QO_i / \sum_t \sum_i D_{it} \tag{1}$$

where $QO_i = \sum_t O_{it}$ and $O_{it} = D_{it} - I_{i,t-1} - \sum_k Q_{ikt}$

If $O_{it} < 0$, let $O_{it} = 0$

(2) Average daily inventory (ADI) reflects the average daily amount of finished products produced before the due date. The inventory of product i on day t (I_{it}) is the initial inventory plus production on day t and minus the demand on day t. The ADI is determined by the total inventory divided by the planning horizon T. It is shown as follow:

$$\text{ADI} = \sum_t \sum_i I_{it}/T \quad (2)$$

$$\text{where } I_{it'} = I_{io} + \sum_{t=1}^{t} \sum_k Q_{ikt} - \sum_{t=1}^{t} D_{it}$$

If $I_{it'} < 0$, let $I_{it'} = 0$

(3) Total Switch Times (TST) of the product is used to check the quality of a product as frequent product switchovers could lead to poor product quality. For everyday t and every line k, product switchover count W_{kt} is 1 if the product scheduled on day t is different from $t-1$. The TST can be defined as:

$$\text{TST} = \sum_k \sum_t W_{kt} \quad (3)$$

where $W_{kt}=1$ if the product scheduled on day t is different from $t-1$.

Several types of production, inventory and quality constraints are considered in scheduling the daily production. They are summarized below.

(1) *Production capacity and capability constraints.* The quantity of product i that can be produced on line k in day t, Q_{ikt} should be less than the capacity C_{ik} of line k. The capacity of line k is a function of product type i. For every i and every k,

$$Q_{ikt} \leq C_{ik} \quad (4)$$

where C_{ik} is 0 if line k is not capable of producing product i.

(2) *Inventory constraints.* Each product is allocated limited storage space in the warehouse. A maximum inventory level IH_i is imposed on each product i.

$$I_{io} + \sum_t \sum_k Q_{ikt} - \sum_t D_{it} \leq \text{IH}_i \quad (5)$$

(3) *Minimum run length constraints.* To achieve the desired quality level, some products have minimum run length requirements as defined by HL_i. For every i, H_{ik} represents the actual run length in days of product i on line k:

$$H_{ik} \geq HL_i \quad (6)$$

(4) *Precedence constraints.* If product i must be produced before product j, then the production starting date of product i on line k (B_{ik}) should be completed before the line starts to process product j.

$$B_{ik} + H_{ik} \leq B_{jk} \quad (7)$$

In our example, product 2SB must be produced prior to product 4B and 4C.

(5) *Group constraints.* Products with similar production conditions should be produced sequentially to ensure product quality. If product i and j are in the same group, it is desired to produce these product back to back on line k.

$$B_{ik} + H_{ik} = B_{jk}, \text{ or } B_{jk} + H_{jk} = B_{ik} \quad (8)$$

For example, Product 17B, 18B, 20 require similar production conditions. Then they should be produced sequentially in the production schedule.

(6) *Resource sharing constraints.* The processing operations of some products require the use of special equipment with capacity constraints. Let RA_r represent the number of tools available and R_r is a set of products required in the production. At any time, no more than RA_r products in R_r can be produced.

$$\sum_{i \in R_r} \sum_k X_{ikt} \leq RA_r \quad (9)$$

where $X_{ikt}=1$ if i is scheduled on k.

For example, no more than two S type products can be produced at one time on all lines as the processing of a S type product requires the use of a special tool 1, and $RA_1=2$.

(7) *Preferable product-line assignment.* Since most of the products can be produced on more than one processing line, priorities on the product-line assignment can be established by the management. The assignment may affect the production efficiency and the quality of the product in some systems. For example, products 2SB and 3SB are preferably produced on line B though they can also be produced on line A.

The number of constraints for this multi-line, multi-product facility is more than one hundred, and the very challenging task is how to satisfy these constraints in the scheduling process while considering the conflicting objectives. This research proposes the development of practical heuristic scheduling algorithms to generate various production schedules based on the demand data and the constraints. With the scheduling tool, the management will be able to generate several feasible production schedules with estimated overdue shipment, daily inventory levels and the number of product switchovers. This information will allow the management to determine the actual daily production schedule that balances the three objectives.

HEURISTIC SCHEDULING ALGORITHMS

The research problem is to construct production schedules that minimize (1) OSR, (2) ADI and (3) TST while satisfying the various production, inventory and quality constraints. With three conflicting objective functions, it is infeasible to develop a global optimization model which optimizes all three functions. Therefore, three heuristic or sequential optimization algorithms are developed in this research to optimize each of the three objectives. The Pull-Backward Heuristic Procedure (PB) first produces an initial production schedule to minimize the OSR. The initial schedule is then adjusted to reduce ADI without affecting the OSR. This procedure is basically a sequential optimization method which optimizes the OSR and ADI in two steps. The Push-Forward Heuristic Procedure (PF) first produces an initial schedule which minimizes the ADI. The initial schedule is then adjusted to reduce OSR without affecting the ADI. Similar to the PB schedule, the PF is a sequential optimization method which optimizes the ADI first then OSR in two steps. Finally, the Reduce Switchover Heuristic Procedure (RS) is used to produce an optimal solution on TST first. Then the OSR is considered next, followed by ADI.

The methodology underlying the three procedures is similar and follows the three basic steps: problem decomposition, product sequencing, and demand shifting. First, the planning horizon is broken into several smaller periods. If the planning horizon is one month, the length of a period can be about one week. All orders of the same product in a period are lumped together to form one product demand. The production for each of the periods is scheduled according to the lumped demand. After all the period schedules are obtained, they are merged together to form the schedule for the whole planning horizon.

In creating a period schedule, two sequencing rules are used to create priorities. The Earliest Due Date (EDD) rule is chosen as the primary rule, where the choice is made based on the product that has the earliest due date. The Shortest Processing Time (SPT) rule, where the choice is made based on the processing operation with the shortest duration, is used as a secondary rule. For example, if 1SB, 1SC and 2SB are produced in line A (see Table 2) in the first 7-day period, 1SB will be produced last as it has an order of 20 due on day 3 compared to the due date of 1 for 1SC (with order quantity 15) and 2SB (with order quantity 20). Since 1SC and 2SB have the same due date, they are sequenced based on SPT rule. The processing priority is given to 2SB because the demand of 35 thousand pounds of 2SB can be produced with a shorter processing time than the time required to produce 55 thousand pounds of 1SC.

When the production cannot be scheduled for on-time delivery due to capacity constraints, the demand can be shifted to another processing line according to line capability. If other lines are also occupied, the demand can be shifted to the next period. To satisfy precedence constraints, group constraints, and resource constraints; the product demand can be shifted to other dates based on the three objectives of the scheduling. Different approaches of shifting are employed for different heuristic procedures. However, the demand of a product is always shifted forward to satisfy the maximum inventory constraints. The procedures are summarized below.

Pull-Backward Heuristic Procedure (PB)

The PB procedure is basically a sequential optimization method which optimizes the OSR and ADI in two steps. If the demand is greater than the production capacity in a given period, then not all production runs can be scheduled before the due dates. In this case, an attempt is made to schedule the production in the previous period to avoid shipment delay. However, the ADI may go up.

- Step 0. Initialization

Adjust the due date of customer order U according to waiting period L to form an adjusted demand date E in planning horizon. Given a product i, let $E_i = U_i - L_i$ for all products with a non-zero waiting period. Break planning horizon T into PN scheduling periods; let v be the length of the period and n be the nth period ($n=1, \dots, \mathrm{PN}$) in the planning horizon, Then $\mathrm{PN} = \frac{T}{v}$. If PN is non-integer, it is rounded down to the nearest integer.

- Step 1. Product-processing line assignment

Assign products to processing lines based on preferable product-line assignment, subject to capability constraints. One product is assigned to one line.

- Step 2. Period schedule generation

Construct a trial production schedule based on demand for period n subject to production capacity constraints and minimum production run length.

- (2.1.) Compute the total demand due in period n by product and by line. Let DD_{ikn} represents this demand for product i assigned to processing line k. For each k, sort due day E_i for all products in ascending order, resulting in $k$sorted lists. If a product has several orders in period n, E_i is computed based on the order with the earliest due date. Products with smaller E_i values have higher priorities to be scheduled for production. If two or more products have the same due day, the product with the smallest processing time (SPT) has a higher priority to be scheduled than other products.
- (2.2.) For each processing line k, sequence products for the line based on the priority established in the sorted list from step 2.1.
- (2.3.) Determine production run length H for each of product scheduled in the period. Run length of a product in a period is determined according to the total quantity of demand, capability and minimum run length constraints:

$$H_{ik} = \begin{cases} DD_{ikn}/C_{ik} & \text{for} \quad H_{ik} > HL_i \\ HL_i & \text{for} \quad H_{ik} \le HL_i \end{cases} \tag{10}$$

where DD_{ikn} is the demand of product i on line k in period n. H_{ik} is rounded up to the next integer if DD_{ink}/C_{ik} is not an integer. Repeat this step for all k. (For example, product 1SC in the first 5-day period in Table 2 has $DD_{ik1}=55{,}000$ and $C_{ik}=25{,}000$. DD_{ink}/C_{ik} is 55,000/25,000, or 2.2. Therefore, H_{ik} would be rounded up to 3.)

- (2.4.) Shift the unfilled demand of a product to another processing line in accordance with capability constraints. If H_{ik} cannot be totally assigned to line k in the period, shift the unscheduled demand to another line which is not fully utilized. If product i has an unfilled demand in period n and cannot be assigned to any other processing line, shift the unfilled demand to period $n+1$.

- Step 3. Period schedule revision
    1. (3.1.) Modify the schedule to meet the precedence constraints. If product j has been scheduled at time t_1 before product i at time t_2 where $[i, j]$ has precedence constraint P, then re-schedule i at t, where $t<t_1$. Adjust production schedule on the processing line accordingly. Repeat this step for all processing lines.
    2. (3.2.) Modify the schedule to meet the group

constraints. If production of a set of products $[i, j]$ in G_g is separated in the schedule and product i is in the earliest position, shift all other products in the same group next to and immediately following the product i. Shift the production of products without group constraints to a later day in the schedule if needed.

- Step 4. Period iteration

Repeat step 1 to step 3 until PN schedules $S_1, S_2, \ldots, S_{PN}$ are obtained for each of the PN periods.

- Step 5. Schedule Combination

Merge period schedules and check constraints.

3. (5.1.) Merge subsets of schedules $S_1, S_2, \ldots, S_{PN}$ to form a complete schedule S.
4. (5.2.) Exchange the position of products scheduled at the same line if precedence constraints are violated.
5. (5.3.) Check resource constraints for the production schedule across all processing lines. If a constraint is violated in day t, shift the demand of the product with lower DD_{ikn} to an earlier day if possible. If the product cannot be produced on an earlier day, shift the demand to day $t+1$. Adjust the production schedule on the same processing line correspondingly.
6. (5.4.) Satisfy the maximum inventory constraints. Assume the production on product i in line k can be shipped in the same day to meet the demand. Determine the inventory level for all products. If the inventory of product is greater than the maximum level allowed, shift that demand to day $t+1$ and then check the resource constraints. The production schedule will be shifted to the future if the new schedule violates the resource sharing constraints.

Push-forward heuristic procedure (PF)

The PF is a sequential optimization method which optimizes the ADI first then OSR in two steps. If the demand is greater than the production capacity in a given period, however, the demand is shifted to the next period to avoid inventory. Based on the previous description, only two steps need to be replaced. They are described below.

1. (3.1.) Modify the schedule to meet the precedence constraints. If product j has been scheduled at time t_1 before product i at time t_2 where $[i,j]$ has precedence constraint P, then re-schedule j at t, where $t > t_2$. Adjust production schedule on the processing line accordingly. Repeat this step for all processing lines.
2. (3.2.) Modify the schedule to meet the group constraints. If production of a set of products $[i,j]$ in G_g is separated in the schedule and product$i$ is in the earliest position, shift all other products in the same group to be produced immediately before the product i. Shift the production of products without group constraints to a later day in the schedule if needed.
3. (5.3.) Check resource constraints for the production schedule across all processing lines. If a constraint is violated in day t, shift the demand of the product with lower DD_{ikn} to day $t+1$. Adjust the production schedule on the same processing line correspondingly.

Reduce Switchover Heuristic Procedure (RS)

The RS procedure focuses on minimizing product switchovers on processing lines which is similar to the current practice of the plant management. It is used to produce an optimal solution on TST first. Then the OSR is considered next, followed by ADI. Based on the current practice, once a processing line is assigned, the production run is continued until the total demand for the planning horizon (normally one month) is satisfied. The advantage of the current practice is that the method is easy to implement. The disadvantage is that both the OSR and ADI are not acceptable.

To implement the RS algorithm, the number of days in a period used in the previous procedures must be increased to allow a longer run. A test on various period lengths shows that the length of the period should be doubled or the number of periods (PN) should be reduced by about 50% in this procedure to observe a statistically significant different number of product switchovers. With RS, all the steps remain unchanged as procedure PB. Only the PN is changed. Other steps remain unchanged as procedure PB.

From the above discussions, it is noted that different production schedules can be obtained by using different heuristic procedures. A computer program written in C language is then implemented to include the three heuristic procedures. Many variables, such as the initial assignment of products to the processing lines and the number of periods in the planning horizon can be changed in the program. Consequently, alternative schedules can be produced for the management to make scheduling decisions.

SIMULATION EXPERIMENTS

Simulation experiments were carried out to evaluate the performance of the three heuristic procedures. The purpose of the experiment was to compare the performance of the procedures by answering the following questions:

- What is the magnitude of the relative performance of one procedure over another?
- If the performances of the two procedures are statistically significant different, then what are the factors related to the problem attributes affecting their differences?

To answer these questions, four factors that may have influence on the performance measures are identified. They are as follows:

1. *Group ratio (G).* The group ratio is expressed as a ratio of the number of products with group constraints to the total number of products produced in the system. For example, if the total number of products is 50, and five of them are involved in group constraints, then the group ratio is 5/50 or 10%. Three levels are tested: 10, 25, and 50%.
2. *Precedence strictness ratio (P).* The precedence constraints specify the products that must be produced

before others. The precedence strictness ratio is defined as the number of precedence requirements among the products over the number of feasible pairs of products. For example, in a system producing five products, the maximum number of precedence constraints is 5!/2!3! or 10. If only one constraint exists for a pair of products, then the precedence strictness ratio is 1/10 or 0.1. Two levels are tested in the experiment: 0.1 and 0.2. These two ratios are relatively low because the type of products in the continuous manufacturing system we studied does not have too many precedence constraints.

3. *Maximum inventory (I).* The maximum finished-goods inventory allowed in the system is basically limited by the space in the warehouse. To reduce the inventory cost, it is desired to reduce the maximum inventory level. Two levels of maximum inventory are tested. A level of 100% represents the actual maximum inventory allowed in a chemical plant, which is about 4500 thousand pounds of products in total. The level of 67% means that the maximum inventory allowed is reduced to 4,500 × 67% or 3000 thousand pounds in total.

4. *Demand (D).* Three sets of real demand data were collected from a chemical plant. They are referred to as D1, D2, and D3. Each data set contains the daily demand for over 60 product types over a month period. The average total daily demands for D1, D2 and D3 are 96.6, 108.1 and 109.4 thousand pounds respectively. The standard deviations of the total daily demand of the three demand patterns are 93.90 for D1, 74.63 for D2, and 75.24 for D3.

Since the production scheduling issues are unique, a well-known data is not available in the literature. Using the three data sets from industry, a factorial experiment was designed and performed to evaluate the performance of the three heuristic procedures and to identify the factors affecting the three objectives scheduling.

Table 3 summarizes the factors and associated levels tested in the experiment. A full factorial experimental design is used, so the number of basic experiments designed for each heuristic procedure is 3 × 2 × 2 × 3 = 36. In order to account for variations in the experiments, five replications are created for each of the basic experiments. Considering three heuristic procedures and five replications per experiment, the total number of experiments would be 36 × 5 × 3 = 540.

Table 3: Experiment factors and levels

Factor	Level
Group (G)	10%; 25%; 50%
Precedence (P)	10%; 20%
Maximum Inventory (I)	100%; 67%
Demand (D)	D1; D2; D3

The method to replicate the experiment is based on a 10% random perturbation on demands. For every customer order in the planning horizon, a uniformly distributed random number between (0, 1] is generated to determine the direction of the perturbation. If the random number is between (0, 0.25], then the value is decreased by 10%. If the random number is between (0.25, 0.5], the order is decreased by 5%. If the random number is between (0.5, 0.75], the corresponding value will be increased by 5%. Finally, the value is increased by 10% if the random number is between (0.75, 1). This allows the demand data of replications to be varied within a 20% range. For example, if the original demand for a product is 25, it may be changed to 22.5 (i.e. 25 × .90), 23.75 (i.e. 25 × 0.95), 26.25 (i.e. 25 × 1.05) or 27.5 (i.e. 25 × 1.10) in the replications.

Three performance measures corresponding to each of the three scheduling objectives are used in the experiments: (1) OSR, (2) ADI and (3) Total switch times (TST). The three performance measures have been defined in Equations (1)–(3).

Based on OSR, ADI and TST, a total cost index (TCI) can be estimated to represent an overall economical performance for a schedule where:

$$TCI = (400\ QO + ADI + 2500\ TST)/100 \tag{11}$$

The quantity QO is the total amount (in thousand pounds) of overdue shipment products, ADI is the ADI in thousand pounds, and TST is the number of times the switch-over occurs in the production schedule. Note that the coefficients of the three parameters are estimated to present a certain relation between parameters, and the values vary in accordance with different production systems.

ANALYSIS OF EXPERIMENTAL RESULTS

Based on the experimental design described in the previous section, a total of 540 experiments were performed. The performance measures of the 180 experiments for each of the heuristic procedures were averaged and presented in Table 4.

Table 4: Summary performance by Heuristic

	Heuristic procedure			% Difference	
Performance measure	PB	PF	RS	PB-RS	PF-RS
OSR	0.09[a]	0.11	0.12	3	1
ADI	73,164	69,883[a]	73,852	1	5
TST	42.40	40.33	36.51[a]	−14	−9
TCI	3032[a]	3233	3211	6	1

[a]Indicates the best performer by performance measure.

Among the three objectives of the scheduling, OSR and ADI can be considered as the primary or key performance measures for comparing different production schedules. As the product quality is critical to the success of the business, every production schedule developed by management must satisfy the run length constraints to insure that the quality of products is acceptable. Therefore, TST is simply used as a benchmark for schedule comparison purpose. In the real world a smaller TST is easier to implement and control, so it can be also used by management to evaluate the feasibility of schedule implementation. With the consideration of the minimum run length constraints in our scheduling problem, all three algorithms proposed can schedule production with acceptable product quality.

Table 4 shows that each heuristic procedure performs best for one of the three objectives as expected. Procedure RS results in an average of 12% OSR for the test problems. If the reduction of OSR is critical, then procedure PB can reduce the rate from 12% to 9%, and procedure PF is also slightly better than RS by 1%. For the ADI, PB is better than RS by 1%, and PF can reduce the inventory by more than 5%. However, the product switchover (TST) for PB and PF will go up by 14 and 9%. Based on the TCI with the coefficients defined in equation (11), procedure PB is the most economical procedure among the three. However, this superiority can be diminished if the coefficients in equation (11) are changed.

The summary results shown in Table 4 reveal that the performance measures of the three heuristic procedures are different. To determine whether the differences are significantly away from 0, t tests were performed. Under the null hypothesis, there should be no difference between the performance measures obtained from each procedure. If the null hypothesis is rejected at a selected level of significance, it is concluded that the difference exists. The assumption of normality on the data was first tested and verified. The test was set at 0.01 level of significance.

Four types of variables were used to measure the relative difference of the heuristic procedures. Variables DOSR(i, j) measures the difference in OSR between two heuristic procedures i and j:

$$\mathrm{DOSR}(i,j) = \mathrm{OSR}(i) - \mathrm{OSR}(j) \tag{12}$$

Variables for measures of ADI, TST and TCI are expressed as percentage difference between two procedures i and j:

$$\mathrm{PADI}(i,j) = [\mathrm{ADI}(i) - \mathrm{ADI}(j)]/\mathrm{ADI}(j) \tag{13}$$

$$\mathrm{PTST}(i,j) = [\mathrm{TST}(i) - \mathrm{TST}(j)]/\mathrm{TST}(j) \tag{14}$$

$$\mathrm{PTCI}(i,j) = [\mathrm{TCI}(i) - \mathrm{TCI}(j)]/\mathrm{TCI}(j) \tag{15}$$

The reason of using only difference in OSR is that the OSR is already a percentage value. Moreover, the value for measure of percentage difference between two procedures sometimes might be infinite since OSR(j) could be near zero.

The difference of the four statistics (OSR, ADI, TST and TCI) and the results of the t tests appear in Table 5. A '*' is marked in the "Significance" column if the hypothesis that the difference is 0 cannot be rejected at the 0.01 level of significance.

Table 5: t-test results by performance measure

Variable	No. Obs	M (%)	SD	Significance
DOSR(PB,RS)	180	−2.24	0.036	0.0001
DOSR(PB,PF)	180	−1.96	0.052	0.0001
DOSR(PF,RS)	180	−0.28	0.064	0.554*
PADI(PF,RS)	180	−5.43	0.017	0.0001
PADI(PF,PB)	180	−4.77	0.019	0.0001
PADI(PB,RS)	180	−0.94	0.015	0.0001
PTST(RS,PB)	180	−14.80	0.081	0.0001
PTST(RS,PF)	180	−9.71	0.032	0.0001
PTST(PF,PB)	180	−7.69	0.192	0.0001
PTCI(PB,RS)	180	−5.17	0.214	0.0002
PTCI(PB,PF)	180	−4.70	0.214	0.0037
PTCI(RS,PF)	180	−4.43	0.428	0.071*

*Indicates that the difference is insignificant at level 0.01.

The t tests show that procedure PB significantly outperforms PF and RS, but the difference between PF and RS is not significant. The OSR generated by procedure PB is about 2% less than that of procedure PF and is 2.24% less than that of procedure RS. Procedure PF can generate the lowest level of ADI, and procedure PB ranks second in ADI. The t tests indicate that all the differences are significant, although the percent deviations are not very large. Procedure RS is the best on TST as the approach gives priority to reduce the number of product switchovers. The differences among the three procedures are statistically significant different. The statistics of the TCI show that procedure PB is significantly superior to the other two procedures for a system with cost coefficients similar to Equation (11). The difference between PF and RS is not significant. The results show that no procedure is absolutely superior to other procedures in every aspect. On the other hand, the experiment shows that the three procedures perform as desired since each of them is given a different priority.

The preceding observation suggests that further statistic analyses can be done to evaluate the effects of various system factors and their interaction effects on the difference between the procedures. To identify important factors affecting the performance of the heuristic procedures, an analysis of variance (ANOVA) was done using the SAS software system to evaluate the effects of the four factors and all their possible interactions. The ANOVA results on the four performance measures are summarized in Tables 6.

Table 6: ANOVA results by performance measure

Variable	Significant effects
DOSR(PB,PF)	G, I, D, G*D, I*D, G*I*D
DOSR(PB,RS)	G, I, D, G*D, I*D
PADI(PB, PF)	G, I, D, G*D, G*I*D
PADI(PB, RS)	G, I, D, G*D
PADI(PF, RS)	G, I, D, G*D, G*I*D
PTST(PB, PF)	G, D, G*D
PTST(PB, RS)	G, D, G*D, I*D
PTST(PF, RS)	G, D, G*D
PTCI(PB, PF)	G, I, D, G*D, I*D, G*I*D
PTCI(PB, RS)	G, G*D, I*D

The variable column identifies the data set tested in the ANOVA model. For example, DOSR(PB, PF) represents the set of experimental data showing DOSR between procedures PB and PF. The level of significance is set at 0.05. All significant effects are listed in the tables.

The observations from ANOVA results are summarized as below:

- Group ratios (G) and demand patterns (D) significantly affect the performance differences in most of the cases.
- The interaction effects of G*D, I*D and G*I*D can also significantly affect the difference in performance measures.
- Three of the four factors (G, I, D) affect the performance. It seems that the performance of the heuristic strongly depends on the operating conditions. Only the precedence constraints (P) do not significantly affect the difference in performance measures.

CONCLUSION

This research presents a multiple objective formulation of production scheduling problem in continuous manufacturing systems and the development of three heuristic procedures to schedule daily production. Each heuristic procedure gives priority to one of three objectives. A computer program was developed to implement the heuristics. Many variables such as assignment of product to processing lines, coefficients of the cost function and the number of periods in the planning horizon can be changed in the program. Consequently, these heuristic procedures can be considered as "decision support" rules because the management will still need to make decisions themselves based on the various schedules produced.

A factorial experiment was designed and performed to evaluate the performance of the three heuristic procedures and to identify the factors that affect the performance differences. The analyses of the results show that no procedure is absolutely superior to the other procedures in all aspects. This is not a surprise because each of the three algorithms produces a local optimal solution. However, both the PB and PF algorithms can generate schedules with better OSR and ADI than the current management approach modeled by the RS algorithm. To provide a composite performance measure, a TCI was suggested. It is noted that the rank of the heuristics can be changed as a function of the cost coefficients in TCI.

Based on the ANOVA results, the differences between procedures are significantly affected by group ratio, maximum inventory and demand pattern. This information is of important value because it suggests that if one of the above factor changes in the manufacturing system, the method to schedule production needs to be re-evaluated. Further research is currently underway to study the sensitivity of the resource availability to the performance measures, and the development of a decision support system based on the heuristics proposed in this research.

REFERENCES

1. Amaro, A. C. S., & Barbosa-Póvoa, A. P. F. D. (1999). Scheduling of industrial distribution manifolds with pre-conditions.*European Journal of Operational Research, 119,*461–478.10.1016/S0377-2217(99)00147-2 ,
2. Anwar, M. F., & Nagi, R. (1997). Integrated lot-sizing and scheduling for just-in-time production of complex assemblies with finite set-ups. *International Journal of Production Research, 35,* 1447–1470.10.1080/002075497195416 ,
3. Askin, R. G., & Standridge, C. R. (1993). *Modeling and analysis of manufacturing systems.* New York, NY: John Wiley.
4. Bose, S., & Pekny, J. F. (2000). A model predictive framework for planning and scheduling problems: A case study of consumer goods supply chain. *Computers & Chemical Engineering, 24,* 329–335.10.1016/S0098-1354(00)00469-5 ,
5. Caramanis, M. C., Anli, O. M., & Paschalidis, I. C. (2003). Supply chain (SC) production planning with dynamic lead time and quality of service constraints. *Proceedings of the IEEE Conference on Decision and Control, 5,* 5478–5485.

6. Daniels, R. (1994). Incorporating preference information into multi-objective scheduling. *European Journal of Operational Research, 77*, 272–286.10.1016/0377-2217(94)90372-7 ,

7. Deepark, B. B., & Grossmann, I. E. (1990). Simultaneous production planning and scheduling in multiproduct batch plants.*Industrial & Engineering Chemistry Research, 29*, 570–580. ,

8. Giannelos, N., & Georgiadis, M. (2003). Efficient scheduling of consumer goods manufacturing processes in the continuous time domain. *Computers & Operations Research, 30*, 1367–1381.10.1016/S0305-0548(02)00076-X ,

9. Gupta, Y. P., Bector, C. R., & Gupta, M. C. (1990). Optimal schedule on a single machine using various due date determination methods. *Computers in Industry, 15*, 245–254.10.1016/0166-3615(90)90023-I ,

10. He, W., & Kusiak, A. (1992). Scheduling manufacturing systems. *Computers in Industry, 20*, 163–175.10.1016/0166-3615(92)90050-W ,

11. Kogan, K., Leu, Y., & Perkins, J. (2002). Parallel-machine, multiple-product-type, continuous-time scheduling: Decomposable cases. *IIE Transactions, 34*, 11–22. ,

12. Lou, P., Liu, Q., Zhou, Z., Wang, H., & Sun, S. X. (2012). Multi-agent-based proactive–reactive scheduling for a job shop. *The International Journal of Advanced Manufacturing Technology, 59*, 311–324.10.1007/s00170-011-3482-4 ,

13. Minton, S., Johnston, M. D., Phillips, A. B., & Laird, P. (1992). Minimizing constraint satisfaction and scheduling problems.*Artificial Intelligence, 58*, 61–205. ,

14. Modarress, B., Ansari, A., & Willis, G. (2000). Controlled production planning for just-in-time short-run suppliers. *International Journal of Production Research, 38*, 1163–1182.10.1080/002075400189077 ,

15. Patterson, J. H., Slowinski, R., Talbit, F. B., & Weglarz, J. (1989). *An algorithm for a general class of research and resource constrained scheduling problems. Advances in Project Scheduling*. Amsterdam: Elsevler Science.

16. Scudder, G. D., Hoffmann, T. R., & Rohleder, T. R. (1993). Scheduling with forbidden early shipments: Alternative performance criteria and conditions. *International Journal of Production Research, 31*, 2287–2305.10.1080/00207549308956858,

17. Seki, Y., & Kogure, K. (1996). Lot scheduling problem for continuous demand. *International Journal of Production Economics,44*, 7–15.10.1016/0925-5273(95)00087-9 ,

18. Tung, L., Lin, L., & Nagi, R. (1999). Multiple-objective scheduling for the hierarchical control of flexible manufacturing systems. *International Journal of Flexible Manufacturing Systems, 11*, 379–409.10.1023/A:1008183330452

19. Viana, A., & de Sousa, J. P. (2000). Using metaheuristics in multiobjective resource constrained project scheduling. *European Journal of Operational Research, 120*, 359–374.10.1016/S0377-2217(99)00163-0 ,

20. Vollmann, T. E., Berry, W. L., & Whybark, D. C. (1992). *Manufacturing planning and control systems.* Boston, MA: Irwin.

21. Wagner, M., & Smits, S. (2004). A local search algorithm for the optimization of the stochastic economic lot scheduling problem. *International Journal of Production Economics, 90*, 391–402.10.1016/S0925-5273(03)00201-9 ,

22. Yim, S., & Lee, D. Y. (1996). Multiple objective scheduling for flexible manufacturing systems using Petri nets and heuristic search. *Proceedings of the IEEE International Conference on Systems, Man and Cybernetics, 4*, 2984–2989.

CHAPTER 9

MGP–CC: A HYBRID MULTIGENE GP–CUCKOO SEARCH METHOD FOR HOT ROLLING MANUFACTURE PROCESS MODELLING

Hossam Faris[a], Alaa F. Sheta[b] & Ertan Öznergiz[c]

[a] King Abdullah II School for Information Technology, The University of Jordan, Amman, Jordan

[b] Computers and Systems Department, Electronics Research Institute (ERI), Giza, Egypt

[c] Marine Engineering Operations Department, Faculty of Naval Architecture and Maritime, Yildiz Technical University, Istanbul, Turkey

ABSTRACT

Maintaining high level of quality in hot rolling manufacturing processes is very challenging problem to keep competitiveness in the iron and steel industrial market. Monitoring the quality of the output product helps enhancing the product outcomes, increase the company profit and improve the economic growth of the country. In this paper, we propose a new hybrid approach based on multigene genetic programming (MGP) and Cuckoo search (CS) algorithms for developing three rigorous models for roll force, torque and slab temperature in the hot rolling industrial process at the Ereg˜li Iron and Steel Factory in Turkey. MGP is a robust variation of the standard genetic programming (GP) algorithm while CS is a new nature-inspired metaheuristic search algorithm. The performance of the developed models is evaluated and compared with those obtained for the standard MGP and GP approaches.

INTRODUCTION

The problem of modelling the hot rolling industrial process is complex and challenging due to the highly dynamic and nonlinear relation between the involved variables in the process such as, friction coefficient, yield stress, disturbances, etc. Moreover, the data sets collected to predict the performance of the process is often noisy and multi-variate (Moussaoui, Selaimia, & Abbassi, 2006). This makes getting accurate measurement of the rolling force/torque very difficult task. However, accurate and reliable models for the rolling force, torque and temperature are vital in designing pass online schedules and maintaining high quality of the production.

Citation: Hossam Faris, Alaa F. Sheta & Ertan Öznergiz (2016) MGP–CC: a hybrid multigene GP–Cuckoo search method for hot rolling manufacture process modelling, Systems Science & Control Engineering, 4:1, 39-49, DOI: 10.1080/21642583.2015.1124032.

Steel market industry has witnesses a high competition every year. Thus, maintaining exceptional level of production quality is essential to reserve a high market share (Cowling, 2003; Liberatore, 1989). In hot rolling strip production in particular, the demand for higher quality products motivates the need for quality control techniques such as reliable and accurate prediction models of the rolling force/torque and temperature (Cowling, 2003; Feng, Liu, Luo, Tang, & Liu, 2008; Kirihata, Maccagno, & Jonas, 1998; Kwak, Kim, Park, Lee, & Hwang, 2000; Schroder, 2003). Typical mathematical models have some limitation since some factors are difficult to be integrated such as the reduction rate and the chemical composition. A typical solution to come around this problem in some rolling processes is to adopt fragmented look-up tables. However, this solution has some disadvantages such as the large size of the generated look-up tables and the lack of effective interpolation capability.

In literature, many authors proposed quality monitoring models for the hot rolling process based on different machine learning techniques such as artificial neural networks (ANN) and fuzzy logic in order to overcome the problems mentioned earlier (Barrios, Torres-Alvarado, & Cavazos, 2012; Bouhouche, Yahi, Hocine, & Bast, 2008; Faris, Sheta, & Öznergiz, 2013; Gorni, 1997; Öznergiz, Özsoy, Delice, & Kural, 2009; Rong, Dan, & Yi, 2005; Sheta, Braik, Öznergiz, Ayesh, & Masud, 2013; Tiensuu, Juutilainen, & Röning, 2011). However, the disadvantage of most of these techniques is the lack of interpritability and considered as black box models. For that reasons authors in Faris et al. (2013) and Sheta and Faris (2014) investigated the application of Symbolic Regression approach using genetic programming (GP) for predicting features of a hot rolling process including roll force, roll torque, and slab temperature.

One of the difficulties and weaknesses in GP is finding the numeric constants that form the GP model along with the set of arithmetic functions (Koza, 1992). In standard GP, numeric constants are generated randomly. Therefore, various studies in the literature proposed different ways to solve this problem. Some of the effective approaches proposed by researchers is the hybridization between GP and metaheuristic algorithms. Two main types of these approaches could be distinguished: the co-evolution approaches and the relay approaches. One example of the co-evolution approaches can be found in Mukherjee and Eppstein (2012) where authors used differential evolution to co-evolution of the coefficients in GP. On the other hand, relay approaches perform optimization of the constants in a separated task when GP finishes evolving the structure of the Symbolic Regression model. An example of this approach is proposed in Alonso, Montaa, and Borges (2009) where evolution strategies is applied to optimize GP constants.

In this paper, we propose a new hybrid relay approach based on combining multigene genetic programming (MGP) and Cuckoo search (CS) algorithm for developing more accurate symbolic regression models for a hot rolling industrial process. MGP and CS are executed separately in sequence. MGP is applied first for finding the structure of the symbolic regression models then CS is used for optimizing the numeric constants in these models. MGP is an evolutionary computation algorithm and a robust variation of the standard GP while CS is a new nature-inspired metaheuristic search algorithm. The objective of the proposed approach is to simulate the behaviour of different features of a hot rolling manufacturing process of the Ereg˜li Iron and Steel Factory in Turkey. Three models for the force, torque and slab temperature in the plate mill are accurately developed and compared to those obtained by MGP and the standard GP approaches.

The paper is organized as follow. Section 2 describes the hot rolling process in Ereg˜li Iron and Steel Factory. Section 4introduces the MGP approach and its main processes. Section 5 gives a brief description of the CS algorithm and explains its application for tuning the generated MGP models. Section 6 describes the data collected for the purpose of the experiments in this study. In Section 7, we list the evaluation criteria used to evaluate the final developed models. Finally, Section 8presents the experiments settings and discusses the obtained results.

PROBLEM DESCRIPTION

The hot-rolling mill plant in Ereg˜li Iron and Steel Factory is shown in Figure 1. The plant consists of two slab furnaces, pre-rolling mill, edger, reversible mill, seven strip rolling stands, a cooling system, a hot leveller, and a shearing system. The plant has also a data acquisition and a computer control system modified by General Electrics. The Ereg˜li Iron and Steel Factory in Turkey has two cold and two hot-rolling rolling mill plants. Cold rolling plants have a total capacity of 2.3 million tons per year and hot-rolling mill plants, 540.000 tons per year, corresponding to a total product capacity of 3 million tons per year.

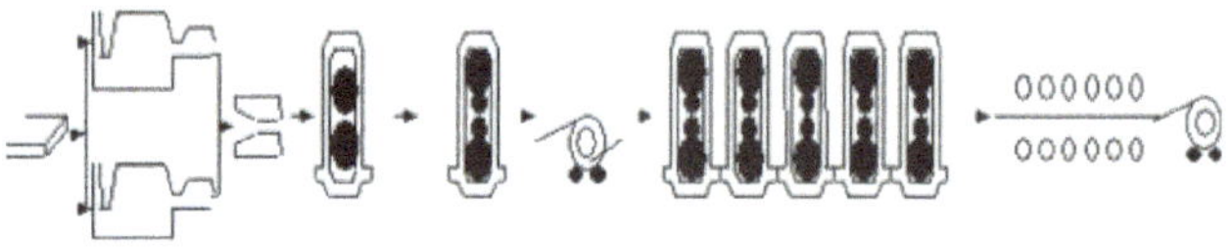

Figure 1: Diagram of the rolling mill plant at the Ereg˜li Iron and Steel Factory in Turkey.

Steel strips with a thickness of 15–16 mm can be produced in the rolling mill plant. In a normal production cycle, each slab passes 5 times in forward and backward directions in the reversible mill. In this plant, the dimensions of slabs are monitored continuously during every passes with an X-ray system, the temperature of slab with a pyrometer, roll force and torque with four load cells placed along the mill. But averages of these measured values for each pass are used for identification. The chemical composition of the low-carbon steels used in this study is given in Table 1.

Table 1: The chemical composition of the low carbon steel (ASTM A53-96 Gr. A)

Carbon	Manganese	Sulphur	Phosphorus	Silicon
0.12%	0.25%	0.2%	0.025%	0.05%

GENETIC PROGRAMMING

GP is an evolutionary algorithm-based methodology for automatically solving problems inspired by biological evolution (Koza, 1991, 1994). GP has many advantages when used to model nonlinear systems (Espejo, Ventura, & Herrera, 2010; Hussein, Sheta, & Abd-Elwahab, 2001; Kotanchek, Smits, & Kordon, 2003; Sheta et al., 2013). They include:

- *Flexibility*: is one of the main advantages of GP over many other heuristic approaches. That is because the general structure of the tree representation of GP individuals can fits a wide range of problems. Moreover, using the terminal set and the function set can do both, represent knowledge and perform computation.
- *Interpretability*: some modelling and prediction techniques are considered as black box methods since they generate models which are hard to understand such techniques traditionally include ANNs, SVMs. Unlike those techniques, tree models generated by GP are more simple, easier to evaluate and their complexity

is controllable by setting the maximum GP tree depth and length.

Therefore, GP has been applied successfully to a large number of manufacturing and industrial processes. Some of these applications can be found in Hussein et al. (2001), Hussian, Sheta, Kamel, Telbany, and Abdelwahab (2000), Faris and Sheta (2013), Faris et al. (2013), Sheta and Faris (2014) and Faris (2013). In particular for the case of hot rolling process, the results in Faris et al. (2013) and Sheta and Faris (2014) show that the developed GP mathematical models are very competitive taking into consideration their simplicity and interpretability.

MULTIGENE GENETIC PROGRAMMING

Multigene GP is a robust variation of the standard GP (Searson, Leahy, & Willis, 2010). Multigene GP can be applied to evolving a population of trees (possible solutions) while minimizing some error criteria. For a system with x input of dimension n × m to produce a model output y with dimension n × 1, we could produce a tree structure which introduce the mathematical relationship:

$$y = f(x), \tag{1}$$

where n is the number of observations taken and m is the number of input variables. In multigene GP, prediction of the output variable $\hat{y}$ is resulted by a weighted output for each trees/genes in the multigene individual plus a bias term. Each tree is a function of zero or more of the N input variables $x_1, \ldots, x_N$. The main processes of the Multigene GP algorithm flow chart is shown in Figure 3.

Initial population and representation

The evolutionary cycle of the algorithm starts by generating random individuals (possible solution or model) each of which consists of M randomly generated trees. A Multigene GP model can be represented in the form shown in Equation (2). A multigene regression model can be written as:

$$\hat{y} = \alpha_0 + \sum_{i=1}^{M} \alpha_i \times \text{tree}_i, \tag{2}$$

where α_0 represents the bias or offset term while $\alpha_1,\ldots,\alpha_m$ are the gene weights and M is the number of genes (i.e. trees) which constitute the available individual. The weights (i.e. regression coefficients) are automatically determined by a least squares (LS) procedure for each multigene individual. In multigene symbolic regression each symbolic model is represented by number of GP trees weighted by linear combination. Each tree is considered as a 'gene' by itself. An example of multigene model is shown in Figure 2. The presented model can be introduced mathematically as given in Equation (3).

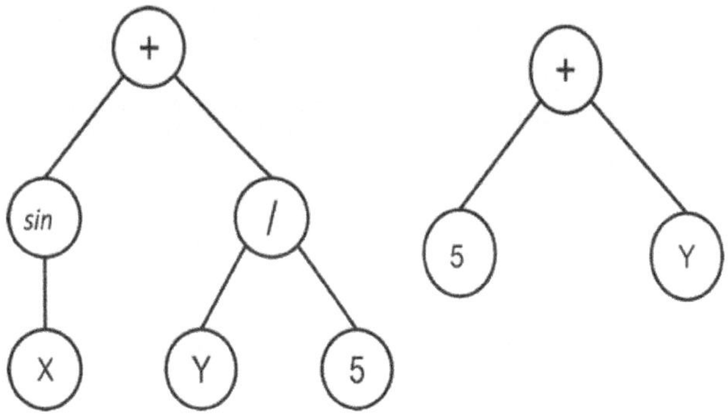

Figure 2: Example of a Multigene symbolic trees.

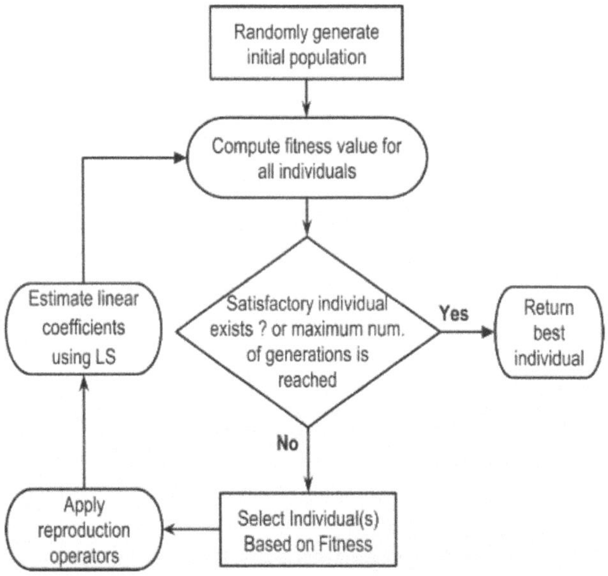

Figure 3: Flow chart of the Multigene GP approach.

$$\alpha_0 + \alpha_1 \left[\sin(X) + \frac{Y}{5}\right] + \alpha_2[5 \times Y]. \tag{3}$$

Each symbolic model will have set of linear coefficients such as $\alpha_0, \alpha_1, \ldots$. These parameters will be estimated ordinary LS techniques. This shows the powerful nature of multigene GP since it combines the classical linear regression theory with the ability to capture nonlinear behaviour based the evolutionary process. It was reported that there are many advantages of multigene symbolic regression than standard GP approach for symbolic regression (Hinchliffe et al., 1996).

A multigene symbolic models are consisted of a function set and a terminal set. The function set is likely to contain of a number of simple arithmetic operators functions and/or and nonlinear mathematical functions such as the ones given in Equation (4).

$$F = \{+, -, \times, /, \sin, \cos, \exp, \log\}. \tag{4}$$

The function F is combined with the terminal set to help the algorithm to develop suitable tree structure which represent a model for the problem. This multigene symbolic model has the advantages that it is linear in the parameters with respect to the coefficients α_0, α_1 and α_2.

Fitness evaluation

The root mean square error (RMSE) is used as a fitness function in order to evaluate GP individuals. RMSE can be described by Equation (5).

$$\text{RMSE} = \sqrt{\frac{1}{n}\sum_{i=1}^{n}(y_i - \hat{y}_i)^2}, \tag{5}$$

where y and $\hat{y}$ are the actual and the estimated values based on proposed model and n is the number of measurements used in the experiments, respectively.

Tournament selection

A selection mechanism consists of two components: a selection scheme and a selection pressure control strategy. This strategy is usually critical in designing a selection mechanism and has been widely studied by the evolutionary computation community (Legg, Hutter, & Kumar, 2004). Tournament selection is a beneficial and robust selection mechanism (Miller, Sutton, & Werbos, 1995). The selection pressure of tournament selection changed according to the tournament size. The more the number of individuals are selected the higher the resulting selection pressure.

Crossover

During the evolutionary process, genes are generated using crossover operator. In our case, we used two point crossover. Crossover makes exchange of genes between individuals. A two point high level crossover adopted in Searson et al. (2010) is used. The example in Equations (6)–(8) shows the operation.

- Individuals for crossover

$$(G_1 G_2 G_3 G_4 G_5) \quad (G_6 G_7 G_8). \tag{6}$$

- Selected genes for crossover

$$(G_1 \mathbf{G_2 G_3} G_4 G_5) \quad (G_6 \mathbf{G_7 G_8}). \tag{7}$$

- Generated individual

$$(G_1 \mathbf{G_7 G_8} G_4 G_5) \quad (G_6 \mathbf{G_2 G_3}). \tag{8}$$

Mutation

Mutation in multigene GP operates in a similar fashion to standard GP evolutionary process. A random gene is selected then a normal sub-tree mutation is performed. The new generated gene replaces the parent gene in the new generation.

Coefficients estimation

In this process, the linear coefficients $\alpha_0, \alpha_1, \ldots, \alpha_M$ of each individual in the population are estimated using ordinary LS techniques based on the training data. In order to save the computation time. This operator is applied on a small fraction of each generation. By default it is set to 5%.

OPTIMIZING MULTIGENE GP MODELS BY CS

CS is one of the recent and powerful metaheuristic optimization algorithms (Yang & Deb, 2009, 2010). CS is inspired by the natural brood of Cuckoo birds and the Lévy flight behaviour of some birds species and fruit flies. The CS algorithm is designed mainly on the following three assumptions (Yang & Deb, 2009):

1. A Cuckoo bird selects randomly a nest and put one egg (solution) in it. Therefore, every nest forms one candidate solution.
2. Some nests that has the best quality solutions will be passed to the next generation.
3. The number of available host nests is fixed, and the ratio that a host discovers a cuckoo egg can be approximated by a fraction P_a of the n nests being neglected and replaced by new random nests (solutions) at their positions.

New solutions x_i^{t+1} are generated using a Lévy flights as shown in Equation (9). Lévy flights are performed by random walks with step-lengths are drawn from a Lévy distribution, a heavy-tailed probability distribution, which has an infinite variance with an infinite mean, Equation (10).

$$x_i^{t+1} = x_i^t + \alpha \oplus \text{Lévy}(\lambda), \tag{9}$$

$$\text{Lévy} \sim \lambda = t^{-\lambda}. \tag{10}$$

In Equation (9), $\alpha > 0$ is the step size which should be related to the scales of the problem of interest. In most cases, is supposed equals 1. The product $\oplus$ denotes entry-wise multiplications.

In this paper, the fitness function used in CS algorithm is the same as MGP which is the RMSE criteria. While nests are encoded as sequence of the coefficients s that connects the trees in the final MGP developed models shown in Figure 4. The goal is to find the best coefficients that minimize the RMSE of the MGP model. CS algorithm enjoys some advantages that make it very competitive to other metaheuristic algorithm. Such advantages include:

- CS has only one parameters to be tuned beside the population size which is the discovery rate P_a (Yang & Deb, 2009).
- In contrast to other metaheuristic approaches that use uniform distributions or Gaussian to generate new explorative moves in the search space, CS performs random walk via Lévy flight which is more efficient in exploring the search space if the search space is large enough (Yang & Deb, 2009).

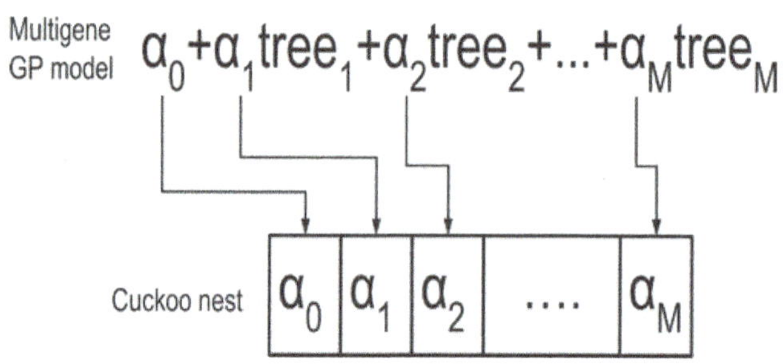

Figure 4: Mapping the constants of Multigene GP model to a Cuckoo nest.

Algorithm 1 Basic steps describing the CS

```
procedure CS
  Objective function f(x), x = (x_1, x_2, ... x_d)^T
  Generate initial population of n host nests x_i (i = 1, 2, ..., n)
  while ( t < Max Generation) or (stop criterion not met)
    Get a cuckoo randomly by Lévy flights evaluate its fitness F_i
    Choose a nest among n (say, j) randomly
    If (F_i > F_j)
      replace j by the new solution
    end if
    A fraction (P_a) of the worse nests are abandoned and new ones are built
    Keep the best solutions/nests
    Rank the solutions/nests and find the current best
    Pass the current best solutions to the next generation
  end while
end
```

The CS algorithm works as described by Algorithm 1 where s is the step size and P_a is the probability of discovering the egg. E_t is derived form Lèvy Flight distribution which is a random walk with step length depending on the current state plus a transition probability (Yang & Deb, 2009).

HOT ROLLING DATA SET

The hot rolling collected data set contains 640 collected from 128 different slabs by a General Electric's data acquisition system. The collected measurements were practically measured for three sub-processes: roll force f, roll torque G and slab temperature T of the rolling process. The thickness and width distribution of the data ranged from 31.68 to 168.6 mm and 948.76 to 1457.26 mm, respectively. In order to compare our obtained results with other reported results from the literature (Öznergiz et al., 2009; Sheta, Öznergiz, Abdelrahman, & Babuska, 2009), same data set partitioning for the training and testing was adopted, 500 for training and 140 for testing.

Each of force f, the torque G and the slab temperature T has six input variables. The inputs and output of the proposed model structures can be summarized as follows:

- Inputs to force f and torque G models: u_1, u_2, u_3, u_4, u_5 and u_6 inputs stand for: entry temperature (T_i), width (W_s), carbon equivalent (C_e), gauge (h_i), draft (i) and roll diameter (R), respectively. The output of each sub-process is stated as M_{Force}, M_{Torque} and given in Equation (11).

$$M_{Force} = f(T_i, W_s, C_e, h_i, i, R),$$
$$M_{Torque} = f(T_i, W_s, C_e, h_i, i, R). \quad (11)$$

- Input to slab temperature T model: u_1, u_2, u_3, u_4, u_5 and u_6 inputs stand for: Entry temperature (T_i), width (W_s), carbon equivalent (C_e), gauge (h_i), torque (G_i), power (E_i), respectively. The output of each sub-process is stated as M_{Temp} and given in Equation (12).

$$M_{Temp} = f(T_i, W_s, C_e, h_i, G_i, E_i). \quad (12)$$

PERFORMANCE EVALUATION

Other performance criteria are used to evaluate the quality of the developed Multigene GP models. The set of criteria used are given as follows:

- Variance-accounted-for (VAF)

$$VAF = \left[1 - \frac{var(y - \hat{y})}{var(y)}\right] \times 100\%. \quad (13)$$

- Mean squares error (MSE)

$$MSE = \frac{1}{n}\sum_{i=1}^{n}(y - \hat{y})^2. \quad (14)$$

- Mean absolute error (MAE):

$$MAE = \frac{1}{n}(\sum_{i=1}^{n} |y_i - \hat{y}_i|), \quad (15)$$

where y and $\hat{y}$ are the actual and the estimated values based on proposed model and n is the number of measurements used in the experiments, respectively.

EXPERIMENTAL RESULTS

In our experiments we use the GPTIPS Tool[1] for developing Multigene GP models for rolling force, torque and slab temperature. The parameters of the application are tuned as listed in Table 2. The best results are obtained using seven genes in each individual. The function set used to combine the terminals in GP individuals is F = $\{+, -, /, *, sqrt, sin, cos\}$.

Table 2: Parameters settings for MGP experiments

Population size	50
Number of generation	100
Selection mechanism	Tournament (size = 3)
Max. tree depth	10
Probability ofcrossover	85%
Probability ofmutation	10%
Elite	2%
Max.no. of genes allowed	7

In Figures 5–7 we show the convergence of MGP toward the best obtained force torque and temperature models, respectively. Figures 8–10 show the actual and estimated force, torque and temperature values based on the developed MGP models.

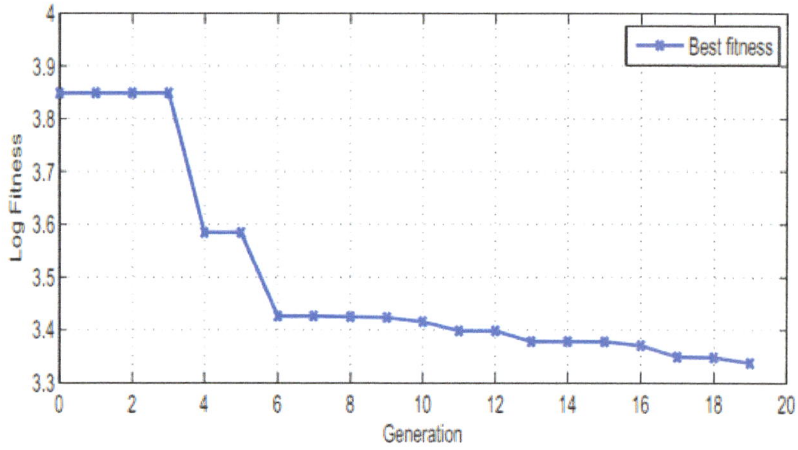

Figure 5: MGP convergence curves for force model.

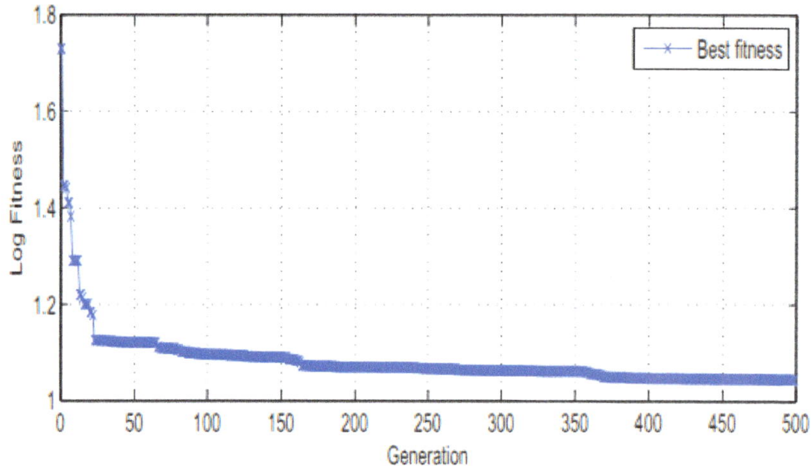

Figure 6: MGP convergence curves for torque model.

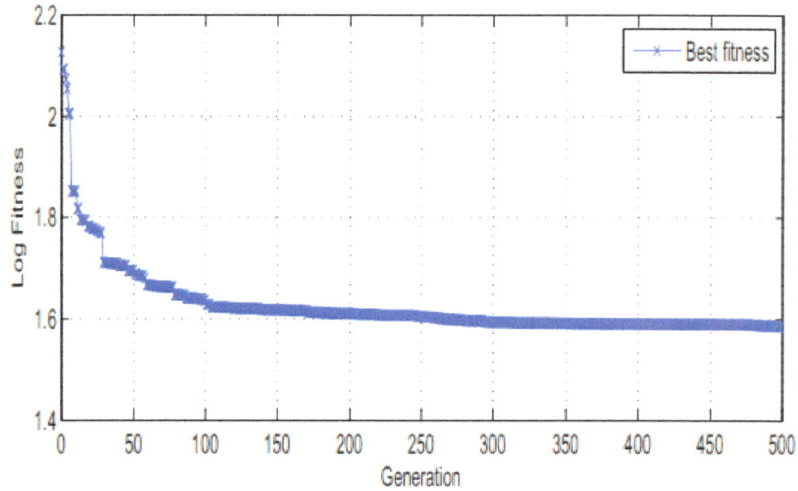

Figure 7: MGP convergence curves for temperature model.

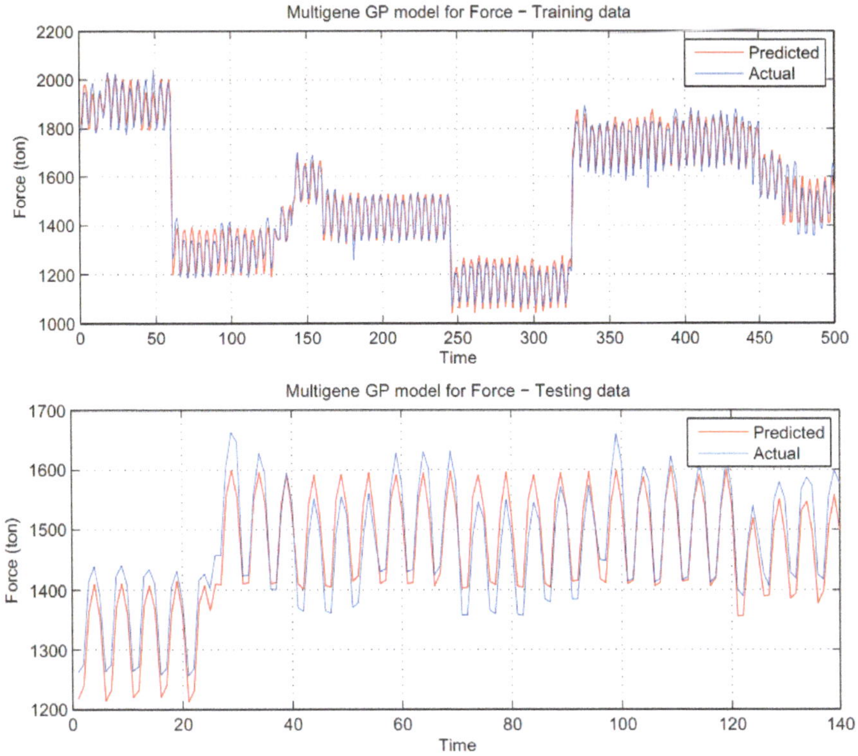

Figure 8: Actual and MGP estimated force in training and testing cases.

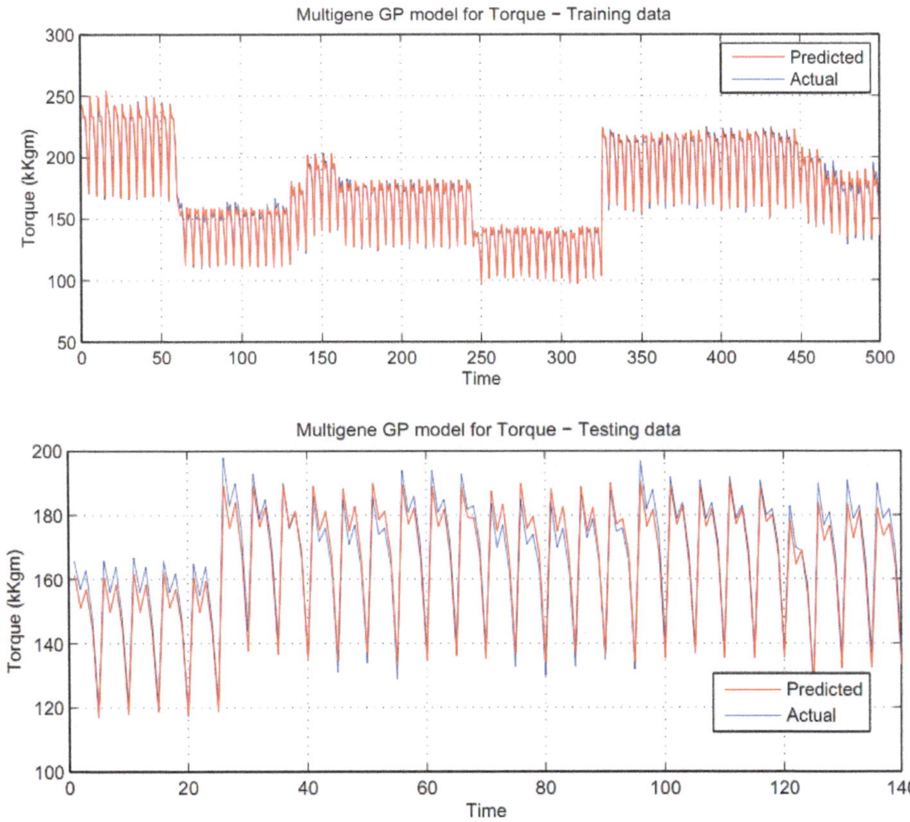

Figure 9: Actual and MGP estimated torque in training and testing cases.

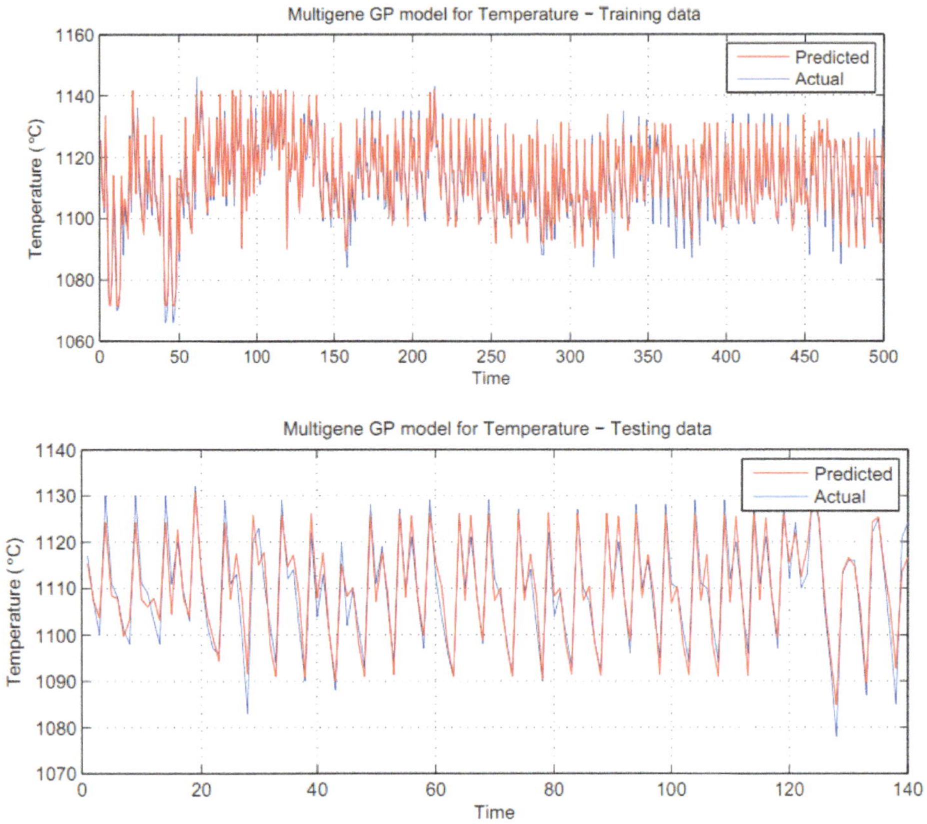

Figure 10: Actual and MGP estimated temperature in training and testing cases.

In order to enhance the accuracy of the developed MGP models, CS algorithm is used to tune the parameters of these models. This approach will be referred to as MGP–CS. CS is applied for 500 cycles. The actual and estimated force, torque and temperature values based on the developed MGP–CS models are shown in Figures 11–13, respectively.

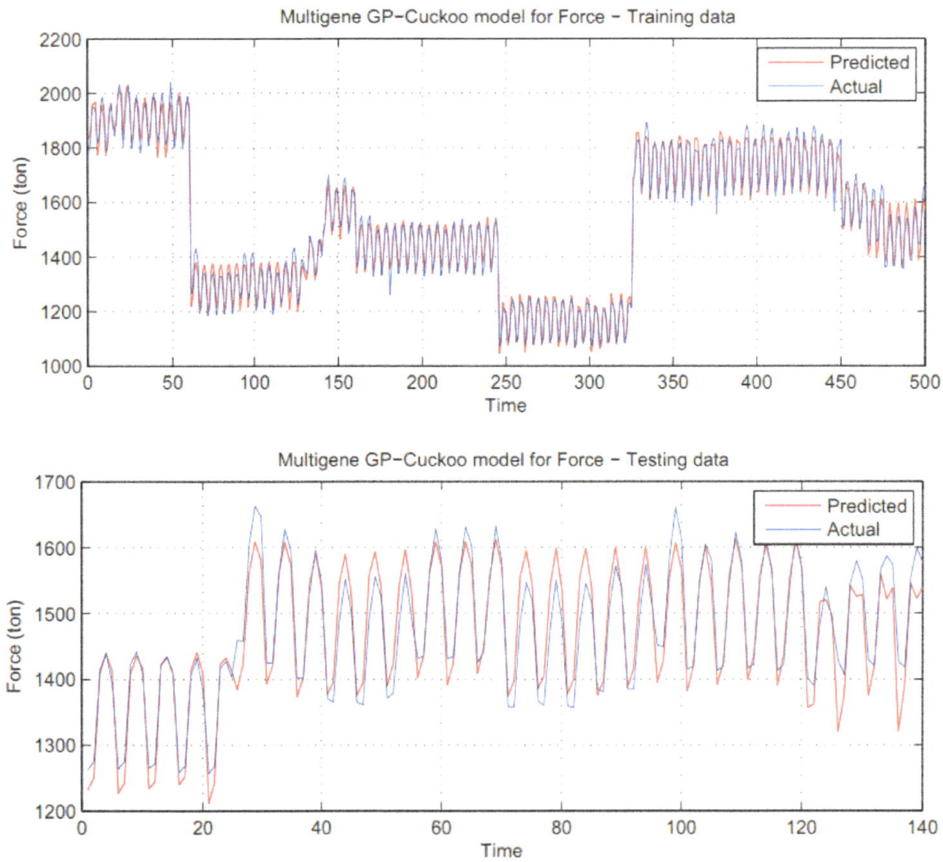

Figure 11: Actual and MGP–CS estimated force in training and testing cases.

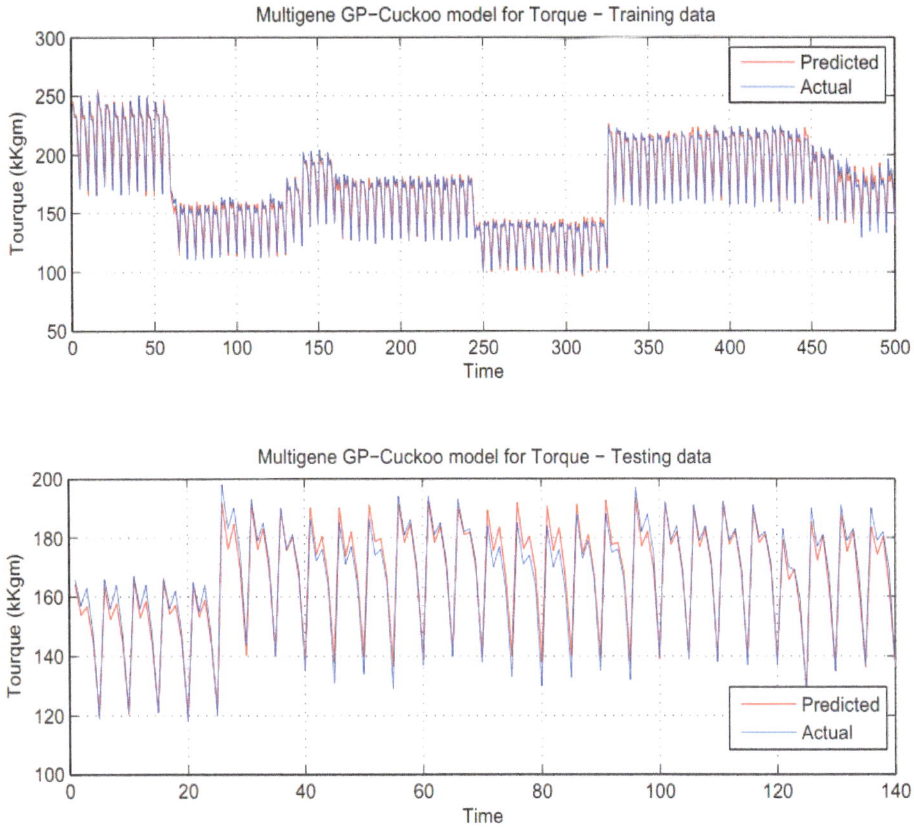

Figure 12: Actual and MGP–CS estimated torque in training and testing cases.

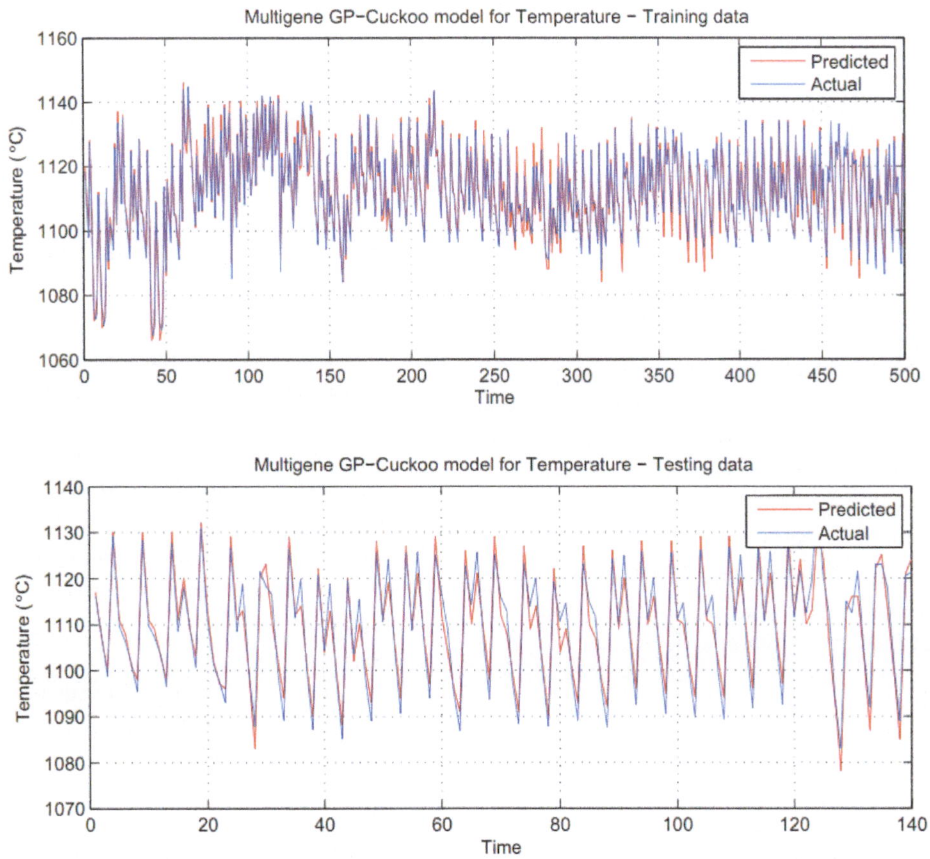

Figure 13: Actual and MGP–CS estimated temperature in training and testing cases.

The developed MGP and MGP–CS models are evaluated using VAF, MSE and MAE; and compared with the previously obtained results for the standard GP approach in Faris et al. (2013) and Sheta and Faris (2014). Based on the comparison results shown in Tables 3–5 we can notice that MGP achieved better results than the standard GP approach in the force and torque cases while there was no improvement in the temperature case. Moreover, combining CS algorithm with MGP in the MGP–CS approach generated the best quality models for the roll force, torque and temperature since it increased the VAF measurement and decreased MAE and MSE for all testing cases.

Table 3: Evaluation results of the developed MGP–CS, MGP and standard GP force models

	MGP–CS		MGP		GP	
	Training	Testing	Training	Testing	Training	Testing
VAF	98.95%	88.86%	98.79%	87.4%	98.54%	84.07%
MAE	20.616	26.075	21.827	33.384	24.04	37.08
MSE	689.74	1127.3	794.3	1451.5	955.06	1987.99

Table 4: Evaluation results of the developed MGP–CS, MGP and standard GP torque models

	MGP–CS		MGP		GP	
	Training	Testing	Training	Testing	Training	Testing
VAF	99.43%	96.518%	99.44%	96.42%	96.24%	97.08%
MAE	2.12	3.1328	2.01	3.74	5.50	15.63
MSE	7.44	14.23	7.35	17.02	49.09	257.80

Table 5: Evaluation results of the developed MGP–CS, MGP and standard GP temperature models

	MGP–CS		MGP		GP	
	Training	Testing	Training	Testing	Training	Testing
VAF	96.43%	92.53%	91.11%	88.88%	88.26%	89.73%
MAE	2.02	2.83	3.64	3.73	4.09	3.46
MSE	7.39	11.16	19.08	19.85	25.14	18.11

In Table 4, there is an enhancement also on the roll torque by decreasing the MAE and MSE values for the testing case. However, in the case of slab temperature which is shown in Table 5 we can notice that the performance is almost the same compared to the standard GP approach.

Finally, it can be concluded that the MGP–CS modelling approach has a promising potential in improving the quality of the roll force and torque prediction models in the hot rolling process. This can be explained by the advantage of Multigene GP that combines the classical linear regression theory with the ability to capture nonlinear behaviour based the evolutionary process.

CONCLUSION

In this paper, MGP is combined with a CS algorithm to develop estimation models for a Hot-Rolling industrial process. Three sub models are proposed for the process; the rolling force, torque and slab temperature. The developed MGP–CS models are evaluated and compared with the standard GP models based on different evaluation criteria. Experimental data collected from the Ereg˜li Iron and Steel Factory in Turkey was used for the verification of the developed models. Compared to MGP and standard GP, MGP–CS models showed better performance results for rolling force, torque and temperature compared to those obtained for the other approaches. This advantage can support implementing higher quality models efficiently in iron and steel manufacturing plants.

ACKNOWLEDGMENTS

Authors would like to thank to Prof. Dr Can ÖZSOY and the Ereg˜li Iron and Steel Factory for providing the required technical assistant and the data for this developed research work.

REFERENCES

1. Alonso, C. L., Montaa, J., & Borges, C. E. (2009). *Evolution strategies for constants optimization in genetic programming*. 21st International Conference on Tools with Artificial Intelligence, 2009, ICTAI'09, IEEE, pp. 703–707.
2. Barrios, J. A., Torres-Alvarado, M., & Cavazos, A. (2012, February). Neural, fuzzy and grey-box modelling for entry temperature prediction in a hot strip mill. *Expert Systems With Applications*, *39*(3), 3374–3384. doi: 10.1016/j.eswa.2011.09.025,
3. Bouhouche, S., Yahi, M., Hocine, B., & Bast, J. (2008). Soft sensor-based artificial neural networks and fuzzy logic: Application to quality monitoring in hot rolling. In *Proceedings of the 10th WSEAS international conference on automatic control, modelling & simulation (ACMOS'08)* (pp. 149–154). Stevens Point, WI: World Scientific and Engineering Academy and Society (WSEAS).
4. Cowling, P. (2003). A flexible decision support system for steel hot rolling mill scheduling. *Computers & Industrial Engineering*,*45*(2), 307–321. doi: 10.1016/S0360-8352(03)00038-X ,
5. Espejo, P. G., Ventura, S., & Herrera, F. (2010). A survey on the application of genetic programming to classification. *IEEE Transactions on Systems, Man, and Cybernetics, Part C: Applications and Reviews*, *40*(2), 121–144. doi:10.1109/TSMCC.2009.2033566 ,
6. Faris, H. (2013). A symbolic regression approach for modeling the temperature of metal cutting tool. *International Journal of Control and Automation*, *6*(4), 431–440.
7. Faris, H., & Sheta, A. (2013). Identification of the Tennessee Eastman chemical process reactor using genetic programming.*International Journal of Advanced Science and Technology*, *50*, 121–140.
8. Faris, H., Sheta, A., & Öznergiz, E. (2013). Modelling hot rolling manufacturing process using soft computing techniques.*International Journal of Computer Integrated Manufacturing*, *26*(8), 762–771. doi: 10.1080/0951192X.2013.766937 ,
9. Feng, X., Liu, Y., Luo, H., Tang, H., & Liu, C. (2008). Numerical simulation during expansion for hot rolling sheet strip. In*Proceedings of the first international*

workshop on knowledge discovery and data mining (WKDD '08) (pp. 310–313). Washington, DC: IEEE Computer Society.

10. Gorni, A. (1997). The application of neural networks in the modeling of plate rolling processes. *Journal of Materials-Electronic*, *49*(4), 1–9.

11. Hinchliffe, M., Hiden, H., McKay, B., Willis, M., Tham, M., & Barton, G. (1996). Modelling chemical process systems using a multi-gene genetic programming algorithm. In J. R. Koza (Ed.), *Late breaking papers at the genetic programming 1996 conference*, Stanford University, July 28–31 (pp. 56–65). Stanford University, CA: Stanford Bookstore.

12. Hussein, A. E.-A., Sheta, A., & Abd-Elwahab, A. (2001). *Modeling of a winding machine using neural networks*. Proceeding of the WSEAS international conference on scientific computation and soft computing, pp. 528–533.

13. Hussian, A., Sheta, A., Kamel, M., Telbany, M., & Abdelwahab, A. (2000). *Modeling of a winding machine using genetic programming*. Proceedings of the congress on evolutionary computation, CEC2000, pp. 398–402.

14. Kirihata, A., Maccagno, T., & Jonas, J. (1998). Mathematical modelling of rolling of multiply-alloyed mean flow stress during medium carbon steels. *ISIJ International-The Iron and Steel Institute of Japan*, *38*(2), 187–195. doi:10.2355/isijinternational.38.187 ,

15. Kotanchek, M., Smits, G., & Kordon, A. (2003). Industrial strength genetic programming. In R. L. Riolo & B. Worzel (Eds.),*Genetic programming theory and practice* (pp. 239–256, Chapter 15). Boston, MA: Springer US.

16. Koza, J. (1991). *Evolving a computer program to generate random numbers using the genetic programming paradigm*. Proceedings of the fourth international conference on genetic algorithms, La Jolla, CA: Morgan Kaufmann.

17. Koza, J. R. (1992). *Genetic programming*. Cambridge, MA: MIT Press.

18. Koza, J. R. (1994). *Genetic programming II*. Cambridge, MA: MIT Press.

19. W. J. Kwak, Y. H. Kim, H. D. Park, J. H. Lee, & S. M. Hwang (2000). FE-based on-line model for the prediction of roll force and roll power in hot strip rolling. *ISIJ International*, 40(10), 1013–1018. doi: 10.2355/isijinternational.40.1013 ,

20. Legg, S., Hutter, M., & Kumar, A. (2004). Tournament versus fitness uniform selection. In *Proc. 2004 Congress on evolutionary computation (CEC-2004)* (pp. 2144–2151). Portland, OR: IEEE.

21. Liberatore, R. L. (1989). Hot rolling mill optimization model. *Computers & Industrial Engineering*, *17*(1–4), 130–135. doi:10.1016/0360-8352(89)90049-1 ,

22. Miller, W., Sutton, R., & Werbos, P. (1995). *Neural networks for control*. Cambridge, MA: MIT Press.

23. Moussaoui, A., Selaimia, Y., & Abbassi, H. A. (2006). Hybrid hot strip rolling force prediction using a Bayesian trained artificial neural network and analytical models. *American Journal of Applied Sciences*, *3*(6), 1885– 1889. doi:10.3844/ajassp.2006.1885.1889

24. Mukherjee, S., & Eppstein, M. J. (2012). *Differential evolution of constants in genetic programming improves efficacy and bloat*. Proceedings of the 14th annual conference companion on genetic and evolutionary computation, ACM, pp. 625–626.

25. Öznergiz, E., Özsoy, C., Delice, I. I., & A. Kural (2009). Comparison of empirical and neural network hot-rolling process models. *Proceedings of the Institution of Mechanical Engineers, Part B: Journal of Engineering Manufacture*, *223*, 305–312. doi:10.1243/09544054JEM1290 ,

26. Rong, Z., Dan, B., & Yi, J. (2005). *A bp neural network predictor model for desulfurizing molten iron*. Proceedings of the First international conference on advanced data mining and applications, ser. ADMA'05, Berlin, Heidelberg: Springer.

27. Schroder, D. K. H. (2003). *A basic understanding of the mechanics of rolling mill rolls*. Austria: ESW & BRC.

28. Searson, D. P., Leahy, D. E., & Willis, M. J. (2010). GPTIPS: An open source genetic programming toolbox for multigene symbolic regression. In *Proceedings of the international multi-conference of engineers and computer scientists 2010 (IMECS 2010)*(pp. 77–80, Vol. 1). Hong Kong.

29. Sheta, A. F., Braik, M., Öznergiz, E., Ayesh, A., & Masud, M., 2013). *Design and automation for manufacturing process: An intelligent business modeling using adaptive Neuro-Fuzzy inference system*. Springer, United Kingdom, Book Chapter. Business Intelligence and Performance Management-Theory, Systems and Industrial Applications Book.

30. Sheta, A., & Faris, H. (2014). Improving production quality of a hot rolling industrial process via genetic programming model. *International Journal of Computer Applications in Technology, special issue on Computational Optimisation and Engineering Applications*, 49(3/4), special Issue on: 'Computational Optimisation and Engineering Applications'.

31. Sheta, A., Öznergiz, E., Abdelrahman, M. A., & Babuska, R. (2009). Modeling of hot rolling industrial process using fuzzy logic. In D. Che (Ed.), *Proceedings of the ISCA 22nd international conference on computer applications in industry and engineering (CAINE)* (pp. 81–86). ISCA.

32. Tiensuu, H., Juutilainen, I., & Röning, J. (2011). *Modeling the temperature of hot rolled steel plate with semi-supervised learning methods*. Proceedings of the 14th international conference on discovery science, ser. DS'11, Berlin, Heidelberg, Springer, pp. 351–364. Retrieved from http://dl.acm.org/citation.cfm?id=2050236.2050264

33. Yang, X.-S., & Deb, S. (2009). *Cuckoo search via Levy flights*. World congress on nature biologically inspired computing, pp. 210–215.

34. X . S. Yang, & Deb, S. (2010). Engineering optimisation by cuckoo search. *International Journal of Mathematical Modelling and Numerical Optimisation*, *1*(4), 330–343. doi: 10.1504/IJMMNO.2010.035430

CHAPTER 10

A LIVELOCK CONTROL POLICY FOR A FLEXIBLE MANUFACTURING SYSTEM MODELING WITH A SUBCLASS OF GENERALIZED PETRI NETS

C.Q. Hou[1], S.Y. Li[1], Y. Cai[1], H.M. Wu[2], A.M. An[3] and Y. Wang[1]

[1] School of Civil Engineering, Lanzhou University of Technology, Lanzhou 730050, People's Republic of China

[2] School of Science, Lanzhou University of Technology, Lanzhou 730050, People's Republic of China

[3] College of Electrical and Information Engineering, Lanzhou University of Technology, Lanzhou 730050, People's Republic of China

ABSTRACT

Livelocks, like deadlocks, can result in the serious problems in running process of flexible manufacturing systems (FMSs) as well. Current deadlock control policies based on the approaches of siphon detection and control, cannot cope with livelocks in a system of sequential systems with shared resources (S^4R), a typical subclass of Petri nets that can model FMSs. On the basis of the mixed integer programming method, this study proposes a livelock control policy (LCP) that can not only solve the new smart siphons (NSSs) associated with livelocks or deadlocks in an S^4R system directly, but also make the solved NSSs max′-controlled by adding the corresponding control places (CPs). As a result, an original S^4R system with livelocks or deadlocks can be turned into the live controlled one in which no NSSs can be found. The related theoretical analysis and several examples are given to demonstrate the proposed LCP. Compared with the existing methods in the literature, the proposed one is more general and powerful.

Public Interest Statement

The paper is appropriate for researchers, graduate students, and engineers who are interested in the control problems arising from flexible manufacturing systems (FMSs) modeling with Petri nets (PNs). Livelocks, like deadlocks, can result in the serious problems in running process of flexible manufacturing systems as well. The research of deadlock and livelock control has become a hot topic in the field of supervisory control theory (SCT). On the basis of the previous results and the mixed integer programming (MIP) method, a livelock control policy (LCP) is proposed for an S^4R system belonged to a subclass of PNs in this paper. By solving the new smart siphons (NSSs) associated with livelocks or deadlocks and making them max′-controlled, an original S^4R system with livelocks or deadlocks can be turned into the live controlled one. The proposed LCP exploit a new path to the deadlock and livelock control problems.

INTRODUCTION

A flexible manufacturing system (FMS) is a novel production mode in which there is some amount of flexibility that allows the system to response to market requirement changes in as timely a manner as possible, no matter predicted or unpredicted.

Citation: C.Q. Hou, S.Y. Li, Y. Cai, H.M. Wu, A.M. An and Y. Wang, A livelock control policy for a flexible manufacturing system modeling with a subclass of generalized Petri nets, DOI: 10.1080/23311916.2014.944766.

An FMS is also a computer controlled configuration in which different operations are executed concurrently, and therefore, have to compete a limited number of resources such as robots, machines, storage devices, fixtures, and conveyors (Li & Zhou, 2009; Wu & Zhou,2009). This competition can cause deadlocks that are highly undesirable phenomena, in which a set of processes keeps waiting indefinitely for other processes in the set to release resources (Chao, 2006; Ezpeleta, Colom, & Martinez, 1995; Huang, Jeng, Xie, & Chung, 2001; Li, Li, & Hu, 2011; Li & Wang, 2011; Li & Zhou, 2004). In a deadlock state, the global or local FMS is blocked or crippled, causing unnecessary cost or catastrophic results. When an FMS is at a livelock state, a special case of deadlocks, some processes continuously change their state, while the other processes are deadlocked (Chao, 2007; Liu & Li, 2010). That is to say, a livelocked system still does some (although not useful) work in which some execution that never makes effective progress cycle infinitely. As a result, deadlocks and livelocks must be considered for designing an FMS.

In order to effectively handle the above problems in an FMS; digraphs, automata, and Petri nets are among the most popular mathematical tools available. Among them, Petri nets have been widely used to model an FMS, because of their own features (Li & Zhou,2009; Wu & Zhou, 2009). They can be well used to describe behavioral properties such as reversibility, boundedness, and liveness. A siphon, as a structural object of a Petri net, is a set of places (where tokens can leak out or inject in) of a Petri net model. Once a siphon has lost all its tokens, i.e. it is emptiable at a reachable marking, output transitions of places in the siphon can never be executed and hence the net is not live. Previous research indicates that siphons are closely related to the deadlocks in Petri net models of FMSs. It is well known that the number of strict minimal siphons (SMSs) in a Petri net grows quickly or may grow exponentially with respect to its size (Li & Zhou,2009) such that it is difficult and time consuming to find out these SMSs. Hence, how to design or select the proper method to solve the siphons is very important for developing new deadlock control policies (DCPs). Many scholars focus on solving and controlling siphons and have developed a large number of siphon-based controlled (SC) policies (Chao, 2006; Huang et al., 2001; Huang, Jeng, Xie, & Chung, 2006; Li, An, Cai, et al., 2013; Li, An, Wang, et al., 2013; Li & Li, 2012a, 2012b; Li, Li, & Al-Ahmari, 2012; Li, Liang, Lun, & Wang, 2004; Li, Hu, & Wang, 2007; Li, Wang, & Wei, 2006; Li & Zhao, 2008; Li & Zhou, 2004, 2006a, 2006b, 2008; Li, Zhou, & Jeng, 2008; Piroddi, Cordone, & Fumagalli, 2009; Uzam & Zhou, 2006; Zhong & Li, 2010).

However, the above DCPs still cannot detect the siphons that cause livelocks and then cannot deal with the livelock problems in OPNs or S^4R. When a system is at a livelock state, the overall states of the system continue to change, while parts of the system are deadlocked. In other words, once a system is at livelock, other processes cannot be processed smoothly. Actually, livelock is a special case of resource starvation although the system may be deadlock-free. As far as the authors know, Chao (2007) first points out that an S^4R without deadly marked siphons (DMSs) that contribute to deadlocks (Park & Reveliotis, 2001) is deadlock-free but may be in a livelock state where both live and dead transitions exist, implying that the representative mixed integer programming (MIP) methods in Chu and Xie (1997) and Park and Reveliotis (2001) are no longer valid to identify DMSs, since the net may have livelocks even though it is deadlock-free. Accordingly, the concept of max′-controlled siphons is proposed in Chao (2007), but the corresponding approach to detect siphons associated with livelocks is not given. Subsequently, based on the max″-controllability condition, Liu and Li (2010) present a more general MIP to detect deadlocks or livelocks caused by siphons that are called extended DMSs, and open a new path to analysis deadlocks and livelocks in view of siphons. The limits of the proposed MIP using max″-controlled siphons in Liu and Li (2010) are non-linear and more complicated, which increases the computational load. Similarly, no corresponding SC policy is presented in Liu and Li (2010).

Considering the shortcomings of the above SC policies, this paper proposes a novel MIP method to directly solve the smart siphons associated with livelocks or deadlocks in Petri net models. By means of the structural analysis of Petri nets and the concept of smart siphons (Li & Li, 2012b), new constrains are generated and then added to the MIP method in Li and Li (2012b), called revised MIP (RMIP). This RMIP can detect smart siphons that cause deadlocks or livelocks in Petri net models directly. Accordingly, a proper control place (CP) (Li et al., 2011; Li & Li, 2012b) is added for each solved smart siphon to make itself max′-controlled (Chao, 2007; Zhong & Li, 2010) during an iterative process. This research claims that if there is no feasible solution (NSF) of an RMIP problem, then the corresponding Petri net system is live. That is to say, livelocks or deadlocks are eliminated in the controlled Petri net system until no new smart siphons (NSSs) can be found by the proposed method in this paper.

The rest of this work is organized as follows. Section 2 reviews preliminaries of Petri nets that are used throughout this work. In Section 3, by adding new constraints to an MIP method in Li and Li (2012b), we present an RMIP method to directly solve NSSs associated with livelocks or deadlocks. By the proposed RMIP method and max′-controlled NSSs, an iterative livelock control policy (LCP) is developed in the same section. In Section 4, three examples are introduced to demonstrate the proposed LCP. We have a discussion in Section 5. Finally, Section 6 concludes this paper and discusses future work.

PRELIMINARIES

Petri nets

A Petri net is a four-tuple N=(P, T, F, W), where P and T are finite and non-empty sets. P is a set of places and T is a set of transitions with $P \cup T \neq \emptyset$ and $P \cap T = \emptyset$. $F \subseteq (P \times T) \cup (T \times P)$ is called the flow relation or the set of directed arcs. $W : (P \times T) \cup (T \times P) \to \mathbb{N}$ is a mapping that assigns a weight to an arc: $W(x, y) > 0$ if $(x, y) \in F$, and $W(x, y) = 0$ otherwise, where $x, y \in P \cup T$ and $\mathbb{N}$ denotes the set of nonnegative integers. N=(P, T, F, W) is called an ordinary net, denoted as N=(P, T, F), if $\forall f \in F$, $W(f)=1$. N = (P, T, F, W) is called a generalized net, if $\exists f \in F$, $W(f)>1$. Given a node $x \in P \cup T$, ${}^{\bullet}x = \{y \in P \cup T \mid (y, x) \in F\}$ is called the preset of x, while $x^{\bullet} = \{y \in P \cup T \mid (x, y) \in F\}$ is called the postset of x. A marking is a mapping $M : P \to \mathbb{N}$. (N, M_0) is called a marked Petri net or a net system, where N is a net and M_0 is an initial marking. The set of markings reachable from M in N is denoted as R(N, M). A string $x_1, \ldots, x_n$ is called a path of N if $\forall i \in \{1, 2, \ldots, n-1\}$, $x_{i+1} \in x_i^{\bullet}$, where $\forall x \in \{x_1, x_2, \ldots, x_n\}$, $x \in P \cup T$. An elementary path from x_1 to x_n is a path whose nodes are all different (except, perhaps, x_1 and x_n). A path $x_1, \ldots, x_n$ is called a circuit if it is an elementary path with $x_1 = x_n$. An ordinary Petri net N is called a state machine if $\forall t \in T, |{}^{\bullet}t| = |t^{\bullet}| = 1$. A state machine component N′=(P′, T′, F′, W′) of a Petri net N=(P, T, F, W) is a state machine and is a subnet of N consisting of places in P′, their presets and postsets, and related arcs. A Petri net is said to be state machine decomposable if it is covered by state machine components (Li & Zhou, 2009).

A P(T)-vector is a column vector I(J) : P(T)→Z indexed by P(T), where Z is the set of integers. I is a P-invariant (P-inv for short) if I≠0 and $I^T [N]=0^T$. P-inv I is said to be a P-semiflow if every element of I is non-negative. $||I||= \{p \in P \mid I(p) \neq 0\}$ is called the support of I. $||I||^+ = \{p \mid I(p)>0\}$ denotes the positive support of I and $||I||^- = \{p \mid I(p)0\}$ denotes the negative support of I. A non-empty set $S \subseteq P$ is a siphon (trap) if ${}^{\bullet}S \subseteq S^{\bullet}$ ($S^{\bullet} \subseteq {}^{\bullet}S$). A siphon is minimal if there is no siphon contained in it as a proper subset. A minimal siphon is called a SMS if it does not contain a trap. Π is used to denote the set of SMSs in a net. Siphon S is inv-controlled by P-inv I under M_0 if $I^T M_0 > 0$ and $\{p \in P \mid I(p) > 0\} \subseteq S$.

A transition t∈T is enabled at a marking M if $\forall p \in {}^{\bullet}t, M(p) \geq W(p, t)$. This fact is denoted as M[t⟩. Given a Petri net (N, M_0), $t \in T$ is live under M_0 if $\forall M \in R(N, M_0), \exists M' \in (N, M), M'[t\rangle$. (N, M_0) is live if ∀t∈T, t is live under M_0 . (N, M_0) is dead under M0 if ∄t∈T, M_0[t⟩. (N, M_0) is deadlock-free (weakly live) if ∀M∈R(N, M_0), ∃t∈T, M[t⟩. A marking M∈R(N, M_0) is a livelock if $\forall M' \in R(N, M_0), M'$, M′ is neither a deadlock nor a final marking, where M′≠M_0 (Li & Zhou, 2009). $N_i = (P_i, T_i, F_i, W_i)$ is said to be a subnet of N=(P, T, F, W), which is generated by X =$P_i \cup T_i$, if $F_i = F \cap (X \times X)$, $\forall f \in F_i$, $W_i(f) = W(f)$, $P_i \subseteq P$, and $T_i \subseteq T$.

Basic definitions and theorems

Definition 2.1 (Li & Zhou, 2008) Let $\{r_1, r_2, \ldots, r_m\} \subseteq P_R (m \geq 2)$ be a set of resources in N. An elementary circuit $C(r_1, t_1, r_2, t_2, \ldots, r_m, t_m)$, denoted as C_R , is called a resource circuit if the following conditions are satisfied: (1) $\forall i \in \mathbb{N}_m = \{1, 2, \ldots, m\}$, $r_i \in {}^{\bullet}t_i$; (2) $\forall i \in \{2, 3, \ldots, m\}, r_i \in t_{i-1}^{\bullet}$; and (3) $r_1 \in t_m^{\bullet}$. Moreover, T_{C_R} is used to denote the set of transitions in C_R.

Definition 2.2 (Li & Zhou, 2009) A well-marked S^4R (Systems of Sequential Systems with Shared Resources) net is a marked Petri net N=(P, T, F, W) with initial marking M_0 such that

(1) $P = P_A \cup P_0 \cup P_R$, where $P_A = \cup_{j=1}^{n} P_{A_j}$ is called the set of operation places such that $P_{A_i} \cap P_{A_j} = \emptyset$, $\forall i, j \in \mathbb{N}_n = \{1, 2, \ldots, n\}, i \neq j, P^0 = \cup_{j=1}^{n} \{p_j^0\}$ is called the set of idle places with $P^0 \cap P_A = \emptyset$, and $P_R = \{r_1, r_2, \ldots, r_m\}$ is called the set of resource places such that $(P^0 \cup P_A) \cap P_R = \emptyset$.

(2) $T = \cup_{j=1}^{n} T_j$ is called the set of transitions, where $\forall i, j \in \mathbb{N}_n, i \neq j, T_i \neq \emptyset$ and $T_i \cap T_j = \emptyset$.

(3) $W = W_A \cup W_R$, where $W_A : ((P_A \cup P^0) \times T) \cup (T \times (P_A \cup P^0)) \rightarrow \{0, 1\}$ such that $\forall i, j \in \mathbb{N}_n, i \neq j, ((P_{A_j} \cup \{p_j^0\}) \times T_i) \cup (T_i \times (P_{A_j} \cup \{p_j^0\})) \rightarrow \{0\}$, and $W_R : (P_R \times T) \cup (T \times P_R) \rightarrow \mathbb{N}$.

(4) $\forall j \in \mathbb{N}_n$, the subnet N_j derived from $P_{A_j} \cup \{p_j^0\} \cup T_j$ is a strongly connected state machine such that every circuit contains idle place p_j^0.

(5) $\forall r \in P_R$, there exists a unique minimal P-semiflow I_r such that $||I_r|| \cap P_R = \{r\}, P^0 \cap ||I_r|| = \emptyset, P_A \cap ||I_r|| \neq \emptyset$, and $\forall p \in ||I_r||, I_r(r) = 1$. Furthermore, $P_A = (\cup_{r \in P_R} ||I_r||) \setminus \{P_R\}$.

(6) N is pure and strongly connected.

(7) $\forall p \in P_A, M_0(p) = 0; \forall r \in P_R, M_0(r) \geq \max_{p \in ||I_r||} I_r(p)$; and $\forall p_j^0 \in P^0, M_0(p_j^0) \geq 1$.

Definition 2.3 (Li & Zhou, 2009) A multiset Ω, over a non-empty set A, is a mapping Ω: A→ℕ, which we represent as a formal sum $\sum_{a \in A} \Omega(a).a$.

In multiset Ω, non-negative integer Ω(a) is the coefficient of element a∈A, indicating the number of occurrences of a in Ω. It is said that a∈A belongs to Ω, denoted by a∈Ω, if Ω(a)>0. It does not belong to Ω, denoted by a∉ Ω, if Ω(a)=0.

Definition 2.4 (Li & Zhou, 2009) $\Omega_1 - \Omega_2 := \sum_{a \in A} (\Omega 1(a) - (\Omega_1 \cap \Omega_2)(a)).a$.

Definition 2.5 (Li & Zhou, 2009) Let r be a resource place and S be an SMS in an S^4R net. The operation places that use r is called the set of holders of r, denoted as H(r). The holders of resource r is defined as the difference of two multisets I_r and r : H(r)=I_r −r. As a multiset, $S = S_A \cup S_R$, $S_R = S \cap P_R$, and $S_A = S \cap P_A$. is called the complementary set of siphon S, where $S = S_A \cup S_R$, $S_R = S \cap P_R$, and $S_A = S \cap P_A$. $||Th(S)|| = \{p \mid p \in \cup_{r \in S^R} H(r), p \notin S\}$ is called the support of complementary set Th(S) of siphon S.

Let $\sum_{p \in ||Th(S)||} (h_s p)p$ denote $Th(S)$. $h_s(p)$ is called the risk coefficient of place p. Clearly, $h_s(p)$ indicates that siphon S loses $h_s(p)$ tokens if the number of tokens in p increases by one.

Definition 2.6 (Zhong & Li, 2010) Let S be a siphon of a well-marked S^4R (N, M_0). S is said to be max′-marked at marking M∈(N, M_0) iff ∃p∈S_A such that M(p)≥1 or ∃p∈S_R such that $M(p) \geq \max_{t \in (p^{\bullet} \cap Th(S)^{\bullet})} \{W(p, t)\}$.

Definition 2.7 (Zhong & Li, 2010) Let S be a siphon of a well-marked S^4R (N, M_0). S is said to be max′- controlled iff S is max′-marked at any reachable marking, that is, ∀M ∈ R(N, M_0), ∃ p ∈S_A such that M(p) ≥ 1 or ∃p∈S_R such that $M(p) \geq \max_{t \in (p^{\bullet} \cap Th(S)^{\bullet})} \{W(p, t)\}$.

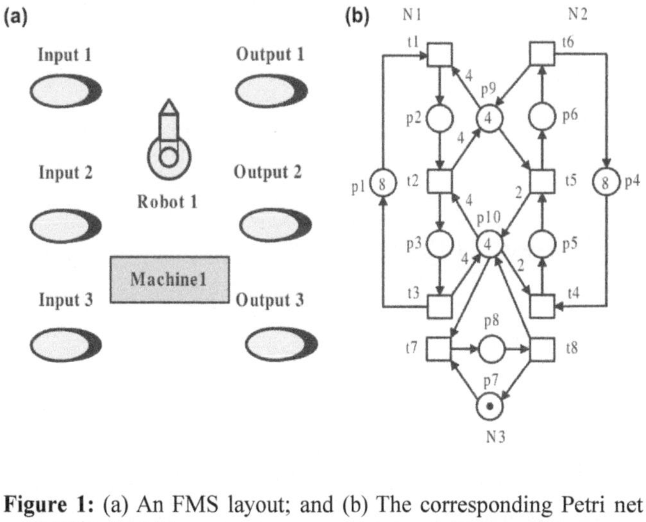

Figure 1: (a) An FMS layout; and (b) The corresponding Petri net model S^4R (N, M_0).

Theorem 2.8 (Zhong & Li, 2010) Let (N, M_0) be a well-marked S^4 R. If all SMSs of the net are max'- controlled, the net is live.

Note that an S^4 R is more general than ES^3 PR, S^3 PR, and L-S^3 PR (Li & Zhou, 2009), i.e. L-S^3 PR⊆S^3 PR⊆ES^3 PR⊆S^4 R. Theorem 2.8 is hence suitable for verifying liveness of ES^3 PR, S^3 PR, and L-S^3 PR.

For example, an FMS layout is shown in Figure 1(a) (Chao, 2007). There are three loading buffers I1–I3 and three unloading buffers O1–O3 to load and unload the FMS corresponding to three raw product types, P1–P3, to be processed by machine M1 and moved by robot R1.

According to the production cycles, a raw product P1 is taken from I1 by R1 and put in M1. After being processed by M1, it is output there as O1. A raw product P2 is taken from I2 and processed by M1 and then moved from M1 to O2 by R1. A raw product P3 is taken from I3, processed by M1 and output from M1 as O3. Figure 1(b) (Chao, 2007) shows its Petri net model S^4 R (N, M_0), where $P^0 = \{p_1, p_4, p_7\}$, $P_A = \{p_2, p_3, p_5, p_6, p_8\}$, $P_R = \{p_9, p_{10}\}$, $|P|=|P^0|+|P_A|+|P_R|=10$, and $|T| = 8$.

By integrated net analyzer (INA, 2003), only an SMS, i.e. $S= \{p_3, p_6, p_8, p_9, p_{10}\}$ can be found. From the above Definitions, $Th(S)=\{p_2, p_5\}$, $CR = \{p_9,t_5, p_{10},t_2\}$, and $TCR = \{t_2,t_5\}$ can be obtained, respectively.

AN ITERATIVE LCP USING THE METHODS OF RMIP AND MAX′-CONTROLLED NSSS

RMIP method

By analyzing the relationship between the states of deadlocks or livelocks and the presence of smart siphons, and then adding new constraints to an MIP method in Li and Li (2012b), the RMIP method is described below.

$$G^{MIP}(M) = \max\left(\sum_{p\in P} v_p + \sum_{p\in P_R} v_p\right) \tag{1}$$

s. t.

$$f_{pt} \geq \frac{M(p) - W(p,t) + 1}{SB(p)}, \quad \forall W(p,t) > 0 \tag{2}$$

$$f_{pt} \geq v_p, \quad \forall W(p,t) > 0 \tag{3}$$

$$z_t \geq \sum_{p\in {}^{\bullet}t} f_{pt} - |{}^{\bullet}t| + 1, \quad \forall t \in T \tag{4}$$

$$v_p \geq z_t, \quad \forall W(t,p) > 0 \tag{5}$$

$$M = M_0 + [N]Y, M \geq 0, Y \geq 0 \tag{6}$$

$$\sum_{p\in P} v_p \leq |P| - 2 \tag{7}$$

$$v_p \geq \frac{M(p)}{SB(p)}, \quad \forall p \in P \tag{8}$$

$$v_r \geq \sum_{t\in r^{\bullet}\cap T_{C_R}} f_{rt} - |r^{\bullet} \cap T_{C_R}| + 1, \quad \forall r \in P_R \tag{9}$$

$$v_p, z_t, f_{pt}, f_{rt} \in \{0, 1\}, \quad \forall p \in P, \forall t \in T \tag{10}$$

where SB(p) =max{|M=M_0+ [N]Y, M≥0, Y ≥0} is the structural bound of place p, f_{rt} is the binary variable for arc (r, t) that is enabled if M(r) is larger than W(r, t). In a live (resp. dead) subnet, all transitions are enabled (resp. disabled) whose z_t=1 (resp. z_t=0). Note that if υ_p=1 for each input place of a transition t in an OPN, then z_t=1. When υ_p=1 but p is unmarked, zt keeps 1, t is said to be pseudo-enabled. For a GPN, arc (p, t) from each input place of a transition t may not be enabled even though p is marked (υ_p=1). That is to say, it is need for p to be sufficiently marked and then υ_p=1 can lead to z_t=1. In order to distinguish the solution of the MIP method in Li and Li (2012b), the solution of the above RMIP method is called NSS.

The RMIP method can be explained as follows: Equation 8 expresses the fact if the number of tokens at a place p is no less than the weight of arc(p, t), then arc(p, t) is enabled and υ_p=1. Physically, when a transition t is enabled and then fired, tokens can flow along arc (t, p) and enter into each output place p. This is consistent with $f\ p_t \geq \upsilon_p$ (Equation 3) where f_{pt}=1 if z_t=1. In addition, if t is an output, rather than input, transition of a place p, f_{pt}=1 does not mean z_t=1; $f\ p_t$=0 means z_t=0 since t is disabled if any input arc is disabled. Equation 9 expresses the fact if all output arcs of resource place r are enabled, then r must be max′-marked and υ_r=1 due to Definition 2.6. Otherwise r is not max′-marked and υ_r=0. The objective function and the other constraints Equations 2–7 and Equation 10 are the same as those in the MIP method in Li and Li (2012b). In summary, this RMIP method can find a set of non-max′-marked places that can form an NSS under a marking M∈R(N, M_0).

Theorem 3.1 Let (N, M_0) be a well-marked S^4 R. If the proposed RMIP method applies to it, then its any feasible solution corresponds to an NSS.

Proof If the feasible solution G^{MIP}(M) of the proposed RMIP method exists, then we obtain an NSS under a certain reachable marking M∈R(N, M_0). By contradiction, suppose that this NSS is not minimal. This implies that the NSS contains a minimal siphon (denoted as S_{min}) as a proper subset. Hence, there exists a reachable dead marking M′ such that υ_p=0 iff p∈S_{min}. It is clear that there exists a new feasible solution G^{MIP}(M′) in which G^{MIP}(M′)>G^{MIP}(M), contradicting the fact that G^{MIP}(M) is maximal among all feasible solutions that meet constraints (Equations 2–10). As a result, Theorem 3.1 holds.

Comparing with the MIP method in Li and Li (2012b), main difference is that this RMIP method can detect a set of non-max′-marked but non-max-marked places that forms a siphon under a marking M there are both live and dead transitions. In other words, the former can only solve a smart siphon causing deadlocks, and the latter can solve an NSS associated with deadlocks and livelocks. In addition, NSF of the proposed RMIP method is obtained, which means no any NSS exists in an S^4 R. Furthermore, an important variable, denoted as U_{z_t}, with respect to the sum of the values of $z_{t_i} (i \in \mathbb{N}^+ = \{1, 2, \ldots, \})$ is introduced to identify the characteristic of the solved NSSs. After executing the above RMIP method, both G^{MIP}(M) and the value of each z_t can be solved. Hence, $U_{z_t} = \sum_{i=1}^{|T|} z_{t_i}$ is obtained accordingly, where |T| denotes the total number of transitions in an S^4 R. From Section 2.1, it is known that if an S4 R is deadlocked under a marking M in which all transitions are disabled, then $z_{t_i}=0$, and $U_{z_t}=0$ is obtained, implying that the characteristic of corresponding NSS is about deadlock. When an S^4 R is livelocked under a marking M in which some transitions are disabled and the rest of transitions are fired, $z_{t_i}\geq 0$ and $0<U_{z_t}<|T|$ are obtained, implying that the characteristic of corresponding NSS is about livelock, denoted as NSS*. Obviously, $z_{t_i}=1$ and $U_{z_t}=|T|$ are true when an S^4 R is live, implying that no any NSS or NSS* exists in an S^4 R.

An iterative LCP combing the RMIP method with max′-controlled NSSs

After solving an NSS (resp. NSS*), its complementary set (denoted as Th(NSS) (resp. Th(NSS) *) can be found by Definition 2.5. Depending on Th(NSS) (resp. Th(NSS) *), two types of CPs in Li and Li (2012b) and Li, An, Wang, et al. (2013) are properly selected and then added for the solved NSS (resp. NSS*). From the above discussion, we add CP for the solved NSS or NSS* to make it max′ not max-controlled. Thus, the key is to determine the initial marking of added CPs (denoted as $M_0(V)$). From Li and Zhou (2009), it is known that the value of $M_0(V)$ is closely related to the value of the control depth variable with respect to the solved NSS (denoted as ξ_{NSS}), i.e. $M_0(V)=M_0(NSS)-\xi_{NSS}$, where $\xi_{NSS}\in\mathbb{N}_m$.

Assume that NSS= $\{p_1, p_2, \ldots, p_k\}$ is a siphon of original net (N_0, M_0), where N_0 = (P_0, T_0, F_0). Add a CP (denoted as V) to make P-vector I be a P-invariant of the augmented net (N* , M*), where ∀ p∈ NSS, I(p) =1, ∀p∈P_0, M* (p) =M_0(p) and I(V)=−1. Thus, $I^T M0 = I^T M^* = M_0(NSS) - M^*(V) > 0$ holds. In order to make the solved NSS max′-controlled, $\xi_{NSS}\geq max_{t\in(r^\bullet\cap Th(NSS)^\bullet)}\{W(p,t)\}$ is determined due to Definitions 2.6 and 2.7. Hence, the definition of ξ_{NSS} is given as follow.

Definition 3.2 Let (N, M_0) be a well-marked S^4 R. ξ_{NSS} is called the control depth variable with respect to an NSS if $\xi_{NSS}\in\mathbb{N}_m$ and $\xi_{NSS}\geq max_{t\in(r^\bullet\cap Th(NSS)^\bullet)}\{W(p,t)\}$, where $\forall r\in NSS_R$, $NSS=NSS_A\cup NSS_R$.

.

Similarly, $\xi_{NSS\star}$ is used to denote the control depth variable with respect to an NSS*, where $\xi_{NSS^*}\in\mathbb{N}_m$ and $\xi_{NSS^*}\geq max_{t\in(r^\bullet\cap_T h(NSS^*)^\bullet)}\{W(p,t)\}$. In general, decreasing ξ_{NSS} (resp. NSS*)intends to relax the control of NSS (resp. NSS*), which may reduce the restriction of the behavioral permissiveness of the controlled system. Moreover, the max′ but not max-controlled condition for NSSs (resp. NSSs*) relaxes the controllability condition and may lead to a live controlled system with more behavioral permissiveness (Liu & Li, 2010; Zhong & Li, 2010).

An NSS or NSS* in a plant net model (N_0 , M_0) can be found by the aforementioned RMIP approach. However, a control policy needs to find all NSSs and NSSs* and then adds the proper CPs for them in the augmented net (N*, M*) such that the final controlled system is live. Hence, an iterative LCP using the proposed RMIP method, which can find a An NSS or NSS* and then add a proper CP for it at each iteration until a live controlled system is obtained, is developed in this subsection and described below, where $|P|=|P_A|+|P^0|+|P_R|$, j ∈$\mathbb{N}^+$, Π, and |V| denote the total number of the places in an original net, the iteration step number, the set of solved NSSs and NSSs* in the augmented net (N_j , M_j), and the number of added CPs in (N_j , M_j), respectively.

The proposed LCP: An iterative LCP combining the RMIP method with max′-controlled NSSs* (NSSs)

Input: a marked S^4 R (N, M_0), where N= ($P_A \cup P^0 \cup P_R$,T,F)

Output: a live controlled net system (N*, M*)

1: preset $\Pi:=\emptyset$, $V:=\emptyset$, and $j:=1$

2: solve an NSS⋆ or NSS by the proposed RMIP method in Section 3.1

3: if NSF for $G^{MIP}(M_1)$ exists then

4: execute NSS_1^* or $NSS_1:=\emptyset$, $N^*:=N$, and $M^*:=M_0$

5: **end if**

6: **while** any feasible solution for $G^{MIP}(M_j)$ exists **do**

7: according to the adding CPs method stated in Li and Li (2012b), Li, An, Wang, et al. (2013), add proper V_j to NSS_j or NSS_j^* in the augmented net (N_j , M_j)

8: execute $\Pi:=\Pi\cup\{NSS_j^*\}\cup\{NSS_j\}$, $V:=V\cup\{V_j\}$, $N^*:=N_j$, $M^*:=M_j$, and $j:=j+1$

9: solve the next NSS_j^* or NSS_j by the proposed RMIP method in Section 3.1

10: end while

11: Output (N*, M*)

At the first iteration, the proposed LCP solves NSS_1^* or NSS_1 and computes $G^{MIP}(M_1)$. NSS_1^* or $NSS_1:=\emptyset$ if NSF for $G^{MIP}(M_1)$ exists, implying that the original Petri net is live. Otherwise, the proposed ICP can obtain NSS_1^* or NSS_1 and then add a proper V_1 to it.

At the jth iteration, the proposed LCP also solves NSS_j^* or NSS_j and then computes $G^{MIP}(M_j)$. Similarly, NSS_j^* or $NSS_j:=\emptyset$ if NFS for $G^{MIP}(M_j)$ holds. Therefore, the corresponding controlled system is live and Policy 1 terminates at this iteration. Otherwise, the proposed LCP solves NSS_j^* or NSS_j and then adds a proper V_j for it. The iteration proceeds until NFS for $G^{MIP}(M_j)$ holds, implying that the corresponding controlled system is live.

Theorem 3.3 Let (N, M_0) be a well-marked S^4 R in which deadlocks and livelocks exist. The proposed LCP is applied to it, which leads to a live augmented net (N* , M*).

Proof Suppose that an NSSj is solved at the jth iteration and the augmented net after the jth iteration is denoted by (N_j, M_j), where $\forall p\in P_A\cup P^0\cup P_R$, $M_{0,j}(p)=M_0(p)$, $M_{0,j}(V)=M_0(NSS_j)-\xi_{NSS_j}$, and $\xi_{NSS_j}\geq max_{t\in(r^\bullet\cap_T h(NSS_j)^\bullet)}\{W(p,t)\}$. By Definition 2.5, Th($NSS_j$) is then computed. Depending on $\xi_{NSS_j}\geq max_{t\in(r^\bullet\cap_T h(NSS_j)^\bullet)}\{W(p,t)\}$. As stated in Li and Li (2012b) and Li, An, Wang, et al. (2013), the addition of both ordinary and general CPs to the original net can make the solved NSSs or NSS* max′-controlled in the iteration. That is to say, an iterative control process can eliminate all existing NSSs and NSS* until NFS for $G^{MIP}(M_j)$ holds, which means that no NSSs and NSS* can be found in the finally augmented net (N*, M*). As a result, the final controlled system (N*, M*) is live owing to Theorem 2.8.

From Chao (2006), Ezpeleta et al. (1995), Huang et al. (2001), Li et al. (2011), Li and Zhou (2004), Li and Zhou (2009), and Zhong and Li (2009), it is known that the number of SMSs in a Petri net grows quickly fast, in the worst case, grows exponentially with respect to its size. Moreover, the computational complexity, structural complexity, and behavior permissiveness are usually used to criticize the different DCPs. Since the proposed LCP can iteratively solve NSSs* (NSSs) that arc belonged to a class of SMSs, this LCP is NP-hard in theory. Due to the fact that any siphon that has more resource places can be composed by those containing less ones (Chao, 2006; Li & Li, 2012b; Li & Zhou, 2008), the solved NSSs* (NSSs) contain the minimal number of resource places. Hence, this LCP may reduce the possibility of adding redundant CPs to some degree. In summary, the proposed LCP has high computational efficiency and can generally obtain a live controlled Petri net system with a simple structure and more behavior permissiveness, which can be verified via two examples in Section 4.

Take an S4 R shown in Figure 1(b) as an illustrative example, where $|P|=|P^0|+|P_A|+|P_R|=10$ and $|T|=8$. By Definition 2.2, $C_R=\{p_9,t_5,p_{10},t_2\}$ and $T_{CR}=\{t_2,t_5\}$ can be obtained, and INA (2003) indicates that this S^4 R has 29⁎ reachable states, where ⁎ denotes maximally permissive behavior.

Table 1: Results of the proposed LCP for a marked S4R shown in Figure 1(b)

j	NSS_j/NSS_j^*	$Th(NSS_j)/Th(NSS_j)^*$	$U_{z,j}$	$^{\bullet}V_j$	$V_j^{\bullet}$	$M_0(V_j)$	$G^{MIP}(M_j)$	$\|P\|+\|V\|$
1	$\{p_3,p_6,p_8,p_9,p_{10}\}^*$	$\{p_2,p_5\}^*$	2				5	10
2	-	-	8	$4t_2, 2t_5$	$4t_1, 2t_4$	4	NFS	11

At the first iteration, $NSS_1^*=\{p_3,p_6,p_8,p_9,p_{10}\}$ with $U_{z,}=2<|T|=8$ is solved by Policy 1. By Definition 2.5, $Th(NSS_1)^*=\{p_2,p_5\}$ with $h_{NSS_1^*}(p_2)=4$ and $h_{NSS_1^*}(p_5)=2$ is computed.

Due to $\xi^*_{NSS_1}\geq \max_{t\in(r^{\bullet}\cap Th(NSS_1^*)^{\bullet})}\{W(p,t)\}=\max\{4,1\}=4, M_0(V_1)=4+4-4=4$ can be determined.

Depending on $h_{NSS_1^*}(p_2)=4$ and $h_{NSS_1^*}(p_5)=2$, a GCP V_1 with $^{\bullet}V_1=\{4t_2,2t_5\}, V_1^{\bullet}=\{4t_1,2t_4\}$, and $M_0(V_1)=4$ is then added.

At the second iteration, Policy 1 obtains NFS for $G^{MIP}(M_2)$, which means that no NSSs or NSS* s can be found and Policy 1 terminates at this iteration. As a result, the finally controlled system is live. The results of Policy 1 are briefly shown in Table 1.

From Table 1, the proposed LCP solves one NSS⋆ and accordingly adds a GCP for it. Policy 1 terminates in the second iteration until NFS for $G^{MIP}(M_2)$ is declared. Finally, the controlled system (N*, M*) is live due to Theorems 2.2 and 3.2. In addition, by INA (2003), (N*, M*) has 29 reachable states and R=100%, where R denotes the ratio of the number of reachable states of the live controlled system (N*, M*) to that of maximally permissive behavior of the original S^4 R (N, M_0).

EXAMPLES

In this section, two examples are presented to demonstrate the proposed ICP in this work. Moreover, its control performance comparison with the existing approaches presented in Zhong and Li (2009) (denoted as ZL), Li et al. (2004) (denoted as LLW⁺), and Li et al. (2006) (denoted as LWW) is also given, where the number of redundant added CPs is obtained by executing the method in Li et al. (2012) and the comparison is carried out on a 2.8GHz Pentium computer with 512MB of RAM under a Windows XP operating system.

Example 1 Figure 2(a) and (b) shows an FMS layout and its Petri net model S^4 R (N, M_0) taken from Zhong and Li (2009), respectively. There are two uploading buffers I1 and I2 and three downloading buffers O1 and O2 to upload and download the FMS corresponding to two raw product types, P1 and P2, to be processed by machine M1 and moved by robot R1 (resp. R2) and R2 (resp. R1).

According to two production cycles, a raw product P1 is taken from I1 by R2 and put in M1. After being processed by M1, it is downloaded by R1 and output to O1. A raw product P2 is taken from I2 by R1 and put in M1 to be processed, and then moved from M1 to O2 by R2. For Figure 2(b), we have

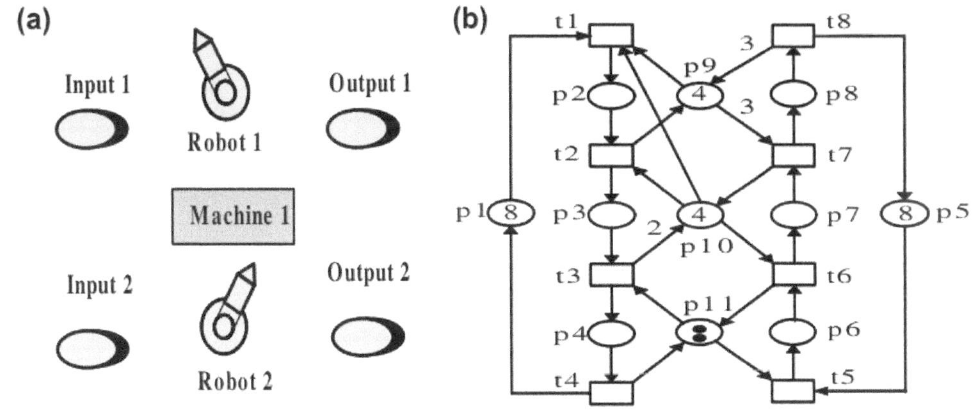

Figure 2: (a) An FMS layout; and (b) The corresponding Petri net model S^4R (N, M_0).

$P^0=\{p_1,p_5\}, P_A=\{p_2,p_3,p_4,p_6,p_7,p_8\}, P_R=\{p_9,p_{10},p_{11}\}, |P|=|P^0|+|P_A|+|P_R|=11$, and $|T|=8$. By Definition 2.1, $C_{R_1}=\{p_9,t_7,p_{10},t_2\}, T_{C_{R_1}}=\{t_2,t_7,\}, C_{R_2}=\{p_{10},t_6,p_{11},t_3\}$, and $T_{C_{R_2}}=\{t_3,t_6\}$ can be obtained. The number of its maximally permissive states is 194⁎ by executing INA (2003). Tables 2 and 3 show the results of the proposed LCP and its control performance comparison with the existing approaches, respectively.

Example 2 Considering another FMS layout and its Petri net model S^4R (N, M_0) depicted in Figure 3(a) and (b) (Zhong & Li, 2009), respectively. This FMS consists of four robots R1–R4, each of which can hold one or three products every time, and three machines M1–M3, each of which can process two or three products every time. There are three loading buffers I1–I3 and three unloading buffers O1– O3 to load and unload the FMS. There are three raw product types, namely P1, P2, and P3, to be processed.

According to three production cycles, a raw product J1 is taken from I1 by R1 and R2 and put in M1. After being processed by M1 it is then moved to M3 by R4. Finally, after being processed by M3, it is processed by M2 and R3 and then moved to O1. A raw product J2 is taken from I2 and processed by M2 and R4, and then processed by M2 only. After being processed by M2 it is then moved from M2 to O2 by R4. A raw product J3 is taken from I3, processed by M3 and R3, and then by M3 and R4. After that, it is then processed by M1. Finally, after being processed by M1, it is then moved from M1 to O3 by R1 and R2 sequentially. For Figure 2(b), we have $P^0 = \{p_8, p_{12}, p_{20}\}$,

$$P_A=\{p_1,p_2,p_3,p_4,p_5,p_6,p_7,p_9,p_{10},p_{11},p_{13},p_{14},p_{15},p_{16},p_{17}\},\ p_{24},p_{25}\},$$

$$P_R=\{p_{18},p_{19},p_{21},p_{22},p_{23},$$

$|P|=|P^0|+|P_A|+|P_R|=25$, and $|T|=19$. By Definition 2.1,

$$C_{R_1}=\{p_{21},t_{16},p_{22},t_9,p_{24},t_8\},\ T_{C_{R_2}}=\{t_8,t_9,t_{16}\},\ C_{R_2}=\{p_{22},t_{14},p_{23},t_5\},$$

$$T_{C_{R_2}}=\{t_5,t_{14}\},\ C_{R_3}=\{p_{24},t_{15},p_{23},t_5,p_{22},t_9\},\ T_{C_{R_3}}=\{t_5,t_9,t_{15}\},$$

$$C_{R_4}=\{p_{24},t_{15},p_{23},t_5,p_{22},t_4\},\ T_{C_{R_4}}=\{t_4,t_5,t_{15}\},\ C_{R_5}=\{p_{21},t_{16},p_{24},t_8\},$$

$$T_{C_{R_5}}=\{t_8,t_{16}\},\ C_{R_6}=\{p_{21},t_{16},p_{24},t_3\},\ T_{C_{R_6}}=\{t_3,t_{16}\}$$

$$C_{R_7}=\{p_{21},t_{16},p_{22},t_9,p_{24},t_3\},\ T_{C_{R_7}}=\{t_3,t_9,t_{16}\},\ C_{R_8}=\{p_{24},t_{12},p_{25},t_5,p_{22},t_4\},$$

$$T_{C_{R_8}}=\{t_4,t_5,t_{12}\},\ C_{R_9}=\{p_{24},t_{12},p_{25},t_5,p_{22},t_9\},\ T_{C_{R_9}}=\{t_5,t_9,t_{12}\},$$

$C_{R_{10}}=\{p_{24},t_{12},p_{25},t_5,p_{22},t_4\}, T_{C_{R_{10}}}=\{t_4,t_5,t_{12}\}$ can be obtained. The number of its maximally permissive states is 8559⁎ by executing INA (2003). Tables 4 and 5 show the results of the proposed LCP and its control performance comparison with the existing approaches, respectively.

From Tables 3 and 5, the number of necessary (redundant) added CPs obtained by this LCP is fewer (equal to zero), leading to the live controlled S^4 R with simple structures. Moreover, this LCP achieves more permissive states compared with the existing methods. That is to say, the proposed LCP can result in a liveness-enforcing Petri net supervisor with a simple structure and more permissive behavior in terms of structural complexity and behavior permissiveness.

Table 2: Results of the proposed LCP for a non-live marked S4R shown in Figure 2(b)

j	NSS_j/NSS_j^*	$Th(NSS_j)/Th(NSS_j)^*$	$U_{z_t,j}$	$^{\bullet}V_j$	$V_j^{\bullet}$	$M_0(V_j)$	$G^{MIP}(M_j)$	$\|P\| + \|V\|$
1	$\{p_4, p_7, p_{10}, p_{11}\}^*$	$\{p_2, 2p_3, p_6\}^*$	4				8	11
2	$\{p_3, p_8, p_9, p_{10}\}^*$	$\{2p_2, p_7\}^*$	4	$2t_3, t_6$	t_1, t_2, t_5	5	10	12
4	-	-	8	$2t_2, t_7$	$2t_1, t_6$	5	NFS	13

Table 3: Control performance comparison

Control criteria	The proposed LCP	ZL
No. of added CPs	2	3
No. of reachable states	189	189
R	97.4%	97.4%
Executing time	≤1 s	≤1 s
Memory occupancy	1.24 M	1.28 M
No. of redundant added CPs	0	1

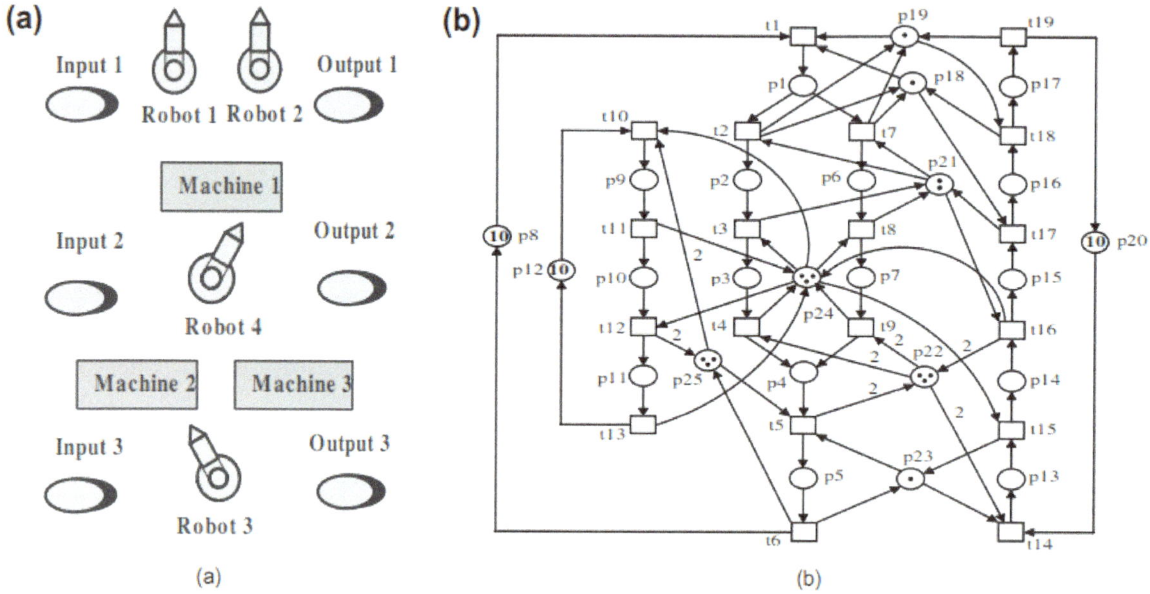

Figure 3: (a) Another FMS layout; and (b) The corresponding Petri net model S^4R (N, M_0).

Table 4: Results of the proposed LCP for another non-live marked S4R shown in Figure 3(b)

j	NSS_j/NSS_j^*	$Th(NSS_j)/Th(NSS_j)^*$	$U_{z_t,j}$	$^{\bullet}V_j$	$V_j^{\bullet}$	$M_0(V_j)$	$G^{MIP}(M_j)$	$\|P\| + \|V\|$
1	$\{p_2, p_6, p_{16}, p_{18}, p_{21}\}^*$	$\{p_1, p_{15}\}^*$	13				25	25
2	$\{p_4, p_9, p_{11}, p_{14}, p_{22}, p_{24}\}^*$	$\{p_3, p_7, 2p_{13}\}^*$	9	t_2, t_7, t_{17}	t_1, t_{16}	2	26	26
3	$\{p_4, p_9, p_{11}, p_{15}, p_{21}, p_{22}, p_{24}\}^*$	$\{p_2, p_3, p_6, p_7, 2p_{13}, 3p_{14}\}^*$	11	$t_4, t_9, 2t_{15}$	$t_1, 2t_{14}$	4	23	27
4	-	-	19	$t_4, t_9, 3t_{16}$	$t_1, 2t_{14}, t_{15}$	6	NFS	28

Table 5: Control performance comparison

Control criteria	The proposed LCP	ZL	LLW*	LWW
No. of added CPs	3	3	6	2
No. of reachable states	7622	4116	2570	2654
R	89.1%	48.1%	30%	31%
Executing time	≤3 s	≤3 s	≤2 s	≤2 s
Memory occupancy	2.74 M	2.38 M	2.14 M	1.97 M
No. of redundant added CPs	0	1	4	0

DISCUSSION

Different types of multiple resources can be requested by different processes in an FMS. As a typical class of generalized Petri nets, an S^4R can model more complicated resource allocation systems with multiple concurrent processes. Thus, an S^4R has better modeling power than ES^3PR, S^3PR, and L-S^3PR (Li & Zhou, 2009). An insufficiently marked siphon in an S^4R can lead to a dead state. Many DCPs (Huang et al., 2006; Li & Li,2012b; Li et al., 2011, 2006; Piroddi et al., 2009; Zhong & Li, 2009) for S^4R are developed based on the concept of max-controlled siphons. However, these policies cannot solve livelock problems that may lead to undesirable results for in running process of an FMS.

As far as the authors know, only limited work of (Chao, 2007; Liu & Li, 2010; Zhong & Li, 2010) is reported on the siphon controllability conditions causing livelock states, i.e. max′ or max″-controlled siphons. But no corresponding SC policies are given and the approach in Liu and Li (2010) is nonlinear and more complicate. By means of the study of Chao (2007), Zhong and Li (2010) and the structural analysis of Petri nets, the presence of a non-max-marked siphon is not a sufficient condition for an S^4 R to be not live. Actually, a set of non-max′-marked places under a marking $M \in R(N, M_0)$ does form a problematic siphon to make an S^4 R be not live. Thus, we add new constrains to the MIP method in Li and Li (2012b) to create a RMIP. The proposed RMIP method in Section 3.1 can solve NSSs in an S^4 R directly and then identify the solved ones causing deadlocks or livelocks depending on the value of U_{zt}. Compared with the method in Liu and Li (2010), this RMIP method is relatively simple and its computation is lower. Moreover, an iterative LCP using the RMIP method and max′-controlled siphon controllability is designed in Section 3.2. Section 4 shows two examples to illustrate that the proposed RMIP method and corresponding LCP are feasible in theory and have potential for development of the theory to cope with both livelocks and deadlocks than the prior theory based on siphon controllability.

CONCLUSIONS

On the basis of the structural analysis of Petri nets and the concept of a set of non-max′-marked places, this research presents an RMIP approach to directly solve NSSs associated with deadlocks or livelocks in an S^4R. Moreover, an LCP adopting the RMIP method and max′-controlled siphon controllability is designed to eliminate deadlocks and livelocks in a non-live S^4R. The results in Section 4 show that the proposed LCP has a certain advantage over the existing policies in terms of behavioral permissiveness and structural complexity of the final live controlled system. The extension of the proposed RMIP and the corresponding LCP to more general classes of Petri nets is the focus of our future research.

ACKNOWLEDGMENTS

The authors would like to thank the editor and anonymous reviewers whose comments and suggestions greatly helped us improve the quality and presentation of this paper.

REFERENCES

1. Chao, D. Y. (2006). Computation of elementary siphons for deadlock control. *The Computer Journal, 49*, 470–479.10.1093/comjnl/bxl019 ,
2. Chao, D. Y. (2007). Max′-controlled siphons for liveness of S^3PGR^2. *IET Control Theory and Applications, 1*, 933–936.10.1049/iet-cta:20060275 ,
3. Chu, F., & Xie, X. L. (1997). Deadlock analysis of Petri nets using siphons and mathematical programming. *IEEE Transactions on Robotics and Automation, Part A, 13*, 793–804.
4. Ezpeleta, J., Colom, J. M., & Martinez, J. (1995). A Petri net based deadlock prevention policy for flexible manufacturing systems. *IEEE Transactions on Robotics and Automation, 11*, 173–184.10.1109/70.370500
5. Huang, Y. S., Jeng, M. D., Xie, X. L., & Chung, D. H. (2006). Siphon-based deadlock prevention policy for flexible manufacturing systems. *IEEE Transactions on Systems, Man, and Cybernetics, Part A, 36*, 1248–1256.10.1109/TSMCA.2006.878953 ,
6. Huang, Y. S., Jeng, M. D., Xie, X. L., & Chung, S. L. (2001). Deadlock prevention policy based on Petri nets and siphons. *International Journal of Production Research,39*, 283–305.10.1080/00207540010002405 ,
7. Integrated Net Analyzer. (2003). *A software tool for analysis of Petri nets* (Version 2.2) [online]. Retrieved July 31, 2003, from http://www.informatik.hu-berlin.de/starke/ina.html
8. Li, S. Y., & Li, Z. W. (2012a). Structure reduction of liveness-enforcing Petri nets using mixed integer programming. *Asian Journal of Control, 14*, 384–399.10.1002/asjc.342 ,
9. Li, S. Y., & Li, Z. W. (2012b). Solving siphons with the minimal cardinality in Petri nets and its applications to deadlock control. *International Journal of Production Research, 50*, 6203–6218.10.1080/00207543.2011.590540 ,
10. Li, S. Y., Li, Z. W., Al-Ahmari, A., & An, A. M. (2012). An extraction algorithm for a set of elementary siphons based on mixed integer programming. *Journal of Systems Science and Systems Engineering, 21*, 106–125.10.1007/s11518-012-5188-z ,
11. Li, S. Y., Li, Z. W., & Hu, H. S. (2011). Siphon extraction for deadlock control in flexible manufacturing systems by using Petri nets. *International Journal of Computer Integrated Manufacturing, 24*, 710–725.10.1080/0951192X.2011.575182 ,
12. Li, S. Y., & Wang, A. R. (2011). A deadlock prevention policy in Petri nets using necessary siphons. *Control Theory & Applications, 28*, 771–780.
13. Li, Y. S., An, A. M., Cai, Y., Hong, C. Q., Han, X. L., & Wang, Y. (2013). A deadlock control policy for a subclass of Petri nets G-system. *Control Theory & Applications, 30*, 1429–1436.
14. Li, Y. S., An, A. M., Wang, Y., Wang, G., Hong, C. Q., & Cai, Y. (2013). Design of liveness-enforcing supervisors with simpler structures for deadlock-free operations in flexible manufacturing systems using necessary

siphons. *Journal of Intelligent Manufacturing, 24*, 1157–1173.10.1007/s10845-012-0647-4 ,

15. Li, Z. W., Hu, H. S., & Wang, A. R. (2007). Design of liveness-enforcing supervisors for flexible manufacturing systems using Petri nets. *IEEE Transactions on Systems, Man, and Cybernetics, Part C, 37*, 517–526.10.1109/TSMCC.2007.897333 ,
16. Li, Z. W., Liang, J. W., Lun, Y., & Wang, A. R. (2004). A deadlock prevention method for FMS with multiple resource acquisitions. In *Proceedings of 8th International Conference on Control, automation, robotics and vision* (pp. 2117–2122). Kunming, China.
17. Li, Z. W., Wang, A. R., & Wei, N. (2006). Liveness-enforcing supervisors for flexible manufacturing systems with multiple resource acquisitions. In *Proceedings of the 2006 IEEE Conference on Networking, sensing and control* (pp. 710–714). Ft. Lauderdale, FL, USA.
18. Li, Z. W., & Zhao, M. (2008). On controllability of dependent siphons for deadlock prevention in generalized Petri nets. *IEEE Transactions on Systems, Man, and Cybernetics, Part A, 38*, 369–384. ,
19. Li, Z. W., & Zhou, M. C. (2004). Elementary siphons of Petri nets and their application to deadlock prevention in flexible manufacturing systems. *IEEE Transactions on Systems, Man, and Cybernetics, Part A, 34*, 38–51.10.1109/TSMCA.2003.820576 ,
20. Li, Z. W., & Zhou, M. C. (2006a). Two-stage method for synthesizing liveness-enforcing supervisors for flexible manufacturing systems using Petri nets. *IEEE Transactions on Industrial Informatics, 2*, 313–325.10.1109/TII.2006.885185,
21. Li, Z. W., & Zhou, M. C. (2006b). Clarifications on the definitions of elementary siphons in Petri nets. *IEEE Transactions on Systems, Man, and Cybernetics, Part A, 36*, 1227–1229.10.1109/TSMCA.2006.878966 ,
22. Li, Z. W., & Zhou, M. C. (2008). On siphon computation for deadlock control in a class of Petri nets. *IEEE Transactions on Systems, Man, and Cybernetics, Part A, 38*, 667–679. ,
23. Li, Z. W., & Zhou, M. C. (Eds.). (2009). *Deadlock resolution in automated manufacturing systems: A novel Petri net approach*. London: Springer.
24. Li, Z. W., Zhou, M. C., & Jeng, M. D. (2008). A maximally permissive deadlock prevention policy for FMS based on Petri net siphon control and the theory of regions. *IEEE Transactions on Automation Science and Engineering, 5*, 182–188.,
25. Liu, G. Y., & Li, Z. W. (2010). General mixed integer programming-based liveness test for system of sequential systems with shared resources nets. *IET Control Theory Applications, 4*, 2867–2878.10.1049/iet-cta.2009.0557 ,
26. Park, J., & Reveliotis, S. A. (2001). Deadlock avoidance in sequential resource allocation systems with multiple resource acquisitions and flexible routings. *IEEE Transactions on Automatic Control, 46*, 1572–1583.10.1109/9.956052 ,
27. Piroddi, L., Cordone, R., & Fumagalli, I. (2009). Combined siphon and marking generation for deadlock prevention in Petri nets. *IEEE Transactions on Systems, Man, and Cybernetics, Part A, 39*, 650–661.10.1109/TSMCA.2009.2013189 ,
28. Uzam, M., & Zhou, M. C. (2006). An improved iterative synthesis method for liveness enforcing supervisors of flexible manufacturing systems. *International Journal of Production Research, 44*, 1987–2030.10.1080/00207540500431321 ,
29. Wu, N. Q., & Zhou, M. C. (Eds.). (2009). *System modeling and control with resource-oriented Petri nets: Control engineering series*. New York, NY: CRC Press.
30. Zhong, C. F., & Li, Z. W. (2009). A deadlock prevention approach for flexible manufacturing systems without complete siphon enumeration of their Petri net models. *Engineering with Computers, 25*, 269–278. ,
31. Zhong, C. F., & Li, Z. W. (2010). Self-liveness of a class of Petri net models for flexible manufacturing systems. *IET Control Theory and Applications, 4*, 403–410.10.1049/iet-cta.2008.0193

CHAPTER 11

A LITERATURE REVIEW ON THE PROGRESSION OF AGILE MANUFACTURING PARADIGM AND ITS SCOPE OF APPLICATION IN PUMP INDUSTRY

V. M. M. Thilak[1] S. R. Devadasan[1] and N. M. Sivaram[2]

[1]Department of Production Engineering, PSG College of Technology, Coimbatore, Tamil Nadu 641004, India

[2]Department of Mechanical Engineering, National Institute of Technology, Tiruchirappalli, Tamil Nadu 620015, India

ABSTRACT

During the recent years, the manufacturing world has been witnessing the application of agile manufacturing paradigm. The literature review reported in this paper was carried out to study this progression. This literature review was carried out in two phases. In the first phase, the literature was reviewed to trace the origin of agile manufacturing paradigm and identify its enablers. Further, during this phase, the applications of agile manufacturing reported in literature arena were reviewed. It was also discernable that certain research works have been initiated to apply agile manufacturing paradigm in pump industry. During the second phase, the researches reported on applying agile manufacturing in pump industry were reviewed. At the end of this review, it was found that so far the implementation of agile manufacturing in pump industry has been examined by the researchers by considering only certain components of pumps. In fact, the holistic implementation of agile manufacturing in the pump industry is yet to be examined by the researchers. In the context of drawing this inference, this paper has been concluded by stating that high scope exists in examining the infusing of agility characteristics in designing and manufacturing of pumps.

INTRODUCTION

The field of manufacturing emerged in the world during ancient days when humankind began to shape the naturally available components to fulfill specific needs. Particularly, the humankind that lived during ancient days would have sharpened the stones for killing animals. Thus, the manufacturing field has ancient day roots. On further development, the manufacturing field expanded to produce little more sophisticated products like utensils and furniture. In order to manufacture these little more sophisticated products, the humankind began to adopt manufacturing paradigms. The first paradigm that the manufacturing world adopted was craft production. Craft production paradigm enabled the humankind to produce products in low volume by employing few people. After the industrial revolution, the manufacturing field witnessed the emergence of mass production paradigm. Mass production paradigm enabled the humankind to produce products in large volume by employing many people. Mass production paradigm dominated the manufacturing scenario till the 1960s. Thereafter, the manufacturing world began to witness competition. In order to face this competition, the companies started adopting lean manufacturing paradigm which focused on eliminating wastes which occurred during the mass

Citation: V. M. M. Thilak, S. R. Devadasan, and N. M. Sivaram, "A Literature Review on the Progression of Agile Manufacturing Paradigm and Its Scope of Application in Pump Industry," The Scientific World Journal, vol. 2015, Article ID 297850, 9 pages, 2015. doi:10.1155/2015/297850.

production. From the 1980s, the world began to experience the intensification of competition. In order to face this kind of intensified competition, the world witnessed the emergence of agile manufacturing paradigm. Agile manufacturing paradigm focuses on manufacturing products based on the dynamic demands of the customers. Furthermore agile manufacturing paradigm enables an organization to supply products according to the choice and specifications of the customer. This capability is addressed by their researchers under the term "mass customization." Today, the companies which have been able to mass-customize the products produced by them through the installation of agile manufacturing paradigm have been able to thrive in the market. Though agile manufacturing has been adding certain industries to face the competition, its application in many other industries is not significantly reported. This situation warrants the need to not only review the literature for studying the state-of-the-art researches on agile manufacturing but also explore its exploitation in an industry in which it is yet to find in-depth application. In this regard, pump manufacturing industry was chosen as the candidate industry for studying the application feasibility of agile manufacturing paradigm. The literature review reported in this paper was carried out in two phases. In the first phase, the state-of-the-art researches on agile manufacturing were studied. In the second phase, the scope applying agile manufacturing concepts in pump industry was studied. The details of carrying out the activities under these two phases are presented in the subsequent sections of this paper.

LITERATURE REVIEW METHODOLOGY

The literature review being reported here was begun by gathering papers reporting researches on agile manufacturing. Then, the meaning, definition, and emanation of agile manufacturing reported in these papers were studied. After reviewing this information, the enablers of agile manufacturing were identified. Subsequently, the different industries where agile manufacturing has been successfully applied were identified and studied. At this stage, the scope of the literature survey being reported here was restricted to studying the research activities on applying agile manufacturing in the pump industry. Subsequently, a search was made in the literature arena to identify the paper reporting research activities in this direction. This search resulted in the identification of three papers reporting the researches on applying agile manufacturing in the pump industry. These papers were studied in detail. The details of these activities are briefly reported in the following sections of this paper.

HISTORY, ORIGIN, AND MEANING OF AGILE MANUFACTURING

The world began to experience globalization from the 1990s. The globalization resulted in the entry of products produced by different countries into the local markets. This scenario fueled the global competition between companies situated in various parts of the world [1]. This global competition enabled the customers to demand products with innovative features at low price and high degree of quality [2–6]. In order to meet the dynamic demands of the customers, few companies in the world acquired agile capabilities to manufacture innovative products within a short period of time to withstand the global competition [7]. A very apt example to be cited to support this claim is the capabilities being demonstrated in contemporary days by the mobile phone manufacturing companies. The modern mobile phone manufacturing companies have been manufacturing and selling several models of mobile phones with amazing innovative features.

In the early period, mobile phone manufacturers produced large sized mobile phones with the provision for telephoning and utilizing other few facilities like calculator and calendar. Today, those manufacturers have brought out mobile phones under different names like tablet and iPad, which are incorporated with facilities like high resolution camera and high speed internet technology. Thus, the mobile phone manufacturers are capable of quickly responding to the dynamic demands of the customers. This is due to the fact that mobile phone manufacturing companies are incorporated with a type of production paradigm that is agile in nature. In the year 1990, a group of researchers coined the term "agile manufacturing" after seeing this kind of developments [8]. Then, researches in the direction of achieving agile manufacturing were pursued by forming agile forum at Iacocca Institute in Lehigh University, USA.

Definitions and Meaning of Agile Manufacturing

The formation of agile forum at Iacocca Institute in Lehigh University resulted in a considerable number of researches in effecting agile manufacturing in different companies. Initially, these researchers began to work on discovering the meaning of agile manufacturing. The common meaning recognized by researchers is that agile manufacturing is a manufacturing paradigm that enables the companies to respond to the dynamic demands of the customers quickly [9]. In the recent times, researches on the procedures to implement agile manufacturing in different manufacturing industries have been developed by the researchers. These efforts have not been successful in all the cases due to certain limitations. The high investments for implementation, rigid nature of the product, and lack of multiskilled employees are few obstacles of successfully implementing agile manufacturing in industries. Thus, smart manufacturing industry like mobile phone manufacturing industry flourishes by either implicitly or explicitly adopting agile manufacturing paradigm, whereas many traditional industries are still struggling to flourish in today's competitive world. However, some manufacturing industries which are yet to implement agile manufacturing have spontaneously imbibed agile manufacturing principles. For instance, the automobile sector has been imbibing agile manufacturing principles spontaneously, as different models of cars are manufactured in this sector to fulfill the different tastes and requirements of customers. Thus, the journey of agile manufacturing which began at Iacocca Institute at Lehigh University advances today through different manufacturing sectors.

After the formation of agile forum at Iacocca Institute at Lehigh University, USA, the researchers made efforts to define agile manufacturing. These efforts of the researchers resulted in several definitions of agile manufacturing in the literature arena. Few authors like Gunasekaran and Yusuf [10] and Jin-Hai et al. [11] have summarized these definitions. A study of these definitions would indicate that the researchers viewed agile manufacturing paradigm as the post-mass production concept which focuses on meeting the global competition by quickly responding to the dynamic demands of the customers [12]. Comprehension of these definitions indicates that agile manufacturing enables a company to be flexible enough to quickly respond to the dynamic demands of the customers and manufacture products with many varieties and innovative features [13]. The manufacturing system reconfigurability and flexibility to produce varieties of products are the critical characteristics of agile manufacturing. In other words, agile manufacturing aids the company to manufacture different varieties of products based on the dynamic changes in customer needs in a shorter period of time.

ENABLERS OF AGILE MANUFACTURING

After developing the definitions of agile manufacturing, researchers began to carry out studies for determining the enablers that would aid the companies with implementing agile manufacturing principles. These researches have been repeatedly reported in literature arena. The agility enablers pinpointed by the authors in these research papers are enumerated in Table 1.

Table 1: Agility enablers reported in literature arena

Paper number	Papers reporting agility enablers	Number of agility enablers identified	Agility enablers
1	Vinodh et al. [23]	2	Computer-Aided Design (CAD), Rapid Prototyping Technology (RPT).
2	Yusuf et al. [24]	4	Core competence management, Virtual Enterprise (VE) formation, capability for reconfiguration, and knowledge driven enterprise.
3	Narasimhan et al. [25]	4	Partnership arrangement, close relationships with suppliers, JIT production, and advanced technologies.
4	Vinodh et al. [26]	5	Organizational structure enabler, manufacturing management enabler, employee agility enabler, technology enabler, and manufacturing strategy enabler.
5	Gunasekaran [27] Bottani [28] Gunasekaran and Yusuf [10]	7	VE formation tools/metrics, physically distributed teams, and manufacturing, rapid partnership formation tools/metrics, Concurrent Engineering (CE), integrated product/production, business information system, rapid prototyping tools, electronic commerce (EC).
6	Vinodh and Kuttalingam [9]	9	CAD, Computer Aided Manufacturing (CAM), Computer Numerical Control (CNC), computer-integrated manufacturing (CIM), Rapid Prototyping (RP), Rapid Tooling (RT), Reverse Engineering (RE), VE, Information Technology (IT).
7	Dowlatshahi and Cao [29]	10	Core competencies, VE, RPT, CE, multiskilled and flexible people, continuous improvement, team working, change and risk management, IT, empowering.
8	Sreenivasa and Devadasan [30]	25	Artificial Intelligence (AI), internet, inter- and intraenterprise activities, Standard for Exchange of Products (STEP), CE, virtual manufacturing, component based distributed shop floor control system, tabu-enhanced genetic algorithm approach, customization, RPT, IT, AI, web technology, innovative total quality function deployment (ITQFD), management responsibility, manufacturing management, employee status, technology and manufacturing strategy, RPT, Computer-Aided Design of Experiments (CADOE), activity based costing (ABC), CAD/CAM, CAD, and RPT.
9	Eshlaghy et al. [31]	25	VE, EC, RPT, improvement, multiskilled and flexible people, team working, CE, change and risk management, integrated information system, continuous improvement, flexible infrastructure, supply chain, improved manufacturing technology, core competence, capability for reconfiguration, knowledge management, innovation, agile strategy, agility process, reward system, culture, reengineering, leadership, collaborative relationships, uncertainty and essential changes, people leverage, and IT.
10	Vázquez-Bustelo et al. [32]	53	Top management support, employee involvement and empowerment, team working, job rotation (multifunctional workforce), training and education, knowledge workers, decentralized decision making, entrepreneurial firm culture, reward schemes to encourage innovation and based on both financial and nonfinancial measures, Enterprise Resource Planning (ERP), Materials Requirement Planning (MRP), Robotics, automated guided vehicle (AGV), automated storage and retrieval systems, CNC machines, CAD/CAM, RPT tools, intranet internet and world wide web, Electronic Data Interchange (EDI), EC, visual inspection, manufacturing cells, virtual reality software, Flexible Manufacturing Systems (FMS), computer-aided process planning (CAPP), group technology (GT), Point of Sales (POS) data collection, bar codes, automatic data collection, real time communication/execution systems, Design for Manufacture/Assembly (DFM/A), strategic alliance/based on core/complementary competencies, virtual firm/organization, rapid partnership formation, integration of functions from purchasing to sales, global supply chain management, customer integrated processes for designing, manufacturing, marketing, and support, strategic relationship with customers and suppliers, internal and external cooperation, Business Process Reengineering (BPR), formation of cross function product development teams, concurrent design of products and processes, multidisciplinary team working environment, intelligent engineering design support system, customer and supplier integrated multidiscipline teams, global access to databases and information, Knowledge Based Systems (KBS), knowledge management systems, sensitive information protection, organization structure that promotes innovation, training and education, learning organization, knowledge acquisition from internal and external resources, and core competence management.

An overview of the contents of Table 1 would indicate that RPT and virtual manufacturing are found to be often mentioned as enablers of agile manufacturing. RPT aids the rapid production of prototypes which enable the companies to reconfigure the products to suit the customers' preferences before the product is actually manufactured in the shop floor. Furthermore, virtual manufacturing enables the viewing of manufacturing procedure in the computer screens before the product is actually manufactured. This capability of virtual manufacturing enables the reconfiguration of manufacturing facilities to suit the manufacturing of the products according to the customers' dynamic demands [14]. An interesting observation made at this juncture was that, during the recent years, researchers have been appraising the power of cloud manufacturing model in enhancing the virtual manufacturing and reconfigurable capabilities of organizations for acquiring agility characteristics. Cloud manufacturing model has the potential to aid the organizations to gather necessary knowledge at low price to become agile companies through the utilization of Information Technology resources [15–19]. Next to RPT and virtual manufacturing, CAD is frequently referred by the researchers as agile enabler. Critical thinking would specify that, without the aid of CAD, the production of prototypes rapidly using RPT and reconfiguring the manufacturing facilities by using virtual manufacturing is not possible. Thus, CAD aids the reconfiguration of products and manufacturing facilities. Hence, the foundational enabler of agile manufacturing is the application of CAD in the case of manufacturing products.

INDUSTRIAL APPLICATIONS OF AGILE MANUFACTURING

Several researchers have endeavored to implement agile manufacturing paradigm in different industries right from its conceptualization. This resulted in appearance of many research papers reporting the experiences of implementing agile manufacturing in various industrial sectors in the literature arena. During the literature review being reported here, a total of 48 research papers on industrial application of agile manufacturing were found. The nature of researches is highlighted in Table 2. The statistical aspects of the paper reviewed are presented in the bar chart shown in Figure 1. As shown, the number of papers reporting the application of agile manufacturing in electronics industry is maximum compared to the number of papers reporting researches on agile manufacturing in other industries. Since various types of researches on agile manufacturing in electronic industry are reported, their nature is described in the next subsection.

Table 2: Papers reporting agile manufacturing research on different manufacturing companies

Serial number	Industry in which agile manufacturing was implemented	Papers reporting the implementation oriented research in agile manufacturing domain	Candidate product	Research reported in the paper
1	Electronic industry	Vinodh et al. [33–45], Deif and ElMaraghy [46], Hooper et al. [47], Khoo and Loi [48], Mondragon et al. [49], Sharifi and Zhang [50], Zandi and Tavana [51]	Several products	Several technologies and models have been applied to achieve agility
2	Pump industry	Vinodh et al. [20], Devadasan et al. [22]	Pump products	Research to infuse agility through the integration of CAD and RPT
3	Aerospace industry	Gunasekaran et al. [52]	Aircraft components	Developed and used an agility audit questionnaire for assessing the agility level of the company
		Mondragon et al. [49]		Studied the impact of information systems in achieving agility
		Sreenivasa et al. [53]		Developed a 30-criterion agility assessment tool
4	Air dryer industry	Sreenivasa et al. [54]	Regenerative air dryer	Reported a research on theoretically mapping air dryer capabilities from agile manufacturing perspectives
		Sreenivasa et al. [55]		Reported a research to enhance the total agility level (TAL) in the company
		Mondragon et al. [49]	Engine and transmission parts	Studied the impact of information systems in achieving agility

5	Automotive industry	Frayret et al. [56]	Motor coach	Demonstrated the effectiveness of a method called "NetMan approach" in acquiring agility characteristics in organisations
		Vinodh and Kuttalingam [9]	Sprocket	Investigated the application of CAD and Computer Aided Engineering (CAE) as enablers of agile manufacturing
6	Machine tool industry	Wang et al. [57]	—	Proposed a multiagent and distributed ruler based approach to carry out production scheduling in agile manufacturing system
		Cheng et al. [58] Cheng et al. [59], Pan et al. [60]	Journal bearings Rolling bearings	Proposed an architecture for implementing agile manufacturing principles by integrating Artificial Intelligence (AI) and internet technology
7	Chemical industry	Guisinger and Ghorashi [61]	—	Described the increasing application of agile manufacturing practices in the specialty chemical industry
8	Casting industry	Yang and Li [62]	—	Researched on evaluating the agility of a company by using the multigrade fuzzy assessment method
9	Household products manufacturing industry	Sharifi and Zhang [50]	Domestic cookers	Designed a conceptual model of agility to enhance the agility of the company
10	Fabrication industry	Fung et al. [63]	Steel frames	Proposed an infrastructure for exercising adaptive production control in an agile manufacturing environment

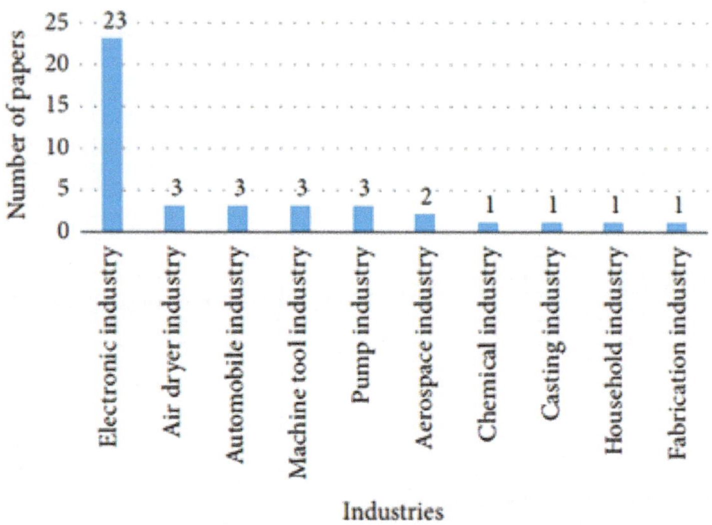

Figure 1: Bar chart of industrial applications of agile manufacturing.

Application of Agile Manufacturing in Electronics Industry

The highlights of the researches reported on implementing agile manufacturing researches in electronic industries are a little more elaborated in Table 3. As shown, the researchers have chosen four products, namely, electronic switch, printed circuit board, computer, and semiconductor while pursuing researches on implementing agile manufacturing. Besides, electronic components were also chosen as the candidate product while pursuing such researches. A careful study of the contents presented in Table 3 would indicate that these researchers have applied several technologies like CAD, CAM and RPT, and e-CRM. Thus, the trace of implementing agile manufacturing in electronic industry has been faster than it is done in the case of other industries. This trend indicates the need to identify the ways of increasing the pace of agile manufacturing in other industries too. As mentioned earlier, in order to initiate research in this direction, the application of agile manufacturing in pump industry is explored in the next subsection.

Table 3: Papers reporting agile manufacturing research in electronics manufacturing industry

Serial number	Papers reporting the research in electronic switch industry	Candidate product chosen for research	Research reported in the paper
1	Vinodh et al. [33], Vinodh et al. [34]	Electronic switch	Designed a model named agile innovative total quality function deployment (agile ITQFD)
2	Vinodh et al. [35]		Designed a tool for quantifying agility in organizations and test its practical compatibility
3	Vinodh et al. [26]		Developed a decision support system named decision support system for quantifying agile criteria (DESSAC)
4	Vinodh et al. [36]		Proposed Computer-Aided Design of Experiment (CADOE) to enable the companies to implement agile manufacturing
5	Vinodh et al. [37]		Research on integrating CAD and CAM to achieve agility in Salzer
6	Vinodh et al. [37]		Research on exploring the interfacing of CAD and CAM to infuse agility in organization
7	Vinodh et al. [38]		Proposed a model named as total agile design system activity based costing model (TADS-ABC)
9	Vinodh [39]		Integrated agility and sustainability to gain business benefits
10	Vinodh et al. [23]		A new concept agile customization program by combining mass customization and agile manufacturing principles
11	Vinodh et al. [40]		Designed and implemented total agile design system (TADS) model
12	Vinodh et al. [41]		Integrated fuzzy logic and analytic hierarchy to identify a group of concepts
13	Vinodh et al. [42]		Assessed agility of an organization using multi-grade fuzzy approach
14	Vinodh et al. [43]		Research on achieving agility by carrying out finite element mold analysis
15	Vinodh et al. [44]		Research on examining the possibilities of acquiring design agility characteristics through the aid of CAD
16	Vinodh et al. [45]		Research on using information technology as an enabler of implementing agile manufacturing
17	Deif and ElMaraghy [46]		Developed a system called agile manufacturing planning and control (agile MPC) system
18	Hooper et al. [47]		Explored the operational issues of agile costing systems by conducting a case study in a PCB manufacturing company
19	Khoo and Loi [48]	Computers	Examined the way of enhancing agility in the manufacturing computers using "product modularity" approach
20	Mondragon et al. [49]		Studied the impact of information systems in achieving agility

21	Sharifi and Zhang [50]	Electronic components	Designed a conceptual model of agility which consisted of three elements, namely, agility drivers, agility capabilities, and agility providers
22	Zandi and Tavana [51]	Semiconductors	A research on evaluating the best electronic-customer relationship management (e-CRM) network for achieving agility

Application of Agile Manufacturing in Pump Industry

A few authors have reported researches on implementing agile manufacturing in pump industry. These researches are briefly described here. Vinodh et al. [20] have carried out a research to analyze the impact of using CAD and RPT in achieving agility in pump industry. These authors chose the impeller of the centrifugal pump as the candidate product and created its new designs using CAD. Then, these new designs were subjected to flow analysis using two software packages, namely, Gambit and Fluent. These authors chose one new design after conducting the analysis and a prototype of that design was created using a rapid prototyping machine. These steps indicated that the new innovative products can be brought out in the industry by using CAD, CAE, and RPT. These authors carried out an activity based costing methodology to determine the costing of manufacturing the impeller. This research proved that CAD and RPT can be used to infuse agility in pump industry.

A case study on assessing the agility level of a traditional pump industry was carried out by Vinodh et al. [21] using a 30-agile-manufacturing-criterion quantification tool. These authors assessed the agility index of the company by collecting the responses from the executives of the company. The score obtained by using the agile manufacturing criteria quantification tool was 529 out of 1000. Since this score is above 500, these authors interpreted that agile manufacturing could be infused in this traditional pump manufacturing industry. Then, the gap analysis was carried out based on the obtained scores. Based on the results obtained, these authors proposed activities to infuse agile characteristics in the pump manufacturing industry.

Devadasan et al. [22] have reported a research on employing Taguchi's offline model in a traditional pump manufacturing company. The literature survey carried out by these authors indicated that the experiments including Taguchi's offline model have been conducted by considering only quantitative features. But in an agile manufacturing environment, the qualitative features should also be considered for carrying out these experiments. Since the cost of conducting actual experiments involves large amount of time and money, these authors preferred conducting of interviews. The interviews conducted were based on Delphi approach. These authors carried out a case study in a submersible pump manufacturing company to assess the success rate of the experiments. In the first phase of the experiment, these authors identified the factors which would aid in infusing agility while designing a new model of pump. Then, using Taguchi's orthogonal array based approach, the experiments containing quantitative and qualitative factors and levels were designed. These authors showed these experiments to two top executives of that company and gathered their feedbacks. In the second phase of the case study, the possibilities of infusing 20 agile criteria were analyzed and it was found that 11 out of 20 criteria 7 are feasible to infuse agility in this company. Since the cost of conducting experiments is more, instead of conducting experiments, these authors interviewed two executives. The response of the executives indicated that there are possibilities of achieving agility in a traditional pump manufacturing industry.

In total, the researches reported in the above papers specified that several agile criteria need to be strengthened in the pump manufacturing industry in several stages. The results of the literature review indicated that the assessment tool which is designed based on agility criteria is useful in determining the gaps to be filled in pump manufacturing industry to adopt agile manufacturing.

CONCLUSION

The literature review reported in this paper was carried out with the motivation behind following the progression of agile manufacturing in the manufacturing world. This literature review was first started by evaluating the looks into investigated agile manufacturing in the literature arena. It was discovered that, to start with, researchers tried exertions to pinpoint the significance and meaning of the term agile manufacturing and its related terms. In the wake of this intersection, the researchers worked towards figuring out the influences of agile manufacturing. During the current phase, the researchers are endeavoring to infuse agility in the manufacturing of traditional products. Throughout the conduct of the literature review reported in this paper, it was observed that just a couple of researchers have worked toward implanting agility in the manufacturing of pumps. Actually, in the researches reported in these papers, just an advance has been made to infuse agile characteristics in the manufacturing of certain components of pumps. Similarly, an exhaustive methodology of infusing agile characteristics while manufacturing pumps has not been addressed in those papers. An interesting observation made while conducting this literature review was that many researchers have attempted to infuse agility characteristics in electronic industry. It is presumed that this has happened so, as the electronic products generally hesitate for digitization. The digitization enables the development of new products in an agile manner through virtual analysis using computers. However, in the case of other products, particularly those developed before digitization technologies emerged in the world, the practice of achieving agility through digitization has been a challenging task. Hence, the modern researchers have begun to make attempts for digitizing these kinds of traditional products. Among these products, the pump fell within the scope of the literature review reported in this paper. Hence, during the conduct of the literature review reported in this paper, the progression of agile manufacturing happening toward infusing agile manufacturing in pump industry was traced. At the end of tracing this progression of agile manufacturing, it is pointed out that there are possibilities of carrying out investigations on infusing agility characteristics while designing and manufacturing of pumps. This sort of research attempts will help the traditional pump manufacturers to generate profit like the way the cell phone manufactures have garnered.

ACKNOWLEDGMENT

The authors would like to thank the anonymous referees for their valuable suggestions.

REFERENCES

1. Y. Yao, T. Kaihara, and N. Fujii, "A study on integration of interdivisional scheduling in dynamically changing manufacturing environment," Journal of Advanced Mechanical Design, Systems and Manufacturing, vol. 6, no. 5, pp. 572–580, 2012.
2. Q. Long, "An agent-based distributed computational experiment framework for virtual supply chain network development," Expert Systems with Applications, vol. 41, no. 9, pp. 4094–4112, 2014.
3. M. Abbasi, R. Hosnavi, and R. Babazadeh, "Agile and flexible supply chain network design under uncertainty," International Journal of Industrial Engineering, vol. 21, no. 4, pp. 190–208, 2014.
4. A. Shekar and J. McIntyre, "Internet user research in product development: rapid and low cost data collection," International Journal of Industrial Engineering, vol. 19, no. 1, pp. 1–13, 2012.
5. A. T. Eiji Morinaga, Y. Sakaguchi, and H. W. A. E. Arai, "Basic study on production scheduling method for highly-distributed manufacturing systems," Journal of Advanced Mechanical Design, Systems, and Manufacturing, vol. 8, no. 5, pp. 1–14, 2014.
6. M. Leber, M. Bastič, and B. Buchmeister, "The trends in usage and barriers of innovation management techniques in new product development," Strojniški Vestnik—Journal of Mechanical Engineering, vol. 60, no. 6, pp. 382–388, 2014.
7. S. Mishra, S. Datta, and S. S. Mahapatra, "Grey-based and fuzzy TOPSIS decision-making approach for agility evaluation of mass customization systems," Benchmarking, vol. 20, no. 4, pp. 440–462, 2013.
8. G. D. Putnik and Z. Putnik, "Lean vs agile in the context of complexity management in organizations,"The Learning Organization, vol. 19, no. 3, pp. 248–266, 2012.
9. S. Vinodh and D. Kuttalingam, "Computer-aided design and engineering as enablers of agile manufacturing: a case study in an Indian manufacturing organization," Journal of Manufacturing Technology Management, vol. 22, no. 3, pp. 405–418, 2011.
10. A. Gunasekaran and Y. Y. Yusuf, "Agile manufacturing: a taxonomy of strategic and technological imperatives," International Journal of Production Research, vol. 40, no. 6, pp. 1357–1385, 2002.
11. L. Jin-Hai, A. R. Anderson, and R. T. Harrison, "The evolution of agile manufacturing," Business Process Management Journal, vol. 9, no. 2, pp. 170–189, 2003.
12. W.-P. Wang, "Toward developing agility evaluation of mass customization systems using 2-tuple linguistic computing," Expert Systems with Applications, vol. 36, no. 2, pp. 3439–3447, 2009.
13. S. A. M. Elmoselhy, "Hybrid lean-agile manufacturing system technical facet, in automotive sector,"Journal of Manufacturing Systems, vol. 32, no. 4, pp. 598–619, 2013.
14. M. K. Lim and Z. Zhang, "A multi-agent system using iterative bidding mechanism to enhance manufacturing agility," Expert Systems with Applications, vol. 39, no. 9, pp. 8259–8273, 2012.
15. D. Wu, M. J. Greer, D. W. Rosen, and D. Schaefer, "Cloud manufacturing: strategic vision and state-of-the-art," Journal of Manufacturing Systems, vol. 32, no. 4, pp. 564–579, 2013.
16. L. Zhang, Y. Luo, F. Tao et al., "Cloud manufacturing: a new manufacturing paradigm," Enterprise Information Systems, vol. 8, no. 2, pp. 167–187, 2014.
17. L. Ren, L. Zhang, F. Tao, C. Zhao, X. Chai, and X. Zhao, "Cloud manufacturing: from concept to practice," Enterprise Information Systems, vol. 9, no. 2, 2015.
18. D. Wu, D. W. Rosen, L. Wang, and D. Schaefer, "Cloud-based design and manufacturing: a new paradigm in digital manufacturing and design innovation," Computer-Aided Design, vol. 59, pp. 1–14, 2015.
19. N. Liu, X. Li, and W. Shen, "Multi-granularity resource virtualization and sharing strategies in cloud manufacturing," Journal of Network and Computer Applications, vol. 46, pp. 72–82, 2014.
20. S. Vinodh, S. R. Devadasan, S. Maheshkumar, M. Aravindakshan, M. Arumugam, and K. Balakrishnan, "Agile product development through CAD and rapid prototyping technologies: an examination in a traditional pump-manufacturing company," International Journal of Advanced Manufacturing Technology, vol. 46, no. 5-8, pp. 663–679, 2010.
21. S. Vinodh, V. U. Kumar, and R. J. Girubha, "Thirty-criteria-based agility assessment: a case study in an Indian pump manufacturing organisation," International Journal of Advanced Manufacturing Technology, vol. 63, no. 9–12, pp. 915–929, 2012.
22. S. R. Devadasan, S. Goshteeswaran, and J. Gokulachandran, "Design for quality in agile manufacturing environment through modified orthogonal array-based experimentation," Journal of Manufacturing Technology Management, vol. 16, no. 6, pp. 576–597, 2005.
23. S. Vinodh, G. Sundararaj, S. R. Devadasan, D. Kuttalingam, and D. Rajanayagam, "Amalgamation of mass customisation and agile manufacturing concepts: the theory and implementation study in an electronics switches manufacturing company," International Journal of Production Research, vol. 48, no. 7, pp. 2141–2164, 2010.

24. Y. Y. Yusuf, M. Sarhadi, and A. Gunasekaran, "Agile manufacturing: the drivers, concepts and attributes,"International Journal of Production Economics, vol. 62, no. 1, pp. 33–43, 1999.

25. R. Narasimhan, M. Swink, and S. W. Kim, "Disentangling leanness and agility: an empirical investigation," Journal of Operations Management, vol. 24, no. 5, pp. 440–457, 2006.

26. S. Vinodh, G. Sundararaj, S. R. Devadasan, R. Maharaja, D. Rajanayagam, and S. K. Goyal, "DESSAC: a decision support system for quantifying and analysing agility," International Journal of Production Research, vol. 46, no. 23, pp. 6759–6780, 2008.

27. A. Gunasekaran, "Agile manufacturing: enablers and an implementation framework," International Journal of Production Research, vol. 36, no. 5, pp. 1223–1247, 1998.

28. E. Bottani, "Profile and enablers of agile companies: an empirical investigation," International Journal of Production Economics, vol. 125, no. 2, pp. 251–261, 2010.

29. S. Dowlatshahi and Q. Cao, "The relationships among virtual enterprise, information technology, and business performance in agile manufacturing: an industry perspective," European Journal of Operational Research, vol. 174, no. 2, pp. 835–860, 2006.

30. C. G. Sreenivasa and S. R. Devadasan, "Identification of arenas unconquered by agile manufacturing researches and practices through a literature survey," International Journal of Services and Operations Management, vol. 9, no. 2, pp. 227–244, 2011.

31. A. T. Eshlaghy, A. N. Mashayekhi, A. Rajabzadeh, and M. M. Razavian, "Applying path analysis method in defining effective factors in organisation agility," International Journal of Production Research, vol. 48, no. 6, pp. 1765–1786, 2010.

32. D. Vázquez-Bustelo, L. Avella, and E. Fernández, "Agility drivers, enablers and outcomes: empirical test of an integrated agile manufacturing model," International Journal of Operations & Production Management, vol. 27, no. 12, pp. 1303–1332, 2007.

33. S. Vinodh, G. Sundararaj, S. Devadasan, D. Rajanayagam, and R. Murugesh, "Agile innovative total quality function deployment and its financial accountability in an Indian company," International Journal of Indian Culture and Business Management, vol. 1, no. 1-2, pp. 244–263, 2007.

34. S. Vinodh, G. Sundararaj, S. R. Devadasan, S. Rajanayagam, and I. Edinbarough, "Agile ITQFD and its financial viability: a pilot project approach," The TQM Journal, vol. 20, no. 5, pp. 520–534, 2008.

35. S. Vinodh, G. Sundararaj, S. R. Devadasan, and D. Rajanayagam, "Quantification of agility: an experimentation in an Indian electronics switches manufacturing company," Journal of Engineering, Design and Technology, vol. 6, no. 1, pp. 48–64, 2008.

36. S. Vinodh, G. Sundararaj, S. R. Devadasan, D. Kuttalingam, and D. Rajanayagam, "Computer-aided design of experiments: an enabler of agile manufacturing," International Journal of Advanced Manufacturing Technology, vol. 44, no. 9-10, pp. 940–954, 2009.

37. S. Vinodh, G. Sundararaj, S. R. Devadasan, D. Kuttalingam, J. Jayaprakasam, and D. Rajanayagam, "Agility through the interfacing of CAD and CAM," Journal of Engineering, Design and Technology, vol. 7, no. 2, pp. 143–170, 2009.

38. S. Vinodh, G. Sundararaj, S. R. Devadasan, and D. Rajanayagam, "TADS-ABC: a system for costing total agile design system," International Journal of Production Research, vol. 47, no. 24, pp. 6941–6966, 2009.

39. S. Vinodh, "Improvement of agility and sustainability: a case study in an Indian rotary switches manufacturing organisation," Journal of Cleaner Production, vol. 18, no. 10-11, pp. 1015–1020, 2010.

40. S. Vinodh, G. Sundararaj, and S. R. Devadasan, "Measuring organisational agility before and after implementation of TADS," International Journal of Advanced Manufacturing Technology, vol. 47, no. 5–8, pp. 809–818, 2010.

41. S. Vinodh, S. G. Gautham, R. Anesh Ramiya, and D. Rajanayagam, "Application of fuzzy analytic network process for agile concept selection in a manufacturing organisation," International Journal of Production Research, vol. 48, no. 24, pp. 7243–7264, 2010.

42. S. Vinodh, S. R. Devadasan, B. V. Reddy, and K. Ravichand, "Agility index measurement using multi-grade fuzzy approach integrated in a 20 criteria agile model," International Journal of Production Research, vol. 48, no. 23, pp. 7159–7176, 2010.

43. S. Vinodh, G. Sundararaj, S. R. Devadasan, D. Kuttalingam, and D. Rajanayagam, "Achieving agility in manufacturing through finite element mould analysis: an application-oriented research," Journal of Manufacturing Technology Management, vol. 21, no. 5, pp. 604–623, 2010.

44. S. Vinodh, S. R. Devadasan, and C. Shankar, "Design agility through computer aided design," Journal of Engineering, Design and Technology, vol. 8, no. 1, pp. 94–106, 2010.

45. S. Vinodh, K. R. Arvind, and D. Rajanayagam, "Development of digital product catalogue for enabling agility in a manufacturing organisation," Journal of Engineering, Design and Technology, vol. 9, no. 2, pp. 143–156, 2011.

46. A. M. Deif and W. H. ElMaraghy, "Agile MPC system linking manufacturing and market strategies,"Journal of Manufacturing Systems, vol. 26, no. 2, pp. 99–107, 2007.

47. M. J. Hooper, D. Steeple, and N. Clive, "Costing customer value: an approach for the agile enterprise,"International Journal of Operations and Production Management, vol. 21, no. 5-6, pp. 630–644, 2001.

48. L. P. Khoo and M. Y. Loi, "A tabu-enhanced genetic algorithm approach to agile manufacturing," The International Journal of Advanced Manufacturing Technology, vol. 20, no. 9, pp. 692–700, 2002.

49. A. E. C. Mondragon, A. C. Lyons, and D. F. Kehoe, "Assessing the value of information systems in supporting agility in high-tech manufacturing enterprises," International Journal of Operations and Production Management, vol. 24, no. 12, pp. 1219–1246, 2004.
50. H. Sharifi and Z. Zhang, "Agile manufacturing in practice—application of a methodology," International Journal of Operations & Production Management, vol. 21, no. 5-6, pp. 772–794, 2001.
51. F. Zandi and M. Tavana, "A fuzzy group quality function deployment model for e-CRM framework assessment in agile manufacturing," Computers and Industrial Engineering, vol. 61, no. 1, pp. 1–19, 2011.
52. A. Gunasekaran, E. Tirtiroglu, and V. Wolstencroft, "An investigation into the application of agile manufacturing in an aerospace company," Technovation, vol. 22, no. 7, pp. 405–415, 2002.
53. C. G. Sreenivasa, S. R. Devadasan, and R. Murugesh, "Thirty criteria-based assessment of agility in a pneumatic enabling products manufacturing company," International Journal of Services and Operations Management, vol. 11, no. 2, pp. 201–221, 2012.
54. C. G. Sreenivasa, S. R. Devadasan, N. M. Sivaram, and S. Karthi, "A case study on mapping air dryer capabilities from agile manufacturing perspectives," International Journal of Services and Operations Management, vol. 16, no. 1, pp. 86–104, 2013.
55. C. G. Sreenivasa, S. R. Devadasan, and R. Murugesh, "Enhancing total agility level through assessment and product mapping: a case study in the manufacturing of refrigeration air dryer," International Journal of Advanced Manufacturing Technology, vol. 60, no. 5-8, pp. 421–436, 2012.
56. J.-M. Frayret, S. D'Amours, B. Montreuil, and L. Cloutier, "A network approach to operate agile manufacturing systems," International Journal of Production Economics, vol. 74, no. 1–3, pp. 239–259, 2001.
57. Y. H. Wang, C. W. Yin, and Y. Zhang, "A multi-agent and distributed ruler based approach to production scheduling of agile manufacturing systems," International Journal of Computer Integrated Manufacturing, vol. 16, no. 2, pp. 81–92, 2003.
58. K. Cheng, D. K. Harrison, and P. Y. Pan, "Implementation of agile manufacturing—an AI and Internet based approach," Journal of Materials Processing Technology, vol. 76, no. 1–3, pp. 96–101, 1998.
59. K. Cheng, P. Y. Pan, and D. K. Harrison, "The Internet as a tool with application to agile manufacturing: a web-based engineering approach and its implementation issues," International Journal of Production Research, vol. 38, no. 12, pp. 2743–2759, 2000.
60. P. Y. Pan, K. Cheng, and D. K. Harrison, "A web-based agile system for rolling bearing design," Integrated Manufacturing Systems, vol. 14, no. 6, pp. 518–529, 2003.
61. A. Guisinger and B. Ghorashi, "Agile manufacturing practices in the specialty chemical industry: an overview of the trends and results of a specific case study," International Journal of Operations & Production Management, vol. 24, no. 5-6, pp. 625–635, 2004.
62. S. L. Yang and T. F. Li, "Agility evaluation of mass customization product manufacturing," Journal of Materials Processing Technology, vol. 129, no. 1–3, pp. 640–644, 2002.
63. R. Y. K. Fung, F. Liang, Z. Jiang, and T. N. Wong, "A multi-stage methodology for virtual cell formation oriented agile manufacturing," International Journal of Advanced Manufacturing Technology, vol. 36, no. 7-8, pp. 798–810, 2008.

CHAPTER 12 COMPARISON AND EVALUATION OF DEADLOCK PREVENTION METHODS FOR DIFFERENT SIZE AUTOMATED MANUFACTURING SYSTEMS

Emad Abouel Nasr[1,2] Abdulaziz M. El-Tamimi[1] Abdulrahman Al-Ahmari[1] andHusam Kaid[1]

[1]Industrial Engineering Department, College of Engineering, King Saud University, Riyadh 11421, Saudi Arabia

[2]Mechanical Engineering Department, Helwan University, Cairo 11792, Egypt

ABSTRACT

In automated manufacturing systems (AMSs), deadlocks problems can arise due to limited shared resources. Petri nets are an effective tool to prevent deadlocks in AMSs. In this paper, a simulation based on existing deadlock prevention policies and different Petri net models are considered to explore whether a permissive liveness-enforcing Petri net supervisor can provide better time performance. The work of simulation is implemented as follows. (1) Assign the time to the controlled Petri net models, which leads to timed Petri nets. (2) Build the Petri net model using MATLAB software. (3) Run and simulate the model, and simulation results are analyzed to determine which existing policies are suitable for different systems. Siphons and iterative methods are used for deadlocks prevention. Finally, the computational results show that the selected deadlock policies may not imply high resource utilization and plant productivity, which have been shown theoretically in previous publications. However, for all selected AMSs, the iterative methods always lead to structurally and computationally complex liveness-enforcing net supervisors compared to the siphons methods. Moreover, they can provide better behavioral permissiveness than siphons methods for small systems. For large systems, a strict minimal siphon method leads to better behavioral permissiveness than the other methods.

INTRODUCTION

Effectively designing and operating an automated manufacturing system can be of assistance to manufacturers to adapt the variety in the market in order to maintain and confirm competitiveness. Different types of parts in AMS enter the system at discrete points of time and are produced concurrently; these parts have to share some common resources, such as robots, machine tools, buffers, automated guided vehicles, and fixtures. The deadlocks can occur in an AMS during its operation by shared resources competition, which are highly unwanted phenomena that often cause low use of some expensive and critical resources and long downtime and can lead to calamitous results in automated manufacturing systems. Therefore, it is required to develop a successful AMS control policy to prevent deadlocks occurrence in AMS.

There are several mathematical tools to solve deadlock problems in AMSs including automata, graph theory, and Petri nets. Recently, Petri nets have been known as one of the most robust mathematical tools for modelling, analyzing, and controlling deadlocks in resource allocation systems including

Citation: Emad Abouel Nasr, Abdulaziz M. El-Tamimi, Abdulrahman Al-Ahmari, and Husam Kaid, "Comparison and Evaluation of Deadlock Prevention Methods for Different Size Automated Manufacturing Systems,"Mathematical Problems in Engineering, vol. 2015, Article ID 537893, 19 pages, 2015. doi:10.1155/2015/537893.

AMS [1]. To overcome the deadlocks in AMS, there are methods which have been derived from a Petri net tool either to forbid the deadlock occurrences by preventing some necessary condition or to detect and resolve a deadlock when it occurs. In general, these methods can be classified into three strategies: deadlock detection and recovery, deadlock avoidance, and deadlock prevention [1, 2].

The researchers proposed three criteria for designing and evaluating a supervisor for a manufacturing system to be controlled. These criteria include behavioral permissiveness, which leads to increase in the resources utilization of system, structural complexity that leads to design supervisor with a few number of monitors to decrease the software and hardware costs, and computational complexity that means a deadlock control policy can be implemented in huge systems [1]. Therefore, the objectives of many researchers are to develop deadlock prevention policies which can provide liveness-enforcing supervisors with maximal permissiveness, a structural complexity, and computational complexity [1].

From previous studies, there are mainly two techniques of analysis for deadlock prevention in AMSs using Petri net: structural analysis [3, 4] and reachability graph analysis [5–7]. Oftentimes structural analysis is applied for structural objects of Petri nets, such as resource transition circuits and siphons. In this case, control steps are simple, and each empty minimal siphon requires a monitor to be added to prevent itself from being not emptied, but the disadvantages of this technique are the fact that the resulting controlled system can be suboptimal and linearly the number of control places is dependent on the size of a net [8].

The reachability graph analysis suffers from a state explosion problem because it requires listing of all or a part of reachable markings. The reachability graph of a Petri net has been classified into two zones: live zone (LZ) and deadlock zone (DZ). In DZ, there is first-met bad markings (FBMs) which are extracted from it; the FBMs are representing the very first entry from LZ to DZ. In this case, a control place is designed and added to forbid the FBM from being reached. This process requires iterations to forbid all FBMs and single out each FBM. The process is terminated if the resulting net is a live [9]. There are several methods and algorithms used for deadlocks prevention; some of them are siphon control methods, a theory of region, and iterative methods [4,6, 9–11].

Most of the researchers have used structural analysis and reachability graph analysis to design deadlocks prevention policies for different scales of AMSs under three criteria that are maximal permissiveness, structural complexity, and computational complexity. The comparison between these policies has been done according to the three aforementioned criteria. It is taken for granted in the literature that a maximally permissive supervisor in the logical level usually leads to the better time performance in an AMS such as productivity and resource utilization, which looks intuitive, as an optimal supervisor will permit more permissive behavior. However, there is a shortage in the quantitative evaluation of time performance of these policies in terms of maximal permissiveness and productivity. Simulation is a powerful tool for performance analysis of complex manufacturing systems.

The contribution of this paper is using a simulation tool based on existing deadlock prevention policies and different types of Petri net models to explore whether a permissive liveness-enforcing Petri net supervisor can provide better time performance (maximal permissiveness).

The work of simulation is implemented in the following steps. (1) Assign the time to the controlled Petri net models, which leads to timed Petri nets. (2) Build the Petri net model using MATLAB software (Petri Net Toolbox (PN Toolbox)). (3) Run and simulate the model to obtain the system time performance. (4) Finally, the simulation results will be analyzed to determine which existing deadlock prevention policies are convenient for different types of manufacturing systems. Different control methods (siphon control methods and iterative methods) are used for deadlocks prevention. Moreover, this paper aims to evaluate the performance of the selected methods such as utilization of resources, plant throughput, and the number of monitors, the number of arcs, and the number of reachable states. This paper focuses on the performance evaluation of automated manufacturing systems by using Petri nets under different deadlock prevention policies. The considered Petri net models belong to M-nets [12] that are a general class of manufacturing-oriented net models in the literature. In an M-net, there are a number of structural properties thanks to its definition [12].

Including this introductory section, the paper is organized as follows. Section 2 provides a literature review of previous researchers works related to deadlock prevention. The basics of Petri nets are discussed in Section 3. The deadlock prevention methods and policies are presented in Section 4. Sections 5 and 6 show the analysis of the selected methods for four different automated manufacturing systems. Finally, Section 7 presents the conclusion and future work.

LITERATURE REVIEW

Deadlock prevention policies have been widely achieved for AMS and led to a huge number of results [1, 11]. In this section, deadlock prevention strategies are reviewed by using Petri nets and developed based on different techniques such as structural analysis and reachability graph analysis.

Structural-Analysis Methods

Ezpeleta et al. [13] used a class of Petri nets, called System of Simple Sequential Processes with Resources (S^3PR) and proposed an algorithm for resource allocation in flexible manufacturing systems (FMSs). The proposed algorithm added new places to the net to impose certain restrictions that forbid the presence of unmarked siphons. The works of Huang et al. [14] and Huang et al. [15] present a new deadlock prevention policy for the class of Petri nets, where deadlocks are related to unmarked siphons. Two types of control places are added to the original model for flexible manufacturing system called ordinary control place and weighted control place to prevent siphons from being unmarked.

The study by Li and Zhou [11] presents a Petri net model which consists of 26 places, 20 transitions, and 18 strict minimal siphons. In [16], elementary siphons concept is introduced to design a liveness-enforcing Petri net supervisor for the same Petri net model, and six elementary siphons are obtained; therefore, only six monitors are added to control the 18 strict minimal siphons. Nevertheless, the proposed algorithm by Ezpeleta et al. [13] adds monitors for all 18 strict minimal siphons.

A new methodology is proposed by Huang [17] to synthesize supervisors for resource allocation in FMSs; the class of Petri net, namely, S^3PR, is considered, where deadlocks are related to unmarked minimal siphons; all minimal siphons should be controlled by adding control places. In this study, the number of control places is reduced by using the concept of the elementary siphon. Based on elementary siphons and P-invariants concepts in Petri nets, a deadlock prevention algorithm is introduced by Li and Wei for a specific class of Petri nets which can properly model several FMSs [16], siphons in a Petri net model are classified into dependent and elementary siphons, and control places are added for all elementary siphons, so that the siphons are invariant-controlled. Therefore, the controllability of dependent siphons is guaranteed by adjusting the control requirements of its related elementary siphons. A novel deadlock prevention algorithm is presented by Chen et al. [1], and the concept of siphon extraction is used to compute unmarked siphons for the Petri net model. First, a siphon extraction algorithm obtains a maximal unmarked siphon, divides the places in it, and determines a necessary siphon from the divided places; this process is carried out for all unmarked siphons. Then, the proposed deadlock prevention algorithm designs a convenient monitor for each necessary siphon to be marked until the controlled Petri net model is a live.

Most of the deadlock control algorithms have been developed for reliable automated manufacturing systems (AMSs). Liu et al. [18] proposed a variety of deadlock control algorithms for AMSs with unreliable resources, monitors and recovery subnets are designed for strict minimal siphons (emptied siphon) and unreliable resources, respectively, and two types of arcs which are normal and inhibitor are used to connect monitors with recovery subnets. In order to obtain a supervisor with structural complexity, elementary siphons extracted from all strict minimal siphons are obviously controlled. The problems of structural complexity, behavioral permissiveness, and computational complexity are considered in the study of Wang et al. [19] to design liveness-enforcing supervisors for a class of Petri nets, namely, S^3PR, without ξ-resources. They used two steps to obtain a supervisor for the model. In the first step, an algorithm is proposed to extract unmarked strict minimal siphons from a given unmarked siphon by using loop resource subsets. In the second step, a siphon-based deadlock prevention algorithm is introduced to provide a maximally permissive liveness-enforcing supervisor with small numbers and no weighted monitors.

Reachability Graph-Based Approaches

Viswanadham et al. [20] presented static resource allocation methods for eliminating deadlocks; in this study, a reachability graph of the Petri net (PN) model is used to reach static resource allocation method. A deadlock prevention algorithm is implemented for a small size manufacturing system consisting of a machine and an automated guided vehicle (AGV). The authors observed that the proposed algorithm can be applied effectively only for small size manufacturing systems. The forbidden state problem of Petri nets (PN) with uncontrollable transitions and liveness requirement is addressed by Ghaffari et al. [5]. An approach is presented to compute a maximally permissive Petri net controller that consists of two steps. In the first step, an ordinary Petri net is used to build the system model. In the second step, control places are designed and added to the original model by using a theory of regions method. The work of Uzam [7, 9] proposes and improves method based on a theory of regions to design an optimal Petri net supervisor. The size of the reachability graph of Petri net model is a major problem to apply the deadlock prevention policy to very large Petri net model. Therefore, they proposed a reduction algorithm to simplify very large Petri nets models in order to make necessary computation easily. Based on the theory of regions, Uzam and Zhou [6] developed an iterative deadlock prevention policy for FMS. In their study, the reachability graph of a Petri net model is divided into two parts: deadlock-free zone (live zone, LZ) and the deadlock zone (DZ). The live zone is developed since the maximal strongly connected component contains the initial marking. The deadlock zone contains markings from which the initial marking is unreachable. FBM is defined as a marking in the deadlock zone. The deadlocks can be eliminated by forbidding the firing of the enabled transitions at FBM. In their work, the presented approach has two problems. First, an optimal supervisor cannot be guaranteed in general even if such an optimal supervisor exists. Second, at each iteration full reachability graph computation is required to check whether the markings in the deadlock zone are reachable. This approach is easy to use and simple if the reachable graph of a system is small but cannot ensure the supervisor behavioral optimality. The redundant monitors in Petri net supervisor may exist when the supervisor is designed by an iterative siphon control approach. Therefore, Uzam et al. [21] introduced an approach to identify and eliminate the redundant monitors by computing the reachability graph of a controlled Petri net model. If the controlled Petri net model does not lose liveness when removing redundant monitors, then the redundant monitors can be eliminated from the supervisor. The reachability graph and liveness check are the major problems in designing liveness-enforcing supervisor. To overcome a full state enumeration, Li and Hu [22] proposed two approaches to eliminating control places from a Petri net supervisor. The first approach is based on the concept of implicit places [23, 24]. The implicity of a monitor is computed by solving a linear programming problem (LPP) that can be implemented in polynomial time. The second approach is extracted from the Mixed Integer Programming- (MIP-) based deadlock detection approach. If the optimal solution to an MIP problem does not lose optimality when removing control place, then the control place is implicit and may provide permissive behavior while liveness is guaranteed. Chen et al. [1] developed a novel and computationally effective approach to design optimal monitors and an iterative approach that can compute the reachability graph to get an optimal supervisor of the flexible manufacturing system, which is demonstrated by a set of control places. The vector covering approach is used to compute the minimal sets of FBMs and legal markings. Iteratively, an FBM is selected from the minimal set of FBMs, an integer linear programming problem (ILPP) is used to design a place invariant (PI) that prevents the FBM from being reached, and all minimal sets of legal markings are not forbidden. The process is terminated if all FBMs cannot be reached and the resulting net is a live. This study is improved in its computational efficiency in [25], and the objective function of the ILPP is maximizing the number of FBMs that are prohibited by the PI; in other words, a control place is designed such that as many FBMs as possible are forbidden and all minimal sets of legal markings are not prohibited. If the ILPP has no solution, it means that there is no maximally permissive Petri net supervisor for the Petri net model.

BASICS OF PETRI NETS

Petri nets are a mathematical and graphical tool suitable for several systems. It is a major tool for studying and describing manufacturing system operations that are characterized as being asynchronous, concurrent, parallel, and distributed.

A Petri net or place/transition net can be defined as a four-tuple $N = (P, T, F, W)$, where P is a finite nonempty set of places, T is a finite nonempty set of transitions, P and T are disjoint, $P \cup T$ are called nodes with $P \cup T \neq \emptyset$ and $P \cap T = \emptyset$, and P and T are represented by circle and bar, respectively. $F \subseteq (P \times T) \cup (T \times P)$ is called a flow relation or directed arcs of the net that can be represented by arcs with arrows from places (transitions) to transitions (places).

$: (P \times T) \cup (T \times P) \rightarrow \mathbb{N}$ is a mapping that assigns a weight to an arc: if $(a, b) \in F$, $(a, b) > 0$; otherwise, $W(a, b) = 0$, where $a, b \in P \cup T$ and $\mathbb{N} = \{0, 1, 2, \ldots\}$. N is called an ordinary net, if $(a, b) = 1$, $\forall (a, b) \in F$. Consider a node $a \in P \cup T$, and ${}^{\bullet}a = \{b \in P \cup T \mid (b, a) \in F\}$ is a preset of a. Postset of a is $a^{\bullet} = \{b \in P \cup T \mid (a, b) \in F\}$. A marking M of N is a mapping from $P \rightarrow \mathbb{N}$. (p) denotes the number of tokens in place p.

Usually, markings and vectors can be described using formal sum. Therefore, vector M is denoted by $\sum_{p \in P} (p)$. For example, a marking that puts two tokens in place p_2, five tokens in place p_4, and no tokens in other places in a net is denoted by $2p_2+5p_4$. (N, M_0) is a net system or marked net and is denoted by $N = (P, T, F, W, M_0)$, where M_0 is an initial marking of N. For a Petri net modeling M_0 represents the different raw parts that are to be synchronously processed in the system, and the state of resources, such as machines and robots. A net is self-loop free or pure if for all $a, b \in P \cup T$, $(a, b) > 0$ and $(b, a) = 0$. Incidence matrix $[N]$ in a net N is an integer matrix that consists of $|P|$ rows and $|T|$ columns with $[N](p, t) = W(t, p) - W(p, t)$.

The enabled rule for each $t \in T$ at marking M if $\forall p \in {}^{\bullet}t$, $M(p) \geq W(p, t)$. As a result, enabled rule is denoted by $[t\rangle$. When a transition t fires, it removes $W(p, t)$ tokens from each place $p \in {}^{\bullet}t$ and deposits $W(t, p)$ tokens in each place $p \in t^{\bullet}$. Therefore, it yields a new marking M', denoted by $[t\rangle M'$, where $M'(p) = M(p) - W(p, t) + W(t, p)$. M' is reachable from M if there exists a fireable finite transition sequence $\delta = \{t_1, t_2, t_3, \ldots, t_n\}$, and markings M_1, M_2, $M_3, \ldots$, and M_{n-1} so that $M[t_1\rangle M_1[t_2\rangle M_2[t_3\rangle M_2, \ldots, M_{n-1}[t_n\rangle M'$, which can be denoted by $M[\delta\rangle M'$ and satisfied the state equation $M + [N]\vec{\delta}$, where $\vec{\delta} : T \rightarrow \mathbb{N}$ is a mapping t in T to the number of occurrences of t in δ, called a Parikh vector or a firing count vector. Obviously, $M_0[\,\rangle$ is the set of all reachable markings of a net N from initial marking $M0$ and is denoted by $R(N, M_0)$. The reachable markings (N, M_0) can be involved in the enumeration of all reachable markings of a net N and graphically expressed by a reachability tree. The reachability tree is denoted by (N, M_0), which is a directed graph and consists of nodes and arcs, nodes are markings in the reachable markings (N, M_0), and arcs are labeled by the transitions of a net N. For instance, an arc from M_2 to M_3 is labeled by t_1 if $M_2[t_1\rangle M_3$. A net N with initial marking M_0 is to be k-bounded if, for all $M \in (N, M_0)$, $(p) \leq k$ $(k = \{1, 2, 3, \ldots\})$. In other words, for each reachable marking "M" from the initial marking $M0$, the number of tokens in each place "p" never exceeds a finite number "k." A net is called safe if all its places are safe, and the number of tokens in each place "p" does not exceed one.

In other words, a net is k- safe if it is 1-bounded. A Petri net with (N, M_0) is a live if all its transitions are a live at M_0. A transition $t \in T$ is a live if, for all $M \in (N, M_0)$, $\exists M \in (N, M)$, there is a firing sequence M' $[t\rangle$. A transition is dead at M_0 if $\nexists t \in T$, $M_0[t\rangle$. P-vector (place vector) and T-vector (transition vector) are column vectors. The former $I{:}P \rightarrow Z$ is catalogued by P and called a P-invariant or place invariant if $I \neq 0$ and $I^T \cdot [N] = 0T$. The latter $J{:}T \rightarrow Z$ is catalogued by T and called a T-invariant or transition invariant if $J \neq 0$ and $[N] \cdot J = 0$, where Z is the integers set. If each element of I is nonnegative the P-invariant I is called a P-semiflow or place semiflow. Suppose that I is a place invariant of a Petri net with (N, M_0) and M is a reachable marking from the initial marking M_0. Then, $I\ TM{=}I^T M_0$. Let $\|I\| = \{p \mid (p) \neq 0\}$ be a support of place invariant I and it can be classified into three parts. Firstly, $\|I\|^+$ is a positive support of place invariant I, and $\|I\|^+ = \{p \mid I(p) > 0\}$. Secondly, $\|I\|^-$ is a negative support of place invariant I, and $\|I\|^- = \{p \mid I(p) < 0\}$. Finally, I is a minimal place invariant if $\|I\|$ is not a superset of the support of any other one and its components are mutually prime. Let l_i's be the coefficients of place invariant I if, for all $p_i \in P$, $l_i = I(p_i)$.

DEADLOCK PREVENTION METHODS AND POLICIES

Siphon Control Methods

This section presents two methods that are motivated by Ezpeleta et al. [13] and Li and Zhou [11]. These methods are called a strict minimal siphons control method and an elementary siphons control method, respectively.

Strict Minimal Siphons (SMSs) Control Method

A siphon in a Petri net is a set of places $S = \{p_1, \ldots, \}$. The very important property of siphon is that, for a given marking, the siphon is unmarked.

Definition 1. Let N be a net. Any set $S \subseteq P$, $S = \emptyset$ with ${}^{\bullet}S \subseteq S^{\bullet}$ is called a siphon. If a siphon does not properly contain another siphon is called a minimal siphon. A minimal siphon S is called a strict if ${}^{\bullet}S \subsetneq S^{\bullet}$.

Siphons play a significant role in the liveness analysis of a Petri net, especially in ordinary ones. If a siphon S in a net has no tokens, then no transitions in $S^{\bullet}$ are enabled and all the transitions connected to S cannot be fired; as a result, the transitions are dead leading to the loss of liveness of the Petri net. In [13], a monitor is added to each strict minimal siphon in order to achieve the liveness of the system. The proposed approach is simple and ensures a success. Nevertheless, it leads to a complex Petri net model than the originally model because the added places are equal to the strict minimal siphons in the target net model and the added arcs are larger than that of places added.

The SMS control based on complementary sets of siphons [S] is used to design the control places. A complementary set of a siphon S is a set of operation places which are the holders of the resources in S but do not belong to S. In addition, [S] ∪ S is the support of a place invariant, which means the operation places in [S] will compete for the resources in S with the operation places belonging to S. S will be unmarked when all tokens in S flow into [S]. The following notation will be used in the establishment of the control policy.

(1) S denotes the set of strict minimal siphons which does not contain the support of any place semiflow (i.e., siphons that can be emptied).

(2) Given the fact that S is a siphon we have $S=S_A \cup S_R$, $S_R = S \cap P_R$, $S_A = S \setminus S_R$, where S_A is the operation places and S_R is the resource places. [S] denotes the following set of state places: $[S] = (U_{r\in S}(r)) \setminus S_A$, where $H(r) = (\|I_r\|^+ \setminus r)$, $\|I_r\|^+ = \{p \mid I(p) > 0\}$ indicates the positive support of place invariant I, $\forall i, j \in \{1, 2, \ldots, n\}$ $i \neq j$, $H(r_i) \cap H(r_j) = \theta$. For a given siphon S,[S]is the set of holders, corresponding to resources in S that do not belong to S. [S] is called S's complementary set.

(3) Given the fact that [S] is a complementary set of S, add a control place for [S] and the initial marking of control place is $M_0(V_S)=M_0(S) - 1$.

Based on the concept of strict minimal siphon, the proposed deadlock prevention algorithm developed by Ezpeleta et al. [13] is stated as follows.

Algorithm 2 (SMS based policy).

Input. Petri net model (N, M_0) of an AMS, where $N = (P^0 \cup P_A \cup P_R, T, F, W)$.

Step 1. For a given Petri net (N, M_0), compute all SMSs.

Step 2. For each siphon SMS, add a monitor V_S such that

(i) the V_S output arcs are connected to the source transitions leading to the sink transitions of S and all arcs weights are ones;

(ii) the V_S input arcs are connected from the stealing places of S and all arcs weights are ones;

(iii) $M_{0A}(V_S) = M_0(S) - 1$, where $M_{0A}(V_S)$ is an initial marking of a monitor.

Step 3. Repeat Step 2 until all existing SMSs are covered.

Step 4. Add all monitors into (N, M_0) and the resulting net is denoted by (N_1, M_1).

Step 5. Output (N_1, M_1).

Step 6. End.

For instance, consider the AMS shown in Figure 1(a). Its Petri net model is shown in Figure 1(b). It consists of one robot R1 that can hold a part at a time, two machines M1 and M2, where each machine can process a part at a time, two loading buffers (I1 and I2), and unloading buffers (O1 and O2). Two parts types are considered in the system: PA and PB. PA moves to M1 and PB moves to M2. The robot R1 reaches the machine that finishes its operation first and grips and loads the part to the next machine 1 or machine 2. The Petri net model consists of 11 places and 8 transitions. The places have the following set partition: $P^0 = \{P_1, P_8\}$, $P_R = \{P_9, P_{10}, P_{11}\}$, and $P_A = \{P_{2,\ldots,7}\}$, where P_0 is the input places, P_R is the resources places, and P_A is the operation places. It has 8 minimal siphons, 3 of which are strict minimal siphons and 20 reachable markings. The three SMSs are $S_1 = \{P_4, P_7, P_9, P_{10}, P_{11}\}$, $S_2 = \{P_4, P_6, P_{10}, P_{11}\}$, and $S_3 = \{P_3, P_7, P_9, P_{10}\}$. Table 1 shows the required monitors using Algorithm 2 for Figure 1(b). Figure 2 illustrates the controlled system of the Petri net model in Figure 1(b) after adding monitors by using Algorithm 2.

Elementary Siphons Control Method

The strict minimal siphons (SMSs) are classified into elementary and dependent ones. Π is used to denote the set of strict minimal siphons while Π_E and Π_D denote the sets of elementary and dependent (redundant) ones, respectively. Otherwise, when a strict minimal one is mentioned, it can be called a siphon.

Definition 3. Let an ordinary Petri net (N, M_0) and S is a siphon, $S \subseteq P$. Place vector λ_S is said to be the characteristic place vector of S if for all $p \in S$, $(p) = 1$; otherwise, $\lambda_S(p) = 0$.

Definition 4. Let $N = (P, T, F)$ be an ordinary Petri net with $|P| = m$, $|T| = n$, and we suppose that N has f SMS, S_1, $S_2, \ldots,$ and $S_f, m, n, f \in \mathbb{N}$. Let $\lambda_{Si}(\eta_{Si})$ be the characteristic place (transition) vector of siphon S_i, $i \in \{1, 2, \ldots, n\}$. We define $[\lambda]_{f\times m} = [\lambda_{S1}, \lambda_{S2}, \ldots, \lambda_{Sf}]\, T$ and $[\eta]_{f\times n} = [\lambda]_{f\times m} \cdot [N]_{m\times n}$ are said to be the characteristic place vector and transition vector matrices of the siphons of N, respectively.

Proposition 5. Let N be an ordinary Petri net and let $[\eta]$ be the characteristic transition vector matrix of the siphons of it. The set of elementary siphons in N can be defined as a rank of $[\eta]$.

From Proposition 5, it can easily enumerate all elementary siphons in a Petri net system (N, M_0) given all siphons. First, compute matrices $[\lambda]$ and $[\eta]$. Then independent linear vectors can be determined in $[\eta]$. Finally, the siphons that correspond to these independent linear vectors are called the elementary siphons in the Petri net (N, M_0). Li and Zhou [11] introduced some results and theorems as follows.

Theorem 6. Let N_0 be a marked ordinary Petri net and $S = \{p_{i1}, p_{i2,\ldots,n}\}$ be a siphon in N0. Monitor VS is added to N_0, and the new net system can be denoted by (N_1, M_1), such that (a) $I_1 = (\ldots, 1p_{i1}, 1p_{i2}, \ldots, 1p_{in}, \ldots, -1V_S, \ldots)T$ is a place invariant of N_1 and (b)$M_1(V_S)=M_0(S)-\xi_S$, where ξ_S is said to be the control depth variable of siphon S, which states the least number of tokens that siphon can hold. Evidently is greater than or equal to 1 to fulfil purpose of control $1 \leq \xi_S \leq M_0(S) - 1$, and (c) for all $p \in P_0$, $M_1(p) = M_0(p)$, where P_0 is the set of operation places of N_0, such that S can be called invariantcontrolled.

Table 1: Control places computations by Algorithm 2

SMS	$\|I\|^+$	$H(r_R)$	$[S]$	${}^\bullet V_{Si}$	$V_{Si}^\bullet$	$M_{oA}(V_{Si})$
$S_1 = \{P_4, P_7, P_9, P_{10}, P_{11}\}$, $S_A = \{P_4, P_7\}$, $S_R = \{P_9, P_{10}, P_{11}\}$.	$\|I_{p9}\|^+ = \{p_2, p_7, p_9\}$, $\|I_{p10}\|^+ = \{p_3, p_6, p_{10}\}$, $\|I_{p11}\|^+ = \{p_4, p_5, p_{11}\}$.	$H(r_9) = \{p_2, p_7\}$, $H(r_{10}) = \{p_3, p_6\}$, $H(r_{11}) = \{p_4, p_5\}$.	$[S_1] = \{p_2, p_3, p_5, p_6\}$.	t_3, t_7	t_1, t_5	2
$S_2 = \{P_4, P_6, P_{10}, P_{11}\}$, $S_A = \{P_4, P_6\}$, $S_R = \{P_{10}, P_{11}\}$.	$\|I_{p10}\|^+ = \{p_3, p_6, p_{10}\}$, $\|I_{p11}\|^+ = \{p_4, p_5, p_{11}\}$.	$H(r_{10}) = \{p_3, p_6\}$, $H(r_{11}) = \{p_4, p_5\}$.	$[S_2] = \{p_3, p_5\}$.	t_3, t_6	t_2, t_5	1
$S_3 = \{P_3, P_7, P_9, P_{10}\}$, $S_A = \{P_3, P_7\}$, $S_R = \{P_9, P_{10}\}$.	$\|I_{p9}\|^+ = \{p_2, p_7, p_9\}$, $\|I_{p10}\|^+ = \{p_3, p_6, p_{10}\}$.	$H(r_9) = \{p_2, p_7\}$, $H(r_{10}) = \{p_3, p_6\}$.	$[S_3] = \{p_2, p_6\}$.	t_2, t_7	t_1, t_5	1

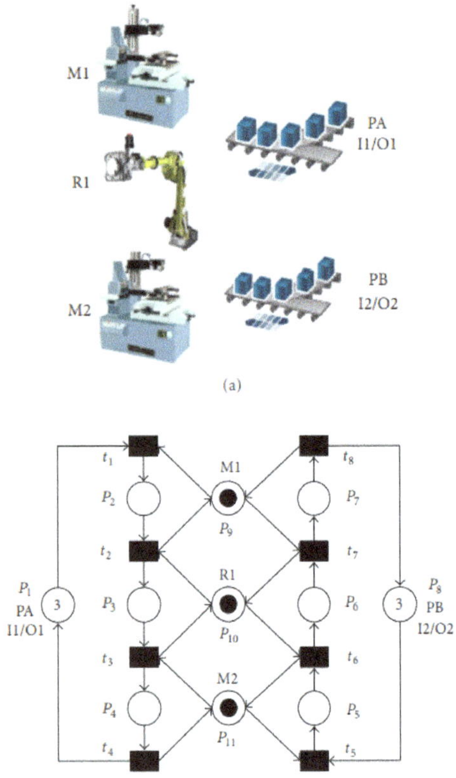

Figure 1: (a) An AMS example and (b) Petri net model of an AMS.

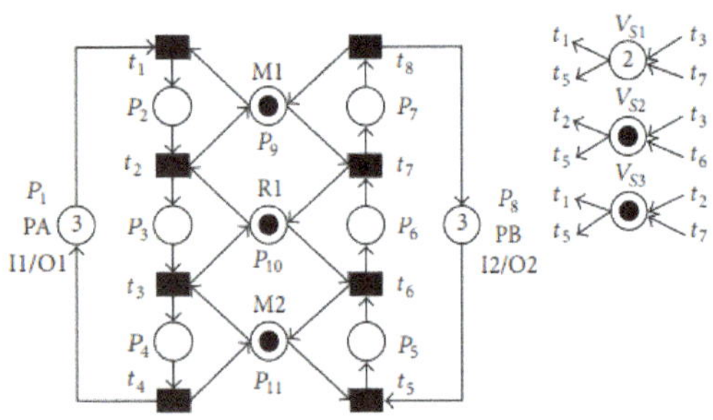

Figure 2: Controlled system of the Petri net in Figure 1(b) by Algorithm 2.

Table 2: Marking relationships between dependent and elementary siphons

Dependent	η relationship	Initial marking relationships, $M_0(S_0) > \sum_{i=1}^{n} a_i \cdot M_0(S_i) - \sum_{i=1}^{n} a_i \cdot \xi_{Si}$	Controlled
S_1	$\eta_1 = \eta_2 + \eta_3$	$M_0(S_1) > (M_0(S_2) + M_0(S_3)) - (\xi_{S2} + \xi_{S3})$ $3 > (2 + 2) - (1 + 1), 3 > 2$	Yes

Theorem 7. Let (N, M_0) be an ordinary Petri net and let S_0 be a generally dependent siphon on elementary siphons S_1, S_2,..., S_n, S_{n+1}, S_{n+2},..., and S_{n+m} that means $\eta_{S0} = \sum_{i=1}^{n} a_i \cdot \eta_{Si} - \sum_{j=n+1}^{n+m} a_j \cdot \eta_{Sj}$. *If* $S_1, S_2, \ldots, S_n, S_{n+1}$, S_{n+2},..., and S_{n+m} are invariant-controlled by adding monitors $V_{S1}, V_{S2}, \ldots, V_{Sn+m}, \ldots$, *and* $M_0(S_0) > \sum_{i=1}^{n} a_i \cdot M_0(S_i) - \sum_{i=1}^{n} a_i \cdot \xi_{Si}$, where $1 \leq \xi_{Si} \leq M_0(S_i) - 1$, S_0 is called controlled. Based on the concept of elementary siphons, the proposed deadlock prevention algorithm developed by Li and Zhou [11] is stated as follows.

Algorithm 8 (elementary siphon-based policy).

Input. Petri net model (N, M_0) of an AMS, where $N = (P^0 \cup P_A \cup P_R, T, F, W)$.

Step 1. For a given Petri net (N, M_0), compute all SMSs.

Step 2. Compute the transition vector matrix $[\eta]$ of the SMS.

Step 3. Compute the elementary siphons of Petri net N. The remaining are the dependent siphons.

Step 4. For each siphon elementary siphon, add a monitor V_S such that

(i) the V_S output arcs are connected to the source transitions leading to the sink transitions of S and all arcs weights are ones;

(ii) the V_S input arcs are connected from the stealing places of S and all arcs weights are ones;

(iii) $M_{0A}(V_S) = M_0(S) - \xi_S$, where $M_{0A}(V_S)$ is an initial marking of a monitor and $1 \leq \xi_S \leq M_0(S) - 1$.

Step 5. Repeat Step 4 until all elementary siphons are covered.

Step 6. Adjust ξ_i, such that each dependent siphon is controlled.

Step 7. Add all monitors to (N, M_0) and the resulting net is denoted by (N_1, M_1).

Step 8. Output (N_1, M_1).

Step 9. End.

For example, there are three SMSs in the Petri net model shown in Figure 1(b), which are $S_1 = \{p_4, p_7, p_9, p_{10}, p_{11}\}$, $S_2 = \{p_4, p_6, p_{10}, p_{11}\}$, and $S_3 = \{p_3, p_7, p_9, p_{10}\}$. One can obtain that $\lambda_{S1} = (0, 0, 0, 1, 0, 0, 1, 0, 1, 1, 1)$, $\lambda_{S2} = (0, 0, 0, 1, 0, 1, 0, 0, 0, 1, 1)$, and $\lambda_{S3} = (0, 0, 1, 0, 0, 0, 1, 0, 1, 1, 0)^T$. Hence, $\eta_{S1} = (-1, 0, 1, 0, -1, 0, 1, 0)$, $\eta_{S2} = (0, -1, 1, 0, -1, 1, 0, 0)$, and $\eta_{S3} = (-1, 1, 0, 0, 0, -1, 1, 0)^T$. The transition vector matrix $[\eta]$ for the Petri net model in Figure 1(b) is as follows. Obviously, the rank of $[\eta]$ is 2 since the first row η_{S1} can be linearly represented by the second and third rows, and we can see that $\eta_{S1} = \eta_{S2} + \eta_{S3}$. As a result, the siphons that correspond to the second and third rows S_2 and S_3 are elementary siphons. Thus, S_1 is a dependent siphon:

$$[\eta] = \begin{pmatrix} -1 & 0 & 1 & 0 & -1 & 0 & 1 & 0 \\ 0 & -1 & 1 & 0 & -1 & 1 & 0 & 0 \\ -1 & 1 & 0 & 0 & 0 & -1 & 1 & 0 \end{pmatrix}. \quad (1)$$

The dependent siphons and their elementary siphons, the initial marking relationships between dependent and elementary siphons, and the controllability of dependent siphons due toTheorem 7 are shown in Table 2. Table 3 shows the required monitors using Algorithm 8 for Figure 1(b). Figure 3 presents the controlled system of the Petri net model in Figure 1(b) after adding monitors by using Algorithm 8.

Iterative Methods

This section presents two techniques that are motivated and developed by [25]. These techniques are used to design a place invariant (PI) that can prohibit as many forbidden bad markings (FBMs) as possible. First, the vector covering method is used to find the minimal covering sets of legal markings and FBMs (see Chen et al. [1]). Then, solving the integer linear programming problem (ILPP) can obtain the coefficients of the place invariant and monitors. An iterative monitor design process is executed. At each iteration, a place invariant is designed to prohibit as many FBMs as possible. All FBMs that are prohibited by the place invariant are eliminated from the minimal covered set of FBMs. This process is terminated when all the minimal covered sets of FBMs are forbidden.

Table 3: Required monitors using Algorithm 8

Siphon	${}^\bullet V_{Si}$	$V_{Si}{}^\bullet$	$M_{oA}(V_{Si})$
S_2	t_3, t_6	t_2, t_5	1
S_3	t_2, t_7	t_1, t_5	1

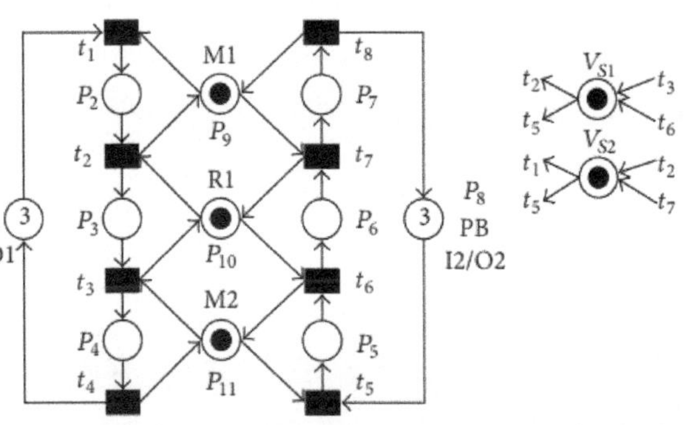

Figure 3: Controlled system of the Petri net in Figure 1(b) by Algorithm 8.

Maximal Number of Forbidding FBM Problem 1

The Maximal number of Forbidding FBM Problem 1 (MFFP1) is a technique used to design a monitor synthesis for FBMs. This technique used ILPP to obtain the coefficients of the place invariant that prohibit as many FBMs as possible. The MFFP1 is shown in the following mathematical model.

MFFP1 is

$$\max \quad f = \sum_{k \in N^*_{\text{FBM}}} f_k \quad (2)$$

subject to

$$\sum_{i \in N_A} l_i \cdot M_l(p_i) \le \beta, \quad \forall M_l \in M^*_L$$

$$\sum_{i \in N_A} l_i \cdot M_k(p_i) \ge \beta + 1 - Q \cdot (1 - f_k),$$

$$\forall M_k \in M^*_{\text{FBM}} \quad (3)$$

$$l_i \in \{0, 1, 2, \ldots\}, \quad \forall i \in \mathbb{N}$$

$$\beta \in \{1, 2, \ldots\}$$

$$f_k \in \{0, 1\}, \quad \forall k \in N^*_{\text{FBM}},$$

where M^*_L and M^*_{FBM} are the minimal covering sets of legal markings and FBMs, respectively. The objective function f is used to maximize the number of FBMs that are forbidden by place invariant. If $f = 0$, we have $f_k = 0$, for all $k \in N^*_{\text{FBM}}$ (N^*_{FBM} is used to denote $\{i \mid M_i \in M^*_{\text{FBM}}\}$); that is to say, no FBMs in M^*_{FBM} can be prohibited by place invariant. In (2), let I be a place invariant, where l_i's ($i \in \mathbb{N}$) are the coefficients of I, β is a positive integer variable, and $M_l(p_i)$ ($\forall M_l \in M^*_L$) are the number of tokens in minimal covering set of legal markings.

All legal markings should be kept after a monitor is added, implying that every marking $M_l \in M^*_L$ cannot be forbidden from being reached, and coefficients l_i's ($i \in \mathbb{N}$) should satisfy (2) that is called the reachability condition.

In (3), Q is a positive integer constant number that must be large enough, $M_k(p_i)$ ($\forall M_k \in M^*_{\text{FBM}}$) are the number of tokens in minimal covering set of FBMs, and f_k's ($f_k \in \{0, 1\}$, $k = 1, 2, \ldots$) are a set of M^*_{FBM} variables. Moreover, in (3), $f_k = 1$ implies that M_k is prohibited by I and $f_k = 0$ implies that this constraint is redundant and M_k cannot be prohibited by I. Based on the concept of MFFP1, the proposed deadlock prevention algorithm developed in [25] is stated as follows.

Algorithm 9 (MFFP1 based policy).

Input. Petri net model (N, M_0) of an AMS, where $N = (P^0 \cup P_A \cup P_R, T, F, W)$.

Step 1. For a given Petri net (N, M_0), find the sets of FBMs M_{FBM} and legal markings M_L.

Step 2. Find the minimal covered sets of FBMs M^*_{FBM} and legal markings M^*_L.

Step 3. $V_M := \emptyset$. /*V_M is used to indicate the set of monitors to be found. */

Step 4. While $\{M^*_{\text{FBM}} \neq \emptyset\}$, do the following.

(i) Design MFFP1.

(ii) Solve MFFP1. Let l_i's ($i \in \mathbb{N}$) and β be the solution if $f^* \neq 0$.
Otherwise, exit, as the solution does not exist.

(iii) Design a place invariant I and a monitor V_S.

(iv) $V_M := V_M \cup \{V_S\}$ and $M^*_{\text{FBM}} = M^*_{\text{FBM}} - F_I$. /* F_I is covered M^*_{FBM}
endwhile.

Step 5. Add all monitors in V_M to (N, M_0) and the resulting net is denoted by (N_1, M_1).

Step 6. Output (N_1, M_1).

Step 7. End.

For the Petri net model shown in Figure 1(b), we have $M^*_{\text{FBM}} = \{p_2+p_5, p_3+p_5, p_2+p_6\}$ and $M^*_L = \{p_2+p_3+p_4, p_5+p_6+p_7\}$. Algorithm 9 is considered. At the first iteration, let

I_1 be the place invariant to be calculated. I_1 should satisfy (2) for the two legal markings in M^*_L: that is, $l_2 \cdot 1+l_3 \cdot 1+l_4 \cdot 1 \leq \beta$ and $l_5 \cdot 1+l_6 \cdot 1+l_7 \cdot 1 \leq \beta$. Therefore, we have the following two constraints:

$$l_2 + l_3 + l_4 \leq \beta,$$
$$l_5 + l_6 + l_7 \leq \beta. \tag{4}$$

We introduce three variables f_1, f_2, and f_3 ($f_1, f_2, f_3 \in \{0, 1\}$) to represent whether I_1 forbids $\text{FBM}_1 = p_2 + p_5$, $\text{FBM}_2 = p_3 + p_5$, and $\text{FBM}_3 = p_2 + p_6$, respectively. Therefore, we have the following three constraints: that is,

$$l_2 + l_5 \geq \beta + 1 - Q \cdot (1 - f_1),$$
$$l_3 + l_5 \geq \beta + 1 - Q \cdot (1 - f_2),$$
$$l_2 + l_6 \geq \beta + 1 - Q \cdot (1 - f_3). \tag{5}$$

Finally, MFFP1 is defined as follows.

MFFP1 is

$$\begin{aligned} \max \quad & f = f_1 + f_2 + f_3 \\ \text{subject to} \quad & l_2 + l_3 + l_4 \leq \beta \\ & l_5 + l_6 + l_7 \leq \beta \\ & l_2 + l_5 \geq \beta + 1 - Q \cdot (1 - f_1) \\ & l_3 + l_5 \geq \beta + 1 - Q \cdot (1 - f_2) \\ & l_2 + l_6 \geq \beta + 1 - Q \cdot (1 - f_3) \\ & l_i \in \{0, 1, 2, \ldots\}, \quad \forall i \in \{2, 3, 4, 5, 6, 7\} \\ & \beta \in \{1, 2, \ldots\} \\ & f_k \in \{0, 1\}, \quad \forall k \in \{1, 2, 3\}. \end{aligned} \tag{6}$$

The optimal solution of the above MFFP1 is $l_2 = 2$, $l_5 = 1$, $l_6 = 1$, β=2, $f_1 = 1$, and $f_3 = 1$, and all the remaining variables are equal to zero. Then, a monitor V_{S1} is designed for $I_1 : 2\mu_2 + \mu_5 + \mu_6 + \mu_{VS1} = 2$. As a result, we have $M_{0A}(V_{S1}) = \beta = 2$, ${}^{\bullet}V_{S1} = \{2t_2, t_7\}$, and $V_{S1}{}^{\bullet} = \{2t_1, t_5\}$. FBM_1 and FBM_3 are prohibited by I_1: that is, $F_{I1} = \{\text{FBM}_1, \text{FBM}_3\}$. Therefore, we have $M^*_{\text{FBM}} := M^*_{\text{FBM}} - F_{I1}$: that is, $M^*_{\text{FBM}} = \text{FBM}_3 = \{p_3 + p_5\}$.

At the next iteration, let I_2 place invariant be calculated. The new MFFP1 can be easily obtained by eliminating constraints that related to FBM_1 and FBM_3 from the first MFFP1. Therefore, the new MFFP1 is stated as follows.

MFFP1 is

$$\begin{aligned} \max \quad & f = f_2 \\ \text{subject to} \quad & l_2 + l_3 + l_4 \leq \beta \\ & l_5 + l_6 + l_7 \leq \beta \\ & l_3 + l_5 \geq \beta + 1 - Q \cdot (1 - f_2) \\ & l_i \in \{0, 1, 2, \ldots\}, \quad \forall i \in \{2, 3, 4, 5, 6, 7\} \\ & \beta \in \{1, 2, \ldots\} \\ & f_2 \in \{0, 1\}. \end{aligned} \tag{7}$$

The optimal solution of the above MFFP1 is $l_3 = 1$, $l_5 = 1$, β=1, and $f_2 = 1$ and all the remaining variables are equal to zero. Then, a monitor V_{S2} is designed for $I_2 : \mu_3 + \mu_5 + \mu_{VS2} = 1$. As a result, we have $M_{0A}(V_{S2}) = \beta = 1$, ${}^{\bullet}V_{S1} = \{t_3, t_6\}$, and $V_{S1}{}^{\bullet} = \{t_2, t_5\}$. Only FBM_2 is prohibited by I_2: , $F_{I2} = \{\text{FBM}_2\}$. Therefore, we have $M^*_{\text{FBM}} := M^*_{\text{FBM}} - F_{I2}$: : that is, $M^*_{\text{FBM}} = \emptyset$.

Algorithm 9 is terminated. Totally, there are two monitors computed for this net. Adding the two monitors to the original Petri net model, the controlled system after addition is as shown in Figure 4. Table 4 shows the two required monitors by using Algorithm 9.

Maximal Number of Forbidding FBM Problem 2

The Maximal number of Forbidding FBM Problem 1 has been modified to improve its efficiency and its number of variables and constraints are slightly reduced. The modified technique is called a Maximal number of Forbidding FBM Problem 2 (MFFP2).The MFFP1 is shown in the following mathematical model.

MFFP2 is

$$\begin{aligned} \max \quad & f = \sum_{k \in N^*_{\text{FBM}}, k \neq j} f_k \\ \text{subject to} \quad & \sum_{i \in N_A} l_i \cdot \left(M_l(p_i) - M_j(p_i)\right) \leq -1, \\ & \forall M_l \in M^*_L \end{aligned} \tag{8}$$

$$\sum_{i \in N_A} l_i \cdot \left(M_k(p_i) - M_j(p_i)\right)$$
$$\geq -Q \cdot (1 - f_k),$$
$$\forall M_k \in M^*_{\text{FBM}},\ k \neq j$$
$$l_i \in \{0, 1, 2, \ldots\},\quad \forall i \in \mathbb{N}$$
$$f_k \in \{0, 1\},\quad \forall k \in N^*_{\text{FBM}},\ k \neq j. \tag{9}$$

The objective function is maximizing f the number of FBMs that are prohibited by place invariant. If no solution has been obtained from MFFP2 for an FBM, then there are not any FBMs in M^*_{FBM} that can be forbidden by that place invariant.

In (8), l_i's ($i \in \mathbb{N}$) are the coefficients of I, (p_i) ($\forall M_l \in M^*_L$) are the number of tokens in minimal covering set of legal markings, and $M_j(p_i)$ are the number of tokens in selected FBM$_j$. In (9), Q is a positive integer constant number that must be large enough, $M_k(p_i)$ ($\forall M_k \in M^*_{\text{FBM}}$) are the number of tokens in minimal covering set of FBMs, and f_k's ($fk \in \{0, 1\}$, $k = 1, 2, \ldots$) are a set of M^*_{FBM} variables. In (9), $f_k = 1$ implies that M_k is prohibited by I and $f_k = 0$ implies that this constraint is redundant and Mk cannot be prohibited by I.

Table 4: Required monitors using Algorithm 9

i	${}^\bullet V_{Si}$	$V_{Si}{}^\bullet$	$M_{oA}(V_{Si})$
1	$2t_2, t_7$	$2t_1, t_5$	2
2	t_3, t_6	t_2, t_5	1

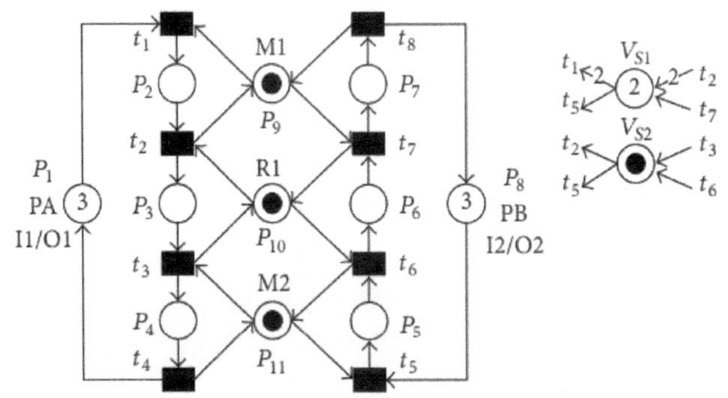

Figure 4: Controlled system of the Petri net in Figure 1(b) by Algorithm 9.

Based on the concept of MFFP2, the proposed deadlock prevention algorithm developed in [25] is stated as follows.

Algorithm 10 (MFFP2 based policy).

Input. Petri net model (N, M_0) of an AMS, where $N = (P^0 \cup P_A \cup P_R, T, F, W)$.

Step 1. For a given Petri net (N, M_0), find the sets of FBMs M_{FBM} and legal markings M_L.

Step 2. Find the minimal covered sets of FBMs $M^\bullet_{\text{FBM}}$ and legal markings $M^\bullet_L$.

Step 3. $V_M := \emptyset$. /$^*V_M$ is used to indicate the set of monitors to be found. */

Step 4. While $\{M^\bullet_{\text{FBM}} \neq \emptyset\}$, do the following.

(i) Design MFF2.

(ii) Solve MFFP2. Let l_i's ($i \in N_A$) and β be the solution if $f^* \neq 0$. Otherwise, exit, as the solution does not exist.

(iii) Design a place invariant I and a monitor V_S $V_M := V_M \cup \{V_S\}$ and $M^\bullet_{\text{FBM}} = M^\bullet_{\text{FBM}} - F_I$. /$^*F_I$ is covered $M^\bullet_{\text{FBM}}$

endwhile.

Step 5. Add all monitors in V_M to (N, M_0) and the resulting net is denoted by (N_1, M_1).

Step 6. Output (N_1, M_1).

Step 7. End.

Consider the Petri net model shown in Figure 1(b) by Algorithm 10. At the first iteration, FBM$_1 = p_2 + p_5$ is chosen. Let I_1 be the place invariant to prohibit FBM$_1$, whose coefficients are l_i's ($i \in \{2, 3, 4, 5, 6, 7\}$). I_1 should satisfy (8) that is called the reachability condition and does not prohibit any marking in M^*_L. Therefore, I_1 should satisfy two constraints:

$$l_2 \cdot (1-1) + l_3 \cdot (1-0) + l_4 \cdot (1-0) + l_5 \cdot (0-1) + l_6 \cdot (0-0) + l_7 \cdot (0-0) \leq -1,$$
$$l_2 \cdot (0-1) + l_3 \cdot (0-0) + l_4 \cdot (0-0) + l_5 \cdot (1-1) + l_6 \cdot (1-0) + l_7 \cdot (1-0) \leq -1 \tag{10}$$

for the two legal markings $p_2 + p_3 + p_4$ and $p_5 + p_6 + p_7$ in M^*_L, respectively. After simplifying the two constraints, we have

$$l_3 + l_4 - l_5 \leq -1,$$
$$-l_2 + l_6 + l_7 \leq -1. \tag{11}$$

Two variables are introduced f_2 and f_3 to represent whether I_1 forbids FBM$_2 = p_3 + p_5$ and FBM$_3 = p_2 + p_6$, respectively. Therefore, we have the two constraints: that is, $l_2 \cdot (0-1) + l_3 \cdot (1-0) + l_4 \cdot (0-0) + l_5 \cdot (1-1) + l_6 \cdot (0-0) + l_7 \cdot (0-0) \geq -Q \cdot (1 - f_2)$ and $l_2 \cdot (1-1) + l_3 \cdot (0-0) + l_4 \cdot (0-0) + l_5 \cdot (0-1) + l_6 \cdot (1-0) + l_7 \cdot (0-0) \geq -Q \cdot (1 - f_3)$, where Q is a positive integer constant that must be big enough and $f_2 = 1$ ($f_3 = 1$) implies that FBM$_2$ (FBM$_3$) is forbidden by I_1 and $f_2 = 0$ ($f_3 = 0$) implies that FBM$_2$ (FBM$_3$) cannot be forbidden by I_1. After simplifying the two constraints, we have

$$-l_2 + l_3 \geq -Q \cdot (1 - f_2),$$
$$-l_5 + l_6 \geq -Q \cdot (1 - f_3). \tag{12}$$

Finally, from the above constrains MFFP2 for FBM$_1$ is stated as follows.

MFFP2 is

$$\begin{aligned}
\max \quad & f = f_2 + f_3 \\
\text{subject to} \quad & l_3 + l_4 - l_5 \le -1 \\
& -l_2 + l_6 + l_7 \le -1 \\
& -l_2 + l_3 \ge -Q \cdot (1 - f_2) \\
& -l_5 + l_6 \ge -Q \cdot (1 - f_3) \\
& l_i \in \{0, 1, 2, \ldots\}, \quad \forall i \in \{2, 3, 4, 5, 6, 7\} \\
& f_k \in \{0, 1\}, \quad \forall k \in \{2, 3\}.
\end{aligned} \tag{13}$$

The optimal solution of the above MFFP2 is $l_2 = 1$, $l_3 = 1$, $l_5 = 2$, and $f_2 = 1$, and all the remaining variables are equal to zero. Then, a monitor V_{S1} is designed for $I_1 : \mu_2 + \mu_3 + 2\mu_5 + \mu_{VS1} = 2$. As a result, we have $M_{0A}(V_{S1}) = 2$, ${}^{\bullet}V_{S1} = \{t_3, 2t_6\}$, and $V_{S1}{}^{\bullet} = \{t_1, 2t_5\}$. FBM_1 and FBM_2 are prohibited by I_1: that is, $F_{I1} = \{FBM_1, FBM_2\}$. Therefore, we have $M^{*}_{FBM} := M^{*}_{FBM} - F_{I1}$: that is, $M^{*}_{FBM} = \{p_2 + p_6\}$. At the second iteration, the last FBM is considered, that is, FBM_3. Let I_2 be the place invariant to prohibit FBM_3. First, the coefficients of I_2 should fulfill the reachability conditions: that is,

$$\begin{aligned}
& l_3 + l_4 - l_6 \le -1, \\
& -l_2 + l_5 + l_7 \le -1.
\end{aligned} \tag{14}$$

Table 5: Required monitors using Algorithm 10

i	${}^{\bullet}V_{Si}$	$V_{Si}{}^{\bullet}$	$M_{0A}(V_{Si})$
1	$t_3, 2t_6$	$t_1, 2t_5$	2
2	t_2, t_7	t_1, t_6	1

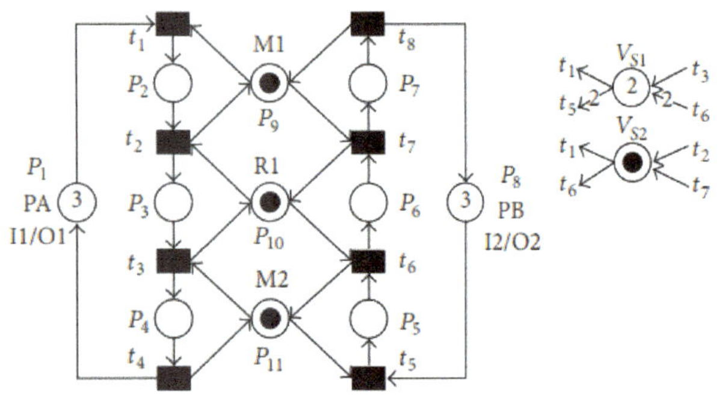

Figure 5: Controlled system of the Petri net in Figure 1(b) by Algorithm 10.

Since FBM_3 is the last marking in M^{*}_{FBM}, no other FBMs in the set require to be prohibited by I_2. By solving the above integer linear system, we have a solution with $l_2 = 1$, $l_6 = 1$, and all the remaining variables are equal to zero. Then, a monitor V_{S2} is designed for $I_2 : \mu_2 + \mu_6 + \mu_{VS2} = 2$. As a result, we have $M_{0A}(V_{S2}) = 1$, ${}^{\bullet}V_{S2} = \{t_2, t_7\}$, and $V_{S2}{}^{\bullet} = \{t_1, t_6\}$. Only FBM_3 is prohibited by I_2: that is, $F_{I2} = \{FBM_3\}$. Therefore, we have $M^{*}_{FBM} := M^{*}_{FBM} - F_{I1}$: that is, $M{*}\ FBM = \theta$.

Algorithm 10 is terminated. Totally, there are two monitors computed for this net. We add the two monitors to the original Petri net model. The controlled system after addition is as shown in Figure 5. Table 5 shows the two required monitors by using Algorithm 10.

CASE STUDIES

This section presents four case studies and the applications of the proposed applied algorithms.

(a) The production sequence of the first AMS case study is shown in Figure 6(a). It consists of two robots (R1 and R2; each one can hold a part at a time) and five machines (Ml, M2, M3, M4, and M5; each one can process one part at a time). There are three loading buffers I1–3 and three unloading buffers O1–3. Three types of parts are considered in the system: PA, PB, and PC. The model of Petri net for this case study is shown in Figure 6(b).

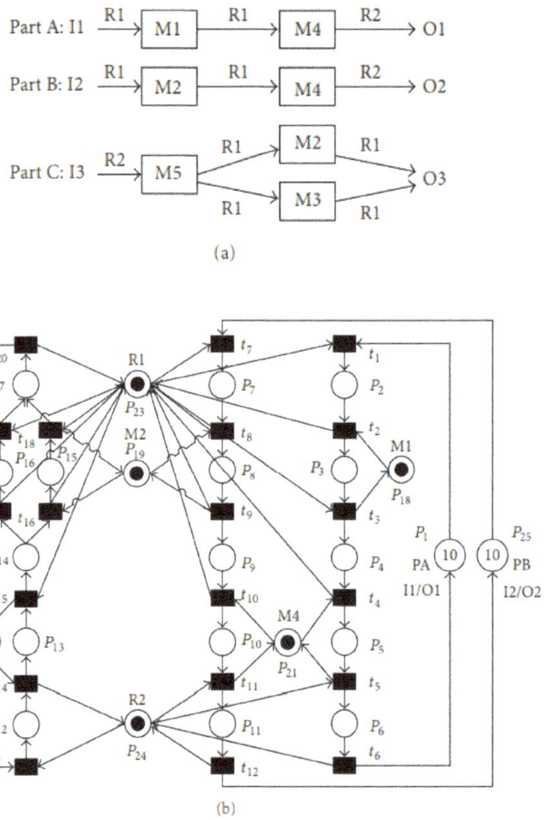

Figure 6: (a) Production sequence and (b) Petri net model of case study 1.

(b) The production sequence of the second AMS case study is shown in Figure 7(a) and several studies have been done in this case study (see Elzpeleta et al. [13] and Chen et al. [25]). The system consists of three robots (Rl, R2, and R3; each one can hold a part at a time) and four machines (Ml, M2, M3, and M4; each machine can process two parts at a time). There are three loading buffers I1–3 and three unloading buffers O1–3. Three types of parts are considered in the system: PA, PB, and PC. The model of Petri net for this case study is shown in Figure 7(b).

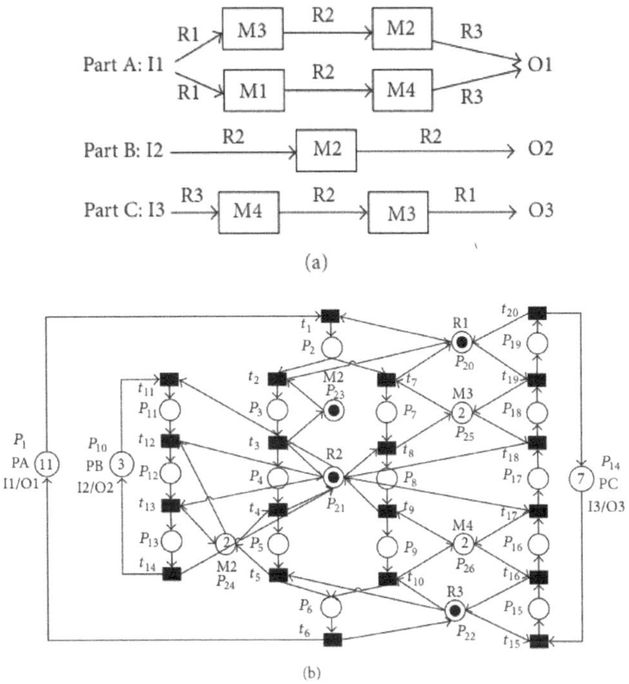

Figure 7: (a) Production sequence and (b) Petri net model of case study 2.

(c) Next, the production sequence of the third AMS case study is shown in Figure 8(a). The cell consists of two robots (R1 and R2; each one can hold a part at a time) and four machines (M1, M2, and M3; each machine can process a part at a time; M4 can process two products at a time) and, for loading and unloading the cell, two loading buffers (I1 and I2) and two unloading buffers (O1 and O2). Two types of parts are considered in the cell: PA and PB. The model of Petri net for this case study is illustrated in Figure 8(b).

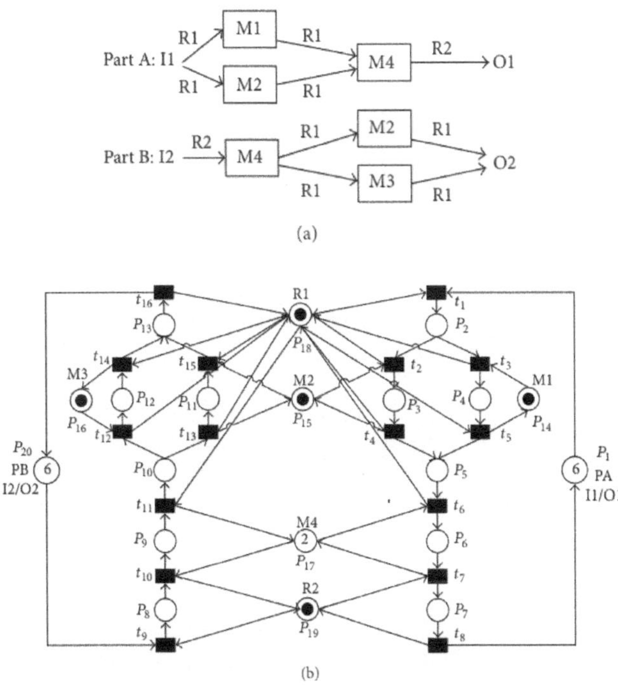

Figure 8: (a) Production sequence and (b) Petri net model of case study 3.

(d) Finally, the production sequence of the fourth AMS case study is shown in Figure 9(a) and several studies have been achieved in this case study (see Uzam [9], Chen et al. [25], Piroddi et al. [26], and Piroddi et al. [27]). The system is composed of two robots (R1 and R2; each robot can hold a part at a time) and four machines (M1, M2, M3, and M4; each machine can process one part at a time). There are two loading buffers (I1 and I2) and two unloading buffers (O1 and O2). PA and PB are considered in the system. The model of Petri net for this case study is shown in Figure 9(b).

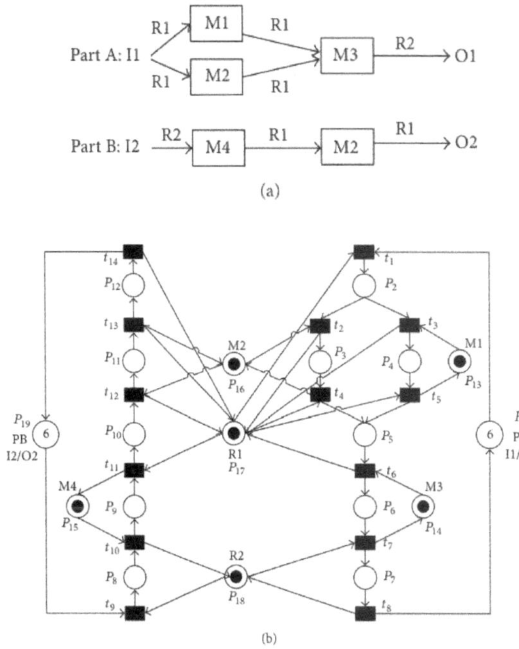

Figure 9: (a) Production sequence and (b) Petri net model of case study 4.

The main parameters of all case studies are presented in Table 6 such as number of places, number of transitions, set partition, SMS, dependent siphons, elementary siphons, reachable markings, FBMs, legal markings, M^*_{FBM}, and M^*_L. . Table 7 illustrates the supervisor performance of applied algorithms for all case studies with respect to the numbers of monitors, arcs, and reachable states.

From Table 7, Algorithms 9 and 10 obtain fewer numbers of monitors than Algorithms 2 and 8 for all case studies while Algorithm 9 provides less or equal number of arcs than the other algorithms in case studies 1, 2, and 3. In case study 4, Algorithm 10 obtains less number of arcs than the other algorithms. As for the reachable states, all algorithms provide the same number of reachable states in case study 1, which is 1072 while both Algorithms 2 and 8 have the same number of reachable states, which is 6287 in case study 2. Moreover, in case study 2, both Algorithms 9 and 10 provide the same number of reachable states, which is 21581. Moreover, in case study 3, both Algorithms 9 and 10 outperform Algorithms 2 and 8 in number of reachable states, which have 656 reachable states. Finally, in case study 4, Algorithms 8, 9, and 10 provide better reachable states than Algorithm 2, which is 205. The required monitors using applied algorithms for all case studied are shown in Tables 8, 9, 10, and 11.

Table 6: Main parameters of all case studies

Parameter	Case study			
	1	2	3	4
Number of places	26	26	20	19
Number of transitions	20	20	16	14
Set partition	$P_0 = \{p_1, p_{25}, p_{26}\}$ $P_R = \{p_{18}–p_{24}\}$ $P_A = \{p_2, \ldots, p_{17}\}$	$P_0 = \{p_1, p_{10}, p_{14}\}$ $P_R = \{p_{20}, \ldots, p_{26}\}$ $P_A = \{p_2, \ldots, p_4, p_6, \ldots, p_{13}, p_{15}, \ldots, p_{19}\}$	$P_0 = \{p_1, p_{20}\}$ $P_R = \{p_{14}, \ldots, p_{19}\}$ $P_A = \{p_2, \ldots, p_{13}\}$	$P_0 = \{p_1, p_{19}\}$ $P_R = \{p_{13}, \ldots, p_{18}\}$ $P_A = \{p_2, \ldots, p_{12}\}$
SMS	11	18	12	5
Dependent siphons	6	9	7	2
Elementary siphons	5	9	5	3
Reachable markings	1492	26750	836	282
FBMs	388	4211	156	54
Legal markings	1072	21581	656	205
M^*_{FBM}	9	34	7	8
M^*_L	92	393	56	26

Table 7: Supervisor performance of applied algorithms for all case studies

Case study	Parameter	Algorithm 2	Algorithm 8	Algorithm 9	Algorithm 10
1	Monitors number	11	5	3	3
	Arcs number	66	23	14	31
	Reachable states number	1072	1072	1072	1072
2	Monitors number	18	9	6	6
	Arcs number	106	50	45	59
	Reachable states number	6287	6287	21581	21581
3	Monitors number	12	5	4	4
	Arcs number	61	25	21	21
	Reachable states number	436	489	656	656
4	Monitors number	5	3	2	2
	Arcs number	21	13	15	12
	Reachable states number	182	205	205	205

Table 8: Required monitors using applied algorithms for case study 1

Algorithm	i	Siphon	${}^\bullet V_{Si}$	$V_{Si}{}^\bullet$	$M_{oA}(V_{Si})$
Algorithm 2	1	S_1	$t_5, t_{11}, t_{18}, t_{19}$	t_1, t_7, t_{13}	6
	2	S_2	$t_5, t_{11}, t_{15}, t_{18}$	t_1, t_7, t_{13}, t_{16}	5
	3	S_3	t_5, t_{11}, t_{15}	t_1, t_9, t_{13}	4
	4	S_4	t_3, t_9, t_{18}, t_{19}	t_1, t_7, t_{15}	3
	5	S_5	$t_5, t_{11}, t_{18}, t_{19}$	t_3, t_7, t_{13}	5
	6	S_6	$t_5, t_{11}, t_{15}, t_{18}$	t_3, t_7, t_{13}, t_{16}	4
	7	S_7	t_5, t_{11}, t_{15}	t_3, t_9, t_{13}	3
	8	S_8	t_9, t_{18}, t_{19}	t_7, t_{15}	2
	9	S_9	t_3, t_9, t_{18}	t_1, t_7, t_{16}	2
	10	S_{10}	t_3	t_1	1
	11	S_{11}	t_9, t_{18}	t_7, t_{16}	1
Algorithm 8	1	S_3	t_5, t_{11}, t_{15}	t_1, t_9, t_{13}	4
	2	S_7	t_5, t_{11}, t_{15}	t_3, t_9, t_{13}	3
	3	S_8	t_9, t_{18}, t_{19}	t_7, t_{15}	2
	4	S_{10}	t_3	t_1	1
	5	S_{11}	t_9, t_{18}	t_7, t_{16}	1
Algorithm 9	1		t_5, t_{11}, t_{15}	t_3, t_9, t_{13}	3
	2		$2t_9, t_{19}, 2t_{18}$	$2t_7, t_{15}, t_{16}$	3
	3		t_3	t_1	1
Algorithm 10	1		$t_5, 5t_{10}, t_{11}, 6t_{18}, 3t_{19}$	$t_4, 6t_7, t_{14}, 2t_{15}, 3t_{16}$	11
	2		$t_2, 11t_4, t_5, t_8, t_{10}, t_{11}, 4t_{14}, t_{15}, t_{16}, t_{17}$	$11t_1, 2t_3, t_7, 2t_9, 6t_{13}$	20
	3		t_5, t_{11}, t_{15}	t_4, t_9, t_{13}	3

Table 9: Required monitors using applied algorithms for case study 2

Algorithm	i	Siphon	${}^{\bullet}V_{Si}$	$V_{Si}{}^{\bullet}$	$M_{oA}(V_{Si})$
Algorithm 2	1	S_1	$t_5, t_{10}, t_{13}, t_{19}$	t_1, t_{11}, t_{15}	10
	2	S_2	$t_5, t_{10}, t_{13}, t_{18}$	t_3, t_7, t_{11}, t_{15}	7
	3	S_3	$t_5, t_{10}, t_{13}, t_{17}$	t_3, t_8, t_{11}, t_{15}	5
	4	S_4	t_2, t_{10}, t_{16}	t_9, t_{15}	2
	5	S_5	t_4, t_9, t_{13}, t_{19}	t_1, t_{11}, t_{16}	9
	6	S_6	t_4, t_9, t_{13}, t_{18}	t_3, t_7, t_{11}, t_{16}	6
	7	S_7	t_4, t_9, t_{13}, t_{17}	t_3, t_8, t_{11}, t_{16}	4
	8	S_8	t_4, t_8, t_{13}, t_{19}	t_1, t_{11}, t_{17}	7
	9	S_9	t_4, t_8, t_{13}, t_{18}	t_3, t_7, t_{11}, t_{17}	4
	10	S_{10}	t_4, t_7, t_{13}	t_3, t_{11}	2
	11	S_{11}	t_3, t_{10}, t_{19}	t_1, t_{15}	8
	12	S_{12}	t_2, t_{10}, t_{18}	t_7, t_{15}	5
	13	S_{13}	t_2, t_{10}, t_{17}	t_8, t_{15}	3
	14	S_{14}	t_3, t_8, t_{19}	t_1, t_{17}	5
	15	S_{15}	t_2, t_8, t_{18}	t_7, t_{17}	2
	16	S_{16}	t_3, t_9, t_{19}	t_1, t_{15}	7
	17	S_{17}	t_2, t_9, t_{18}	t_7, t_{16}	4
	18	S_{18}	t_2, t_9, t_{17}	t_8, t_{16}	2
Algorithm 8	1	S_1	$t_5, t_{10}, t_{13}, t_{19}$	t_1, t_{11}, t_{15}	10
	2	S_3	$t_5, t_{10}, t_{13}, t_{17}$	t_3, t_8, t_{11}, t_{15}	5
	3	S_4	t_2, t_{10}, t_{16}	t_9, t_{15}	1
	4	S_{10}	t_4, t_7, t_{13}	t_3, t_{11}	2
	5	S_{11}	t_3, t_{10}, t_{19}	t_1, t_{15}	8
	6	S_{14}	t_3, t_8, t_{19}	t_1, t_{17}	5
	7	S_{15}	t_2, t_8, t_{18}	t_7, t_{17}	1
	8	S_{17}	t_2, t_9, t_{18}	t_7, t_{16}	4
	9	S_{18}	t_2, t_9, t_{17}	t_8, t_{16}	1
Algorithm 9	1		$17t_5, 50t_{10}, 15t_{13}, 49t_{17}, 2t_{19}$	$3t_1, 14t_3, 44t_8, 3t_9, 15t_{11}, 50t_{15}, t_{18}$	196
	2		$t_5, 24t_8, 3t_{10}, t_{13}, 18t_{18}$	$t_3, 27t_7, t_{11}, 6t_{15}, 3t_{16}, 9t_{17}$	71
	3		$6t_3, t_5, 5t_8, t_9, 6t_{19}$	$6t_1, t_4, 5t_{15}, 4t_{17}, t_{18}$	34
	4		t_4, t_{13}	t_3, t_{11}	2
	5		t_9, t_{17}	t_8, t_{16}	2
	6		t_{10}, t_{16}	t_9, t_{15}	2
Algorithm 10	1		$9t_5, 50t_8, 25t_{10}, 8t_{13}, 74t_{17}, t_{19}$	$2t_1, 7t_3, 73t_7, 8t_{11}, 50t_{15}, 25t_{16}$	196
	2		$7t_5, 20t_9, 19t_{10}, 6t_{13}, 39t_{17}, t_{19}$	$t_1, 6t_3, 38t_8, 6t_{11}, 20t_{15}, 19t_{16}, t_{18}$	71
	3		$7t_5, t_8, 27t_{10}, 6t_{13}, 19t_{16}, 8t_{17}, t_{19}$	$t_1, 6t_3, 27t_9, 6t_{11}, 27t_{15}, t_{18}$	34
	4		t_4, t_{13}	t_3, t_{11}	2
	5		$6t_3, t_5, 23t_8, t_{10}, 18t_{18}, 5t_{19}$	$6t_1, t_4, 17t_7, t_9, t_{15}, 22t_{17}$	2
	6		t_3, t_7, t_{19}	t_1, t_{17}	2

Table 10: Required monitors using applied algorithms for case study 3

Algorithm	i	Siphon	${}^{\bullet}V_{Si}$	$V_{Si}{}^{\bullet}$	$M_{oA}(V_{Si})$
Algorithm 2	1	S_1	t_7, t_{14}, t_{15}	t_1, t_9	6
	2	S_2	t_6, t_{14}, t_{15}	t_1, t_9	5
	3	S_3	t_7, t_{11}, t_{14}	t_1, t_9, t_{12}	5
	4	S_4	t_6, t_{11}, t_{14}	t_1, t_9, t_{12}	4
	5	S_5	t_4, t_5, t_{14}, t_{15}	t_1, t_9	3
	6	S_6	t_4, t_5, t_{14}	t_1, t_{12}	2
	7	S_7	t_7, t_{14}, t_{15}	t_1, t_9	5
	8	S_8	t_7, t_{11}	t_1, t_9	3
	9	S_9	t_7, t_{10}	t_6, t_9	2
	10	S_{10}	t_6, t_{14}, t_{15}	t_1, t_9	4
	11	S_{11}	t_6, t_{11}	t_4, t_5, t_9	2
	12	S_{12}	t_4, t_{14}, t_{15}	t_2, t_{11}	2
Algorithm 8	1	S_5	t_4, t_5, t_{14}, t_{15}	t_1, t_9	3
	2	S_6	t_4, t_5, t_{14}	t_1, t_{12}	2
	3	S_9	t_7, t_{10}	t_6, t_9	2
	4	S_{11}	t_6, t_{11}	t_4, t_5, t_9	2
	5	S_{12}	t_4, t_{14}, t_{15}	t_2, t_{11}	2

Algorithm 9	1	t_4, t_{14}, t_{15}	t_2, t_{11}	2
	2	t_6, t_{14}	t_1, t_{12}	2
	3	$2t_7, t_{10}, t_{11}$	$t_4, t_5, t_6, 2t_9$	5
	4	t_6, t_{11}	t_4, t_5, t_{10}	2
Algorithm 10	1	t_4, t_5, t_{14}	t_1, t_{12}	2
	2	$t_6, t_7, 2t_{11}$	$2t_4, 2t_5, t_9, t_{10}$	5
	3	t_4, t_{14}, t_{15}	t_2, t_{11}	2
	4	t_7, t_{10}	t_6, t_9	2

Table 11: Required monitors using applied algorithms for case study 4

Algorithm	i	Siphon	$^\bullet V_{S_i}$	$V_{S_i}^\bullet$	$M_{0A}(V_{S_i})$
Algorithm 2	1	S_1	t_7, t_{13}	t_1, t_9	5
	2	S_2	t_4, t_5, t_{13}	t_1, t_{11}	2
	3	S_3	t_3, t_7, t_{13}	t_1, t_9	4
	4	S_4	t_7, t_{11}	t_1, t_9	3
	5	S_5	t_4, t_{13}	t_2, t_{11}	1
Algorithm 8	1	S_2	t_4, t_5, t_{13}	t_1, t_{11}	2
	2	S_4	t_7, t_{11}	t_1, t_9	3
	3	S_5	t_4, t_{13}	t_2, t_{11}	1
Algorithm 9	1		$3t_4, 5t_6, t_{12}, 7t_{13}$	$4t_1, 4t_2, t_5, t_9, 7t_{11}$	14
	2		$2t_7, 3t_{11}$	$t_1, t_2, t_5, 3t_9$	9
Algorithm 10	1		$t_4, 2t_4, 2t_{13}$	$t_1, t_{12}, 2t_{11}$	3
	2		$2t_7, 3t_{11}$	$t_1, t_2, t_5, 3t_9$	9

COMPUTATIONAL RESULTS AND ANALYSIS

This section presents that the computational results and analysis of the applied methods are conducted for four AMSs case studies. All cases have been modeled using MATLAB software. The simulation is done for 24 time units. After running and simulating the case studies, the results of MATLAB simulation in terms of the utilization resources and throughput can be summarized as follows. For case study 1, Figures 10 and 11 show the resources utilization and throughput of case study 1, respectively. From the figures, it can be found that all applied Algorithms 8, 9, and 10 obtain the same utilization for all resources, while Algorithm 2 obtains utilization better than Algorithms 8, 9, and 10 at M1, M2, M5, and R1, and same utilization at M4 and R2. For the throughput, Algorithm 2 obtains greater produced number of part A and part C than Algorithms 8, 9, and10. Moreover, Algorithms 8, 9, and 10 provide the same throughput for all part types. For case study 2, Figures12 and 13 show the resources utilization and throughput of the case study 2, respectively.

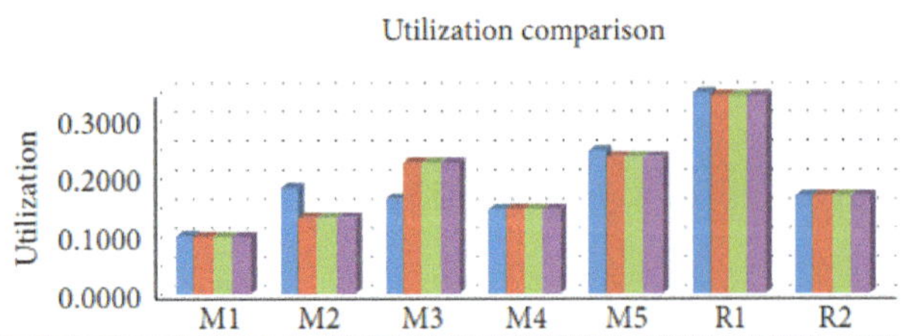

	M1	M2	M3	M4	M5	R1	R2
Algorithm 2	0.1034	0.1858	0.1663	0.1473	0.2497	0.3489	0.1714
Algorithm 8	0.1018	0.1342	0.2285	0.1477	0.2385	0.3437	0.1714
Algorithm 9	0.1018	0.1342	0.2285	0.1477	0.2385	0.3437	0.1714
Algorithm 10	0.1018	0.1342	0.2285	0.1477	0.2385	0.3437	0.1714

Figure 10: Utilization of resources for the Petri net model in case study 1.

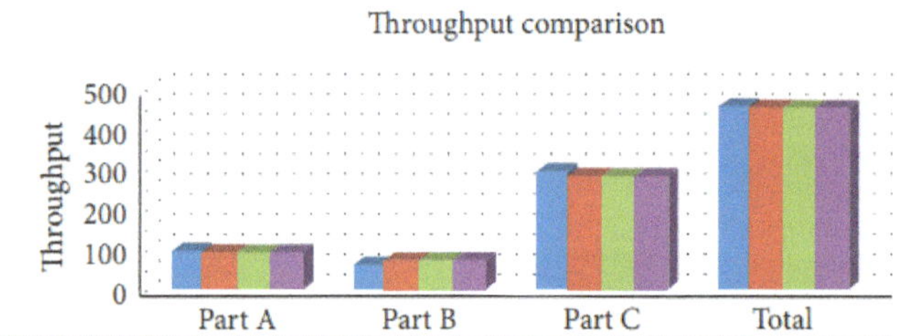

	Part A	Part B	Part C	Total
Algorithm 2	98	64	295	457
Algorithm 8	96	75	284	455
Algorithm 9	96	75	284	455
Algorithm 10	96	75	284	455

Figure 11: Throughput of the system for the Petri net model in case study 1.

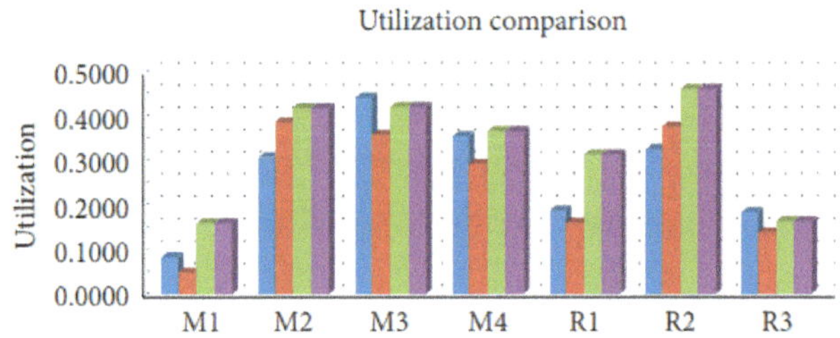

	M1	M2	M3	M4	R1	R2	R3
Algorithm 2	0.0838	0.3063	0.4410	0.3551	0.1877	0.3263	0.1840
Algorithm 8	0.0488	0.3853	0.3577	0.2904	0.1591	0.3762	0.1366
Algorithm 9	0.1595	0.4186	0.4206	0.3662	0.3144	0.4611	0.1631
Algorithm 10	0.1595	0.4186	0.4206	0.3662	0.3144	0.4611	0.1631

Figure 12: Utilization of resources for the Petri net model in case study 2.

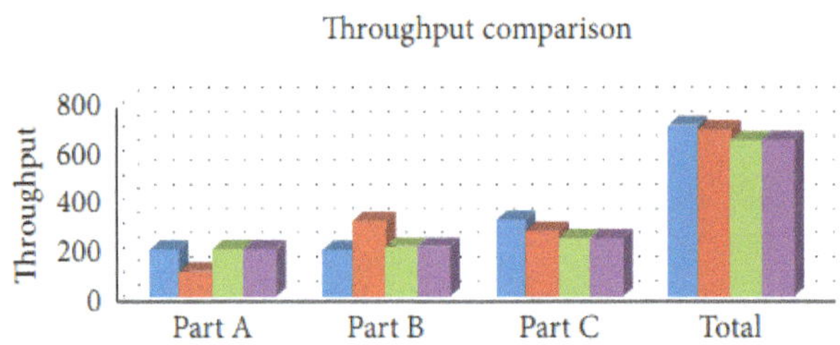

	Part A	Part B	Part C	Total
Algorithm 2	195	195	318	708
Algorithm 8	105	313	272	690
Algorithm 9	197	207	241	645
Algorithm 10	197	207	241	645

Figure 13: Throughput of the system for the Petri net model in case study 2.

The figures indicate that both applied Algorithms 9 and 10 have the same utilization for all resources and have better results than Algorithms 2 and 8 at M1, M2, M4, R1, and R2. Algorithm 2 obtains utilization better than Algorithms 8, 9, and 10 at M3 and R3. For the throughput, Algorithm 2 obtains greater produced number of part C than Algorithms 8, 9, and 10, while Algorithms 9 and 10 have greater produced number of part A than Algorithms 2 and 8. Moreover, Algorithm 8 obtains a greater produced number of part B than Algorithms 2, 9, and 10. For case study 3,

Figures 14 and 15 show the resources utilization and throughput of case study 3, respectively. The figures indicate that both applied Algorithms 9 and 10 have the same utilization of all resources and obtain better results than Algorithms 2 and 8 at M1, M3, and R2. Both Algorithms 2 and 8 obtain the same utilization for all resources and provide utilization better than Algorithms 9 and 10 at M2, M4, and R1. For the throughput, Algorithms 2 and 8 have greater produced number of part B than Algorithms 9 and 10, while Algorithms 9 and 10 obtain greater produced number of part A than Algorithms 2 and 8. Finally, for case study 4, Figures 16 and 17 show the resources utilization and throughput of case study 4, respectively. From the figures, it can be found that both applied Algorithms 9 and 10 obtain the same utilization of all resources and obtain better results than Algorithms 2 and 8 at M1, M2, R1, and R2. Both Algorithms 2 and 8 obtain the same utilization of all resources and obtain utilization better than Algorithms 9 and 10 at M3 and M4. For the throughput, Algorithms 9 and 10 obtain greater produced number of parts A and B than Algorithms 2 and 8.

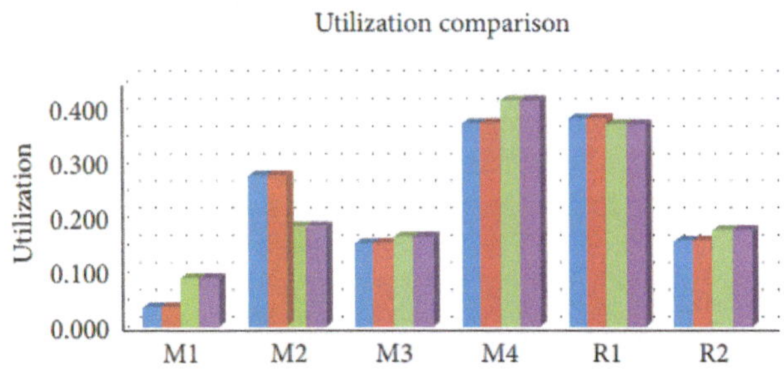

	M1	M2	M3	M4	R1	R2
Algorithm 2	0.0352	0.2770	0.1522	0.3715	0.3804	0.1563
Algorithm 8	0.0352	0.2770	0.1522	0.3715	0.3804	0.1563
Algorithm 9	0.0879	0.1832	0.1647	0.4141	0.3704	0.1761
Algorithm 10	0.0879	0.1832	0.1647	0.4141	0.3704	0.1761

Figure 14: Utilization of resources for the Petri net model in case study 3.

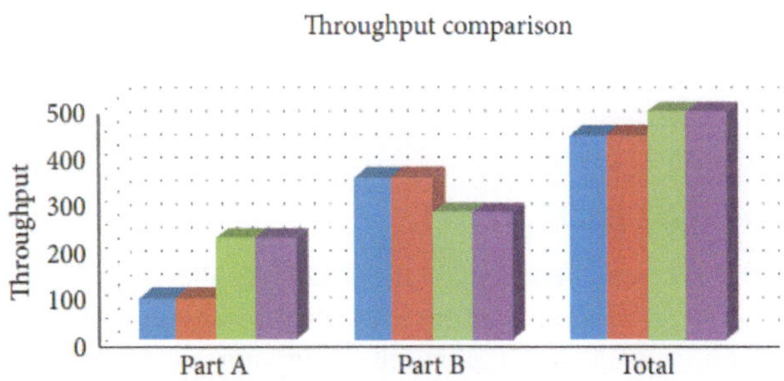

	Part A	Part B	Total
Algorithm 2	91	348	439
Algorithm 8	91	348	439
Algorithm 9	217	271	488
Algorithm 10	217	271	488

Figure 15: Throughput of the system for the Petri net model in case study 3.

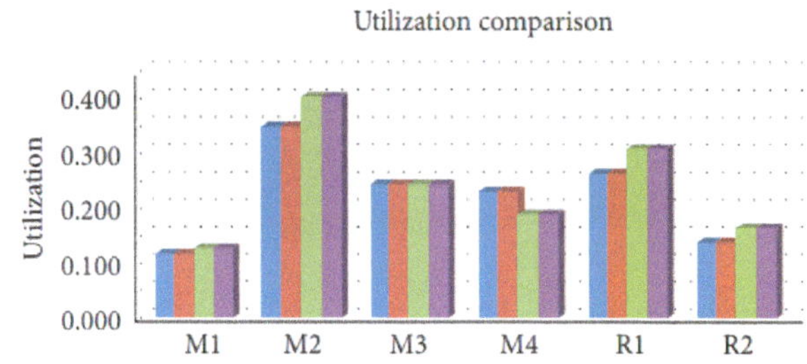

	M1	M2	M3	M4	R1	R2
Algorithm 2	0.1168	0.3504	0.2454	0.2325	0.2651	0.1392
Algorithm 8	0.1168	0.3504	0.2454	0.2325	0.2651	0.1392
Algorithm 9	0.1268	0.4043	0.2452	0.1898	0.3096	0.1652
Algorithm 10	0.1268	0.4043	0.2452	0.1898	0.3096	0.1652

Figure 16: Utilization of resources for the Petri net model in case study 4.

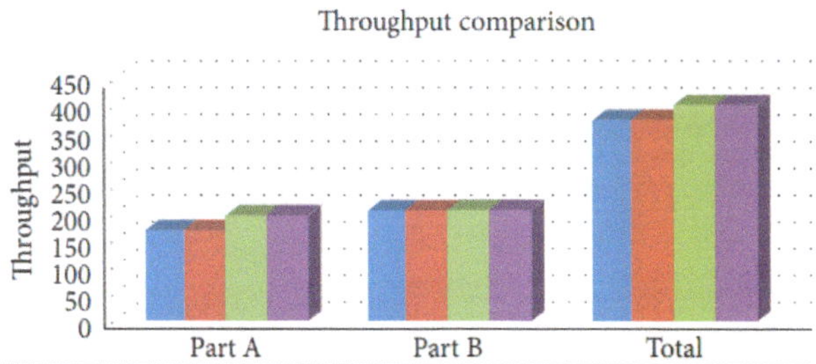

	Part A	Part B	Total
Algorithm 2	169	207	376
Algorithm 8	169	207	376
Algorithm 9	197	208	405
Algorithm 10	197	208	405

Figure 17: Throughput of the system for the Petri net model in case study 4.

Based on the above computational results for all case studies, it is shown that four applied deadlock prevention policies may not imply high resource utilization and plant productivity which has been shown theoretically in previous publications, since the utilization of any resource at case studies does not exceed 50% and there are several resources that have utilization less than 20%. Moreover, it is found that the resource utilization and plant productivity are different for case studies by using applied policies. The main reason of the occurrence of this difference is a conflict situation in places. When a control place enables more than one transition at the same time, only one transition can fire. For instance, in case study 1, the number of control places using Algorithm 2 are 11 and more than one transition is enabled at the same time; that is, several transitions can be fired. While Algorithm 10 reduces the control places into 3 places and enables more than one transition at the same time, few transitions can be fired compared with Algorithm 2. Therefore, the performance is different for different cases by using applied policies. However, we can introduce the comparison between applied policies based on the above computational results, from the viewpoint of behavior permissiveness. Both Algorithms 9and 10 can provide better behavioral permissiveness than Algorithms 2 and 8 for small size systems (case studies 3 and 4). For the large systems size (case studies 1 and 2), Algorithm 2 leads to better behavioral permissiveness than the other algorithms. The supervisor performance of applied algorithms such as the number of monitors, arcs, and reachable states as averages values is summarized. Figures 18, 19, and 20 show the summarized supervisor performance of all algorithms. From the figures, both Algorithms 9 and 10 obtain a supervisor with structural complexity and computational complexity compared to Algorithms 2 and 8.

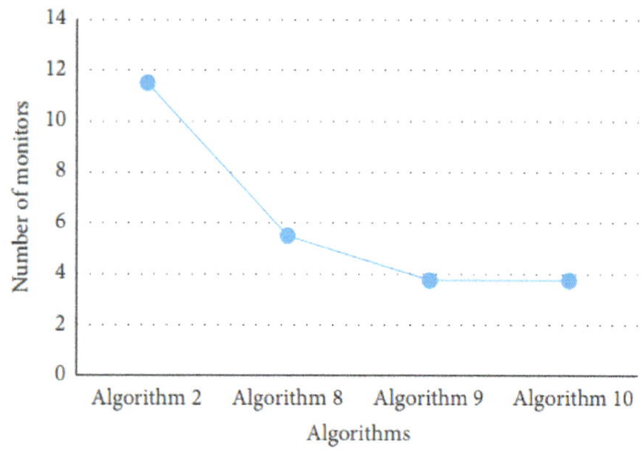

Figure 18: Average of monitors for all cases.

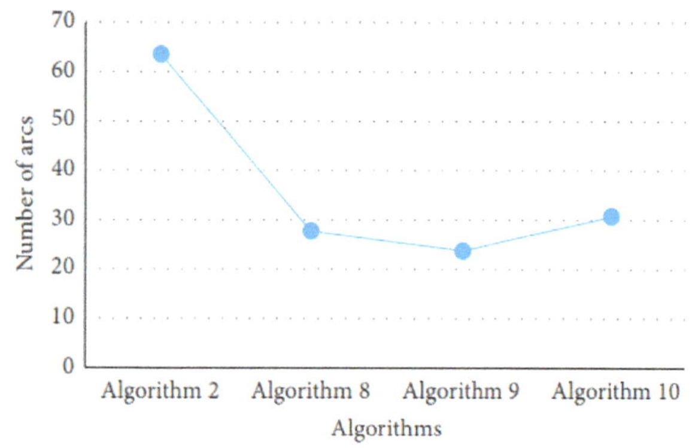

Figure 19: Average of arcs for all cases.

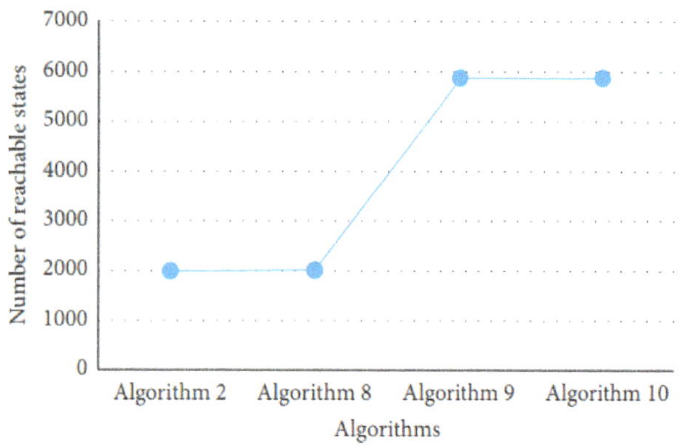

Figure 20: Average of reachable states for all cases.

CONCLUSION AND FUTURE WORK

This paper presents four applied deadlock prevention policies, two and two of which are iterative methods and siphon control methods, respectively. Four applied deadlock control methods are used to design a liveness-enforcing supervisor for four automated manufacturing system case studies. The controlled case studies are modeled and simulated using MATLAB software (Petri net toolbox) to analyze and evaluate the performance of selected methods such as utilization of resources and throughput; the computational results can be summarized as follows.

The applied deadlock prevention policies may not imply high resource utilization and plant productivity which has been shown theoretically in previous publications. Moreover, it is found that the resource utilization and plant productivity are different for case studies by using applied policies. The main reason for this difference is a conflict situation in places. When a control place enables more than one transition at the same time, only one transition can fire. For instance, in case study 1, the number of control places using Algorithm 2 are 11 and more than one transition is enabled at the same time; that is, several transitions can be fired. While Algorithm10 reduces the control places into 3 places and enables more than one transition at the same time, few transitions can be fired compared with Algorithm 2. Therefore, the performance is different for different cases by using applied policies. However, for all automated manufacturing systems, with respect to structural complexity and computational complexity, the selected iterative methods can always lead to structurally complicated and computationally expensive liveness-enforcing net supervisors compared to the siphon control methods. From the aspect of behavioral permissiveness, the selected iterative methods can provide better behavioral permissiveness than siphon control methods for small size systems. For large systems size, strict minimal siphons control method leads to better behavioral permissiveness than the other methods.

Real-world AMSs can suffer from impacts of unreliable resources (resources with failures) under the supervisory control of deadlocks. Therefore, the need for new effective deadlock control policies based on Petri nets becomes increasingly more important. Thus, the future research direction involves designing new effective deadlock prevention policies for automated manufacturing systems with unreliable resources [8]. Then, a simulation will be used to measure the performance of these policies. Different and complex automated manufacturing systems are required and should be considered as a benchmark for these deadlock prevention policies under a variety of control specifications. Optimal deadlock-free scheduling with high time performance will also be interesting.

ACKNOWLEDGMENT

This work was funded by the National Plan for Science, Technology and Innovation (MARRIFAH), King Abdulaziz City for Science and Technology, Kingdom of Saudi Arabia, Award no. 12-ELE2506-02.

REFERENCES

1. Y. Chen, Z. Li, M. Khalgui, and O. Mosbahi, "Design of a maximally permissive liveness-enforcing Petri net supervisor for flexible manufacturing systems," IEEE Transactions on Automation Science and Engineering, vol. 8, no. 2, pp. 374–393, 2011.
2. R. A. Wysk, N. S. Yang, and S. Joshi, "Detection of deadlocks in flexible manufacturing cells," IEEE Transactions on Robotics and Automation, vol. 7, no. 6, pp. 853–859, 1991.

3. D. Y. Chao, "Direct minimal empty siphon computation using MIP," The International Journal of Advanced Manufacturing Technology, vol. 45, no. 3-4, pp. 397–405, 2009.

4. D. Y. Chao, "Improvement of suboptimal siphon-and FBM-based control model of a well-known," IEEE Transactions on Automation Science and Engineering, vol. 8, no. 2, pp. 404–411, 2011.

5. A. Ghaffari, N. Rezg, and X. Xie, "Design of a live and maximally permissive Petri net controller using the theory of regions," IEEE Transactions on Robotics and Automation, vol. 19, no. 1, pp. 137–141, 2003.

6. M. Uzam and M. Zhou, "Iterative synthesis of petri net based deadlock prevention policy for flexible manufacturing systems," in Proceedings of the IEEE International Conference on Systems, Man and Cybernetics (SMC '04), pp. 4260–4265, October 2004.

7. M. Uzam, "The use of the Petri net reduction approach for an optimal deadlock prevention policy for flexible manufacturing systems," The International Journal of Advanced Manufacturing Technology, vol. 23, no. 3-4, pp. 204–219, 2004.

8. K. Lautenbach, "Linear algebraic calculation of deadlocks and traps," in Concurrency and Nets, pp. 315–336, Springer, Berlin, Germany, 1987.

9. M. Uzam, "An optimal deadlock prevention policy for flexible manufacturing systems using Petri net models with resources and the theory of regions," The International Journal of Advanced Manufacturing Technology, vol. 19, no. 3, pp. 192–208, 2002.

10. D. Y. Chao, "Fewer monitors and more efficient controllability for deadlock control in S^3PGR^2 (systems of simple sequential processes with general resource requirements)," The Computer Journal, vol. 53, no. 10, pp. 1783–1798, 2010.

11. Z. Li and M. Zhou, "Elementary siphons of Petri nets and their application to deadlock prevention in flexible manufacturing systems," IEEE Transactions on Systems, Man, and Cybernetics Part A: Systems and Humans, vol. 34, no. 1, pp. 38–51, 2004.

12. Z. Li, G. Liu, H.-M. Hanisch, and M. Zhou, "Deadlock prevention based on structure reuse of Petri net supervisors for flexible manufacturing systems," IEEE Transactions on Systems, Man and Cybernetics, Part A: Systems and Humans, vol. 42, no. 1, pp. 178–191, 2012.

13. J. Ezpeleta, J. M. Colom, and J. Martinez, "A petri net based deadlock prevention policy for flexible manufacturing systems," IEEE Transactions on Robotics and Automation, vol. 11, no. 2, pp. 173–184, 1995.

14. Y. S. Huang, M. D. Jeng, X. L. Xie, and S. L. Chung, "A deadlock prevention policy for flexible manufacturing systems using siphons," in Proceedings of the IEEE International Conference on Robotics and Automation (ICRA ‹01), pp. 541–546, May 2001.

15. Y. S. Huang, M. Jeng, X. Xie, and D.-H. Chung, "Siphon-based deadlock prevention policy for flexible manufacturing systems," IEEE Transactions on Systems, Man, and Cybernetics Part A: Systems and Humans, vol. 36, no. 6, pp. 1248–1256, 2006.

16. Z. Li and N. Wei, "Deadlock control of flexible manufacturing systems via invariant—controlled elementary siphons of petri nets," The International Journal of Advanced Manufacturing Technology, vol. 33, no. 1-2, pp. 24–35, 2007.

17. Y. S. Huang, "Design of deadlock prevention supervisors using Petri nets," The International Journal of Advanced Manufacturing Technology, vol. 35, no. 3-4, pp. 349–362, 2007.

18. G. Y. Liu, Z. W. Li, K. Barkaoui, and A. M. Al-Ahmari, "Robustness of deadlock control for a class of Petri nets with unreliable resources," Information Sciences, vol. 235, pp. 259–279, 2013.

19. S. Wang, M. Zhou, and W. Wu, "Design of a maximally permissive liveness-enforcing supervisor with reduced complexity for automated manufacturing systems," Asian Journal of Control, vol. 17, no. 1, pp. 190–201, 2014.

20. N. Viswanadham, Y. Narahari, and T. L. Johnson, "Deadlock prevention and deadlock avoidance in flexible manufacturing systems using Petri net models," IEEE Transactions on Robotics & Automation, vol. 6, no. 6, pp. 713–723, 1990.

21. M. Uzam, Z. Li, and M. Zhou, "Identification and elimination of redundant control places in petri net based liveness enforcing supervisors of FMS," The International Journal of Advanced Manufacturing Technology, vol. 35, no. 1-2, pp. 150–168, 2007.

22. Z. Li and H. Hu, "On systematic methods to remove redundant monitors from liveness-enforcing net supervisors," Computers & Industrial Engineering, vol. 56, no. 1, pp. 53–62, 2009.

23. F. Garcia-Valles and J. Colom, "Implicit places in net systems," in Proceedings of the 8th International Workshop on Petri Nets and Performance Models, pp. 104–113, Zaragoza, Spain, 1999.

24. L. Recalde, E. Teruel, and M. Silva, "Improving the decision power of rank theorems," in Proceedings of the IEEE International Conference on Systems, Man, and Cybernetics, Computational Cybernetics and Simulation, pp. 3768–3773, October 1997.

25. Y. Chen, Z. Li, and M. Zhou, "Behaviorally optimal and structurally simple liveness-enforcing supervisors of flexible manufacturing systems," IEEE Transactions on Systems, Man, and Cybernetics Part A:Systems and Humans, vol. 42, no. 3, pp. 615–629, 2012.

26. L. Piroddi, R. Cordone, and I. Fumagalli, "Selective siphon control for deadlock prevention in Petri nets,"IEEE Transactions on Systems, Man, and Cybernetics Part A:Systems and Humans, vol. 38, no. 6, pp. 1337–1348, 2008.

27. L. Piroddi, R. Cordone, and I. Fumagalli, "Combined siphon and marking generation for deadlock prevention in Petri nets," IEEE Transactions on Systems, Man, and Cybernetics Part A: Systems and Humans, vol. 39, no. 3, pp. 650–661, 2009.

CHAPTER 13

OPTIMAL PRODUCTION PLANNING FOR MANUFACTURING SYSTEMS WITH INSTANTANEOUS STOCK-DEPENDENT DEMAND AND IMPERFECT YIELDS

Longfei He[1] Huangli Peng[1] Zhanwen Niu[1] Haili Lu[2] and Xiangli Xie[1]

[1]College of Management and Economics, Tianjin University, Tianjin 300072, China

[2]Department of Logistics Management, Wuhan University of Technology, Hubei 430063, China

ABSTRACT

We consider an EPL model like manufacturing system in presence of production imperfectness and stock-demand dependence simultaneously. During the production process, the system can evolve from in-control state into out-of-control state at any random time, after which the defective items will be generated likely causing quantity loss. Meanwhile, the market demand rate is instantaneously dependent on the timely holding inventory. The manufacturer has to determine his production run length and cycle time by taking into account possible imperfect production, stock-dependent demand, and inventory holding capacity bound. We empolder a model to capture this problem and develop computational algorithm to solve it. We further conduct numerical studies to validate our model and solving method. Sensitivity analyses are reported to show the effect of parameters on the system performance.

INTRODUCTION

Managing supply/production uncertainty and changeable demand is always one of the challenging operations since matching them is fairly hard due to instability occurring in two sides, respectively. The firm can suffer managerial issues resulting from both out-of-control production (e.g., Rosenblatt and Lee [1] and Sana [2]) and demand's instantaneous dependence on inventory (e.g., Gupta and Vrat [3] and Urban [4]).

This study involves two aspects of imperfect production and demand-stock dependence in the OM field. Imperfect production property means that defective items can be produced as the production system evolves over time. These imperfect goods will be reworked or deposed directly. As a consequence, the inventory/production planning is spoiled. On the other hand, the demand rate is instantaneously dependent on the timely inventory holding level of quality goods. That means the random production imperfection will eventually affect demand rate through defective items. Setting longer production cycle time to obtain higher inventory can obtain more demand rate resulting in more revenue. However, the longer the production cycle time is, the larger likeliness the imperfect production happens. The annual total cost comprises annual ordering cost and annual inventory holding cost. The production cycle time has opposite effects on these two kinds of cost components. Settling the optimum trade-off between the benefits and corresponding costs for the sake of optimizing the annual total profit is accordingly a crucial problem confronting practitioners.

Citation: Longfei He, Huangli Peng, Zhanwen Niu, Haili Lu, and Xiangli Xie, "Optimal Production Planning for Manufacturing Systems with Instantaneous Stock-Dependent Demand and Imperfect Yields," Mathematical Problems in Engineering, vol. 2015, Article ID 271902, 14 pages, 2015. doi:10.1155/2015/271902

Our study intends to address this problem by considering the production cycle time and pure production run time length as decision variables in the modeling of the manufacturing system existing in imperfect production and demand-inventory dependence. Specifically, we research the following problem: a manufacturer schedule his/her production annually by deciding each cycle time and run length, whereas the manufacturing system may switch from the in-control state to the out-of-control state after some time on the account of continuous run of the machine, which takes place at any random time during the production process, also as presented in Shah and Shukla [5]. Moreover, this is not immediately under observation and finally causes the quantity loss. The market demand is sensitive to the on-hand effective inventory excluding the defective goods.

Although we are concentrating on a new model of production planning contextualized in uncertain yields simultaneously confronting inventory-dependent demand, our modeling processes are to some extent inspired by preceding papers. We give an elucidation on the board literature related to our study below.

The existing relevant literature mainly includes two streams: (1) uncertain production management;(2) inventory/production planning with inventory-dependent demand and their extensions. The production uncertainty incurred by various factors like machine failure, staff changes can always cause the entire production system instable resulting in defective products has inspired extensive studies, such as Rosenblatt and Lee [1], Lee and Rosenblatt [6], Lee and Rosenblatt [7], and a variety of their succeeding papers like Liou et al. [8], Kim and Hong [9], Chung and Hou [10], Guo et al. [11], Hu et al. [12], and so on. Drawing on the celebrated EOQ/EMQ models, early pioneering researches study the production planning optimization considering defection inspecting, restoration cost dependence, detection delay, and maintenance, say, Rosenblatt and Lee [1], Lee and Rosenblatt [6], and Lee and Rosenblatt [7]. Their valuable investigations firstly pull the production planning research closer to the realistic practice where there usually exist in out-of-control states for the production process incurring a lot of associated management problems. Furthermore, inspection error types are proposed to be other factors having effects on production planning in an imperfect EMQ manufacturing system; see Liou et al. [8]. From a different perspective, Kim and Hong [9] link the distinction among three types of deteriorating processes to the optimal production run length seeking. Later, Chung and Hou [10] generalize former literature to allow the shortage in the imperfect system. After those static models in this line of research, Sana [2] addresses a production-inventory model with imperfect production and time-varying demand. Their study substantially extends proceeding researches into dynamic situations. In contrast, some recent researches turn to exploring imperfect production planning in the fuzzy environment, like Guo et al. [11] and Hu et al. [12]. Although the aforementioned literature has done intensive studies on uncertain production incorporating features incurred by itself, that existing research fails to consider the change of demand pattern during the production cycle time. However, they usually assume the demand characteristics are invariable over time, which is just a very special case of our model in this study.

Observation to practice has shown that the real market demand is very complex and it is not a fixed constant, as implied by Levin et al. [13], Silver and Peterson [14], and Gupta and Vrat [3]. Incited by their ground-breaking work, a variety of studies have incorporated stock-dependent rate component into diverse classical inventory models, for instance, by considering items deterioration in Mandal and Phaujdar [15], Datta and Pal [16], and Dixit and Shah [17], involving shelf-space allocation in Baker and Urban [18], allowing stock shortage in Datta and Pal [16], and combining delayed payment schemes in Soni and Shah [19], Min et al. [20], Shah et al. [21], Teng et al. [22], and Zhou et al. [23]. A considerable amount of literature has also addressed relevant problems, like Goh [24], Wee [25], Padmanabhan and Vrat [26], Ray and Chaudhuri [27], Sarker et al. [28], Giri and Chaudhuri [29], Yan and Cheng [30], Mandal and Maiti [31], Chung and Tsai [32], and so on. To learn more about the research of this kind, one is referred to a recent literature review by Urban [4]. Almost all literature mentioned above has only focused on the dependence between demand and inventory with the implicitly obvious a priori prerequisite of completely perfect production, which usually does not fit the practice well. On the contrary, in this paper, we alter that precondition and emphasize the imperfect production frequently occurring in real world.

To the best of our knowledge, few of existing studies have considered the production imperfectness and inventorydemand dependence simultaneously. In contrast, this is just what we concentrate on in the present study. We model the production and demand consumption processes with the aforementioned characteristics as a constrained nonlinear programming system combining different scenarios. Furthermore, our model is distinguished from existing relevant models in several ways: (1) we combine imperfect production and inventory-dependent demand together to explore the optimal production problem; (2) we use optimizing the annual total profit as the objective function instead of annual total cost, which substantially reflects the effect of demandinventory dependence on benefits; and (3) we incorporate the inventory capacity upper boundary into traditional models of imperfect production and inventory holding. We select and develop a suitable computational algorithm to solve the nonlinear programming model with detailed operational steps and procedures. Subsequently, we conduct numerical examples and parameters sensitivity analysis to examine the proposed computational algorithm so as to get some managerial insights (Figures 7, 9, 10, 11, and 13).

The rest of the paper is organized as follows. In Section 2, we sketch the problem features, notations, and its assumptions. In Section 3, we discuss the various scenarios arising from our studied system and derive the associated model. In Section 4, we describe a computational algorithm for solving the constructed model. In Section 5, numerical studies and associated analysis are executed to validate the algorithm proposed. Finally, concluding remarks are given to summarize this study in Section 6.

PROBLEM CHARACTERISTICS, NOTATIONS, AND ASSUMPTIONS

We study a manufacturing system described like a variant of economic production lot (EPL) model, which is distinct from classical EPL model by characterizing its specific production property and demand changing rules, respectively. On the production side, the system may randomly evolve into an out-of-control state in which a proportion of goods will be generated following some probability distribution. On the demand side, the impact of inventory level on demand is considered. The demand rate consists of the constant demand rate and the inventory level dependent demand rate addressed as

$$D(t) = a + bI(t), \quad 0 \le t \le T, \tag{1}$$

where $a > 0$ and $b > 0$.

So the dynamics where the stock $I(t)$ is depleted at time t can be presented as follows:

$$\frac{dI(t)}{dt} = -D(t) = -a - bI(t), \quad 0 \le t \le T. \tag{2}$$

The production process can be described like this: for scheduled production cycle time T, the manufacturer runs the production line from the beginning to time S, that is, $[0, S]$, where $0<S<T$. The production line is closed down in interval $(S, T]$. The manufacturing system may evolve into out-of-control situation incurring defective goods produced randomly. Denoting the production rate V and the production duration S yields aggregate volume VS during the cycle time. It may produce defects after producing continuous τ time and the system failure results in inferior goods in γ proportion of all production during $[\tau, S]$, where τ obeys an exponential distribution with parameter λ. During the whole cycle time, the inventory is being consumed by the demand at a rate varying according to the associated stock level at the same time.

The following notations will be used throughout the paper.

Parameters:

- $I(t)$: inventory level at any time t (units),
- $D(t)$: sales rate at time t with $D(t) = a + bI(t)$ where $a>0$ and $b>0$ (units),
- I_{max}: maximum system inventory capacity (units),
- V: productivity, that is, production rate (units per year),
- τ: random variable of the time moment when production system is changing to uncontrolled state, the probability of which obeys an exponential distribution with parameter λ (year),
- γ: the proportion of defective goods generated after production status changing time τ,
- k: fixed setup cost per order (\$/order),
- c: variable cost per unit product (\$/unit),
- p: the selling price per unit qualified product (\$/unit),
- σ: the salvage value per unit defective product (\$/unit),
- h: holding cost per unit product per unit time (\$/unit item/unit time).

Decision Variables:

- T: the whole production and supplying cycle time (year),
- S: total production duration, $0<S<T$ (year),
- AC: total annual costs, including inventory and ordering costs (\$/year),
- AC_I: annual inventory costs (\$/year),
- AC_o: annual ordering costs (\$/year),
- AR: total annual revenue, including sales income of qualified products and scrapped income of defective (\$/year),
- AR_s: sales income of qualified products (\$/year),
- AR_n: crapped income of defective (\$/year),
- AP: total annual profits, $AR - AC$ (\$/year).

To refine and reflect the essence of our problem, we need assumptions in our model as follows.

(A1)The lead time is zero and the time span is unlimited.

(A2)The production capacity is limited but out-of-stock is not allowed, and productivity is greater than steady section, that is $v > a$..

(A3) After producing continuous τ time, the system enters an uncontrolled state and defective rate of the product obtained is γ. After discovering the defective, it needs be scrap processing timely instead of entering the inventory system, where downtime τ obeys an exponential distribution with parameter λ.

(A4) Production process is independent in different cycles, which is in each cycle, and the duration τ of controlled state is independent.

(A5) During each production cycle, the beginning and ending inventory are both zero, that is, (0) = 0 and (T) = 0.

THE MODEL

For a given production cycle time T, there may be two cases in the whole production process because of the occurrence of imperfect production at time τ. One case is that breakdown time τ does not occur before the production stopping, that is, $\tau \in [S, +\infty]$, and products obtained in during the period are all qualified products. The other is that the breakdown time τ occurs during the production process, that is, $\tau \in [0, S]$, and the defective rate is γ $(0 < \gamma < 1)$ from time τ to time S. Assume (1) the upper bound of the stocking system capacity is I_{max}; (2) inventory levels at the beginning and the ending times of the production period are both zero, namely, $I(0) = 0$ and $I(T) = 0$. We plot the inventory level changing trajectory of qualified products in Figure 1 with two cases, where S_1 is the stopping time of the continuous production with system under control completely, and so is S_2 when system out of control occurs.

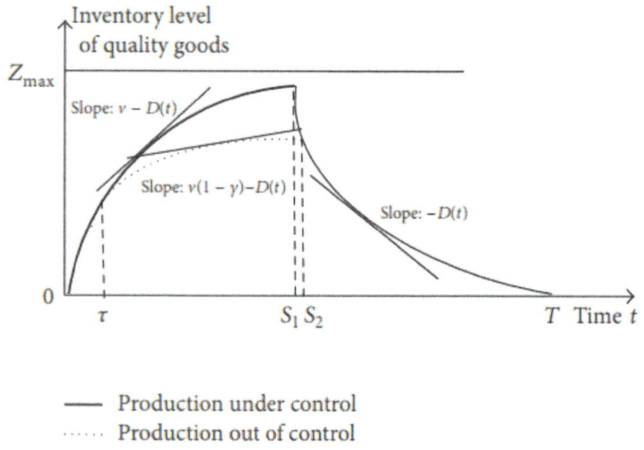

Figure 1: Comparison of effective inventory level changing trends in two scenarios.

System Under Control Completely before Production Stopping

Figure 2 shows the inventory changes in case 1.

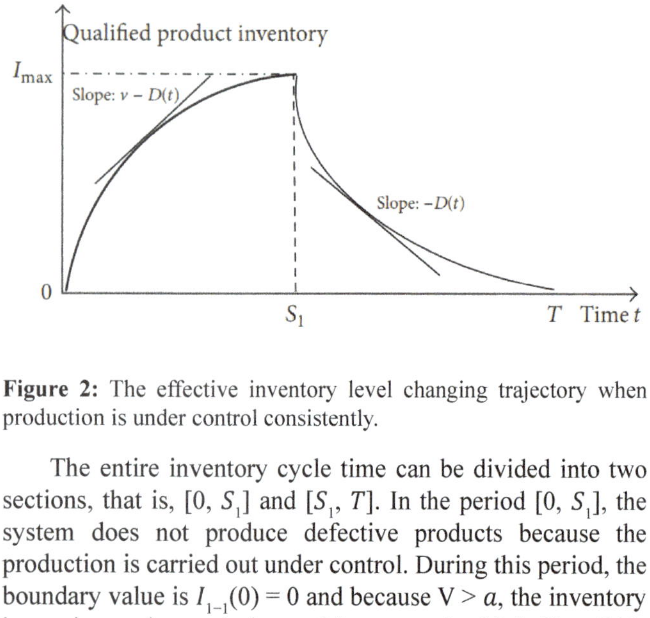

Figure 2: The effective inventory level changing trajectory when production is under control consistently.

The entire inventory cycle time can be divided into two sections, that is, $[0, S_1]$ and $[S_1, T]$. In the period $[0, S_1]$, the system does not produce defective products because the production is carried out under control. During this period, the boundary value is $I_{1-1}(0) = 0$ and because $V > a$, the inventory keeps increasing and slope of inventory $I_{1-1}(t)$ is $V - D(t)$, which is given by

$$\frac{dI_{1-1}(t)}{dt} = v - a - bI_{1-1}(t), \quad 0 \le t \le S_1. \tag{3}$$

In the period $[S_1, T]$, production ceases. The inventory consumption comes from the fixed demand and the needs with the inventory changing. During this period, the boundary value is $I_{1-2}(T) = 0$ and the slope of inventory $I_{1-2}(t)$ is $-D(t)$, which is given by

$$\frac{dI_{1-2}(t)}{dt} = -a - bI_{1-2}(t), \quad S_1 \le t \le T. \tag{4}$$

Solving differential equations (4) and differential equations (4), the inventory level in case 1 is $I_1(t)$:

$$I_1(t) = \begin{cases} \dfrac{(a-v)\left(e^{-bt}-1\right)}{b}, & 0 \le t \le S_1, \\ \dfrac{a\left[e^{b(T-t)}-1\right]}{b}, & S_1 < t \le T. \end{cases} \tag{5}$$

The inventory function $I_1(t)$ is a continuous function. So where $t=S_1$, there is $I_{1-1}(S_1)=I_{1-2}(S_1)$; that is,

$$\frac{(a-v)\left(e^{-bS_1}-1\right)}{b} = \frac{a\left[e^{b(T-S_1)}-1\right]}{b}. \tag{6}$$

Simplify (6) and then obtain the relationship between T and S_1, which is given by

$$S = \frac{1}{b}\ln\frac{ae^{bT} - a + v}{v}. \tag{7}$$

Total annual costs AC_1 include annual inventory costs ACI_1 and annual ordering cost AC_{o1}; that is, $AC_1 = ACI_1 + AC_{o1}$, in which, annual ordering cost AC_{o1} consists of two parts: one is the preparation costs for each production and the other is production costs which change with the production volume:

$$\begin{aligned} AC_{I1} &= \frac{h}{T}\left[\int_0^{S_1} I_{1-1}(t)\,dt + \int_{S_1}^{T} I_{1-2}(t)\,dt\right] \\ &= \frac{h}{Tb^2}\left[(v-a)e^{-bS_1} + vbS - v - abT + ae^{b(T-S_1)}\right], \end{aligned}$$

$$AC_o = \frac{1}{T}(cvS + k),$$

$$\begin{aligned} AC_1 = {} & \frac{h}{Tb^2}\left[(v-a)e^{-bS} + vbS - v - abT + ae^{b(T-S)}\right] \\ & + \frac{1}{T}(cvS + k). \end{aligned} \tag{8}$$

Total annual revenue AR_1: there are no defects in case 1; total annual revenue AR_1 equals ARs, the sales income of qualified products. Consider

$$AR_1 = AR_s = \frac{pvS}{T}. \tag{9}$$

Total annual profit AP_1 is total annual revenue minus total annual costs, that is, $AP_1 = AR_1 - AC_1$, which is given by

$$\begin{aligned} AP_1 = {} & \frac{1}{T}\left[(p-c)vS - k\right] \\ & - \frac{h}{Tb^2}\left[(v-a)e^{-bS} + vbS - v - abT + ae^{b(T-S)}\right] \end{aligned} \tag{10}$$

Substituting formula (7) into formula (10), then we can get total annual profits function which depends on T:

$$AP_1 = \frac{1}{T}\left[\left(p - c - \frac{h}{b}\right)\frac{v}{b}\ln\frac{ae^{bT} - a + v}{v} - k\right] + \frac{ah}{b}. \tag{11}$$

For any fixed inventory cycle T, the ending inventory of system equals zero, that is, $(T) = 0$. This paper will denote $(1/b)\ln((ae^{bT} - a + V)/V)$ as a critical point of τ, denoted by $\tau 0$. Where $\tau > (1/b)\ln((ae^{bT} - a + V)/V)$, the production processes will follow case 1. And where $\tau > (1/b)\ln((aebT - a + V)/V)$, there will be defective goods in $[\tau, S_2]$. For a detailed analysis, see case 2.

System Running Out of Control before the End of Production

To maintain the ending inventory $(T) = 0$, S_2 needs to change with τ. Figure 3 shows the qualified inventory changes.

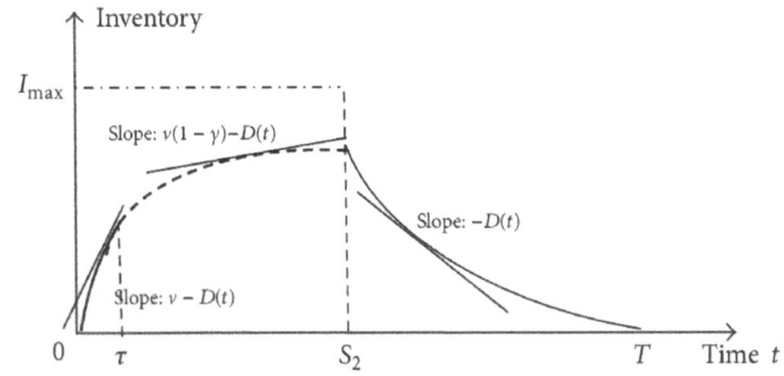

Figure 3: The trajectories of change of the qualified inventory level when the system is out of control.

In this case, the entire inventory cycle is divided into three sections, that is, $[0, \tau]$, $[\tau, S2]$, and $[S2, T]$. In the period $[0, \tau]$, the system will never produce defective products because the production is carried out completely under control. And during this period, the boundary value is $I_{2-1}(0) = 0$ and the slope of inventory $I_{2-1}(t)$ is $V - D(t)$; that is,

$$\frac{dI_{2-1}(t)}{dt} = v - a - bI_{2-1}(t), \quad 0 \le t \le \tau. \tag{12}$$

In the period $[\tau, S_2]$, production turns into an uncontrolled state. So there will be defective products and the defect rate is γ. Where $t=\tau$, we have $I_{2-1}(\tau) = I_{2-2}(\tau)$ and the slope of the qualified inventory $I_{2-2}(t)$ is $V(1 - \gamma) - D(t)$; that is,

$$\frac{dI_{2-2}(t)}{dt} = v(1-\gamma) - a - bI_{2-2}(t), \quad \tau \le t \le S_2. \tag{13}$$

In the period $[S_2, T]$, the inventory is consumed continuously until it reduces to zero.The boundary values $I_{2-3}(T) = 0$ and the slope of inventory $I_{2-3}(t)$ are $-D(t)$; that is,

$$\frac{dI_{2-3}(t)}{dt} = -a - bI_{2-3}(t), \quad S_2 \le t \le T. \tag{14}$$

Solve differential equations (12), (13), and (14) and then get the expression of inventory levels $I_2(t)$ in case 2:

$$I_2(t) = \begin{cases} \dfrac{(a-v)\left(e^{-bt} - 1\right)}{b}, & 0 \le t \le \tau, \\ \dfrac{v\gamma e^{b\tau} - (v-a)}{b} e^{-bt} + \dfrac{v(1-\gamma) - a}{b}, & \tau \le t \le S_2, \\ \dfrac{a\left[e^{b(T-t)} - 1\right]}{b}, & S_2 < t \le T. \end{cases} \tag{15}$$

Obviously, $I_2(t)$ is also a piecewise function. It is easy to prove, where $\tau \in [0, (1/b)\ln((V - a)/V\gamma)]$, in the period $\tau \le t \le S_2$, the inventory $I_{2-2}(t)$ is an increasing function. And where $\tau \in ((1/b)\ln((V - a)/V\gamma), S_2]$, in the period $\tau \le t \le S_2$, the inventory $I_{2-2}(t)$ is a decreasing function. For details see appendix, Proof 1.

Due to the continuity of the inventory function, where $t=S_2$, we have $I_{2-2}(S_2)=I_{2-3}(S_2)$; that is,

$$-\frac{a - v(1-\gamma)}{b} + \frac{v\gamma - (v-a)e^{-b\tau}}{be^{-b\tau}} e^{-bS_2} = \frac{a\left[e^{b(T-S_2)} - 1\right]}{b}. \tag{16}$$

Simplify (16) and then obtain the relationship among T,, and S_2:

$$S_2 = \frac{1}{b} \ln \frac{ae^{bT} + v - a - v\gamma e^{b\tau}}{v(1-\gamma)}. \tag{17}$$

In case 2, total annual costs AC_2 also consist of annual inventory costs AC_{I2} and annual ordering costs AC_{o2}, that is $AC_2 = AC_{I2} + AC_{o2}$. Consider

$$AC_{I2} = \frac{h}{T}\left[\int_0^{\tau} I_{2-1}(t)\,dt + \int_{\tau}^{S_2} I_{2-2}(t)\,dt + \int_{S_2}^{T} I_{2-3}(t)\,dt\right]$$

$$= \frac{h}{Tb^2}\left\{-v + bvS_2 - bv\gamma(S_2 - \tau) + v\gamma - v\gamma e^{-b(S_2-\tau)} + (v-a)e^{-bS_2} + a\left[e^{b(T-S_2)} - bT\right]\right\},$$

$$AC_{o2} = \frac{1}{T}(cvS_2 + k),$$

$$AC_2 = \frac{h}{Tb^2}\begin{Bmatrix} -v + bvS_2 - bv\gamma(S_2 - \tau) + v\gamma - v\gamma e^{-b(S_2-\tau)} \\ + (v-a)e^{-bS_2} + a\left[e^{b(T-S_2)} - bT\right] \end{Bmatrix} + \frac{1}{T}(cvS_2 + k). \tag{18}$$

In case 2, defects occur during production, so total annual revenue AR_2 consists of two sections: sales income of qualified products ARS_2 and scrapped income of defective annual revenue ARN_2:

$$AR_{S2} = \frac{p\left[vS_2 - \gamma v(S_2 - \tau)\right]}{T},$$

$$AR_{N2} = \frac{\sigma\gamma v(S_2 - \tau)}{T},$$

$$AR_2 = AR_{S2} + AR_{N2} = \frac{pvS_2 + (\sigma - p)\gamma v(S_2 - \tau)}{T}. \tag{19}$$

Total annual profits are equal to the annual revenue minus annual costs (Figure 5), that is, $AP_2 = AR_2 - AC_2$, which is given by

$$AP_2 = \frac{1}{T}\left[(p-c)vS_2 + (\sigma - p - \omega)\gamma v(S_2 - \tau) - k\right] - \frac{h}{Tb^2}\left\{-v + bvS_2 - bv\gamma(S_2 - \tau) + v\gamma - v\gamma e^{-b(S_2-\tau)} + (v-a)e^{-bS_2} + a\left[e^{b(T-S_2)} - bT\right]\right\}. \tag{20}$$

Substituting (17) into (20), then simplify to

$$AP_2 = \frac{1}{T}\left[\begin{array}{l}\left(p-c-\frac{h}{b}\right)\frac{v}{b}\ln\frac{ae^{bT}+v-a-v\gamma e^{b\tau}}{v(1-\gamma)} \\ +\left(\sigma-p+\frac{h}{b}\right) \\ \cdot\gamma v\left(\frac{1}{b}\ln\frac{ae^{bT}+v-a-v\gamma e^{b\tau}}{v(1-\gamma)}-\tau\right)-k\end{array}\right] + \frac{ha}{b}. \quad (21)$$

Optimizing Annual Total Profit

Objective function: expected annual profits .

In this paper, the inventory model is divided into case 1 and case 2, according to whether there will be defects in the production. This paper assumes that τ, the moment of appearing defect in $(0, +\infty)$, obeys an exponential distribution with parameter λ. Therefore the expected annual profits:

$$\begin{aligned}AP &= E[AP_1] + E[AP_2] \\ &= \int_{\tau_0}^{+\infty} AP_1\lambda e^{-\lambda\tau}d\tau + \int_0^{\tau_0} AP_2\lambda e^{-\lambda\tau}d\tau \\ &= \frac{1}{T}v\left[\left(p-c-\frac{h}{b}\right)+\gamma\left(\sigma-p+\frac{h}{b}\right)\right] \\ &\quad\cdot\int_0^{(1/b)\ln((ae^{bT}-a+v)/v)}\lambda e^{-\lambda\tau}\frac{1}{b} \\ &\quad\cdot\ln\frac{ae^{bT}+v-a-v\gamma e^{b\tau}}{v(1-\gamma)}d\tau \\ &\quad+\frac{1}{T}\left\{\gamma v\left(\sigma-p+\frac{h}{b}\right)\right\} \\ &\quad\cdot\left[\left(\frac{1}{\lambda}+\frac{1}{b}\ln\frac{ae^{bT}-a+v}{v}\right)\left(\frac{ae^{bT}-a+v}{v}\right)^{-\lambda/b}-\frac{1}{\lambda}\right] \\ &\quad+\frac{1}{T}\left[\left(p-c-\frac{h}{b}\right)\frac{v}{b}\ln\frac{ae^{bT}-a+v}{v}\right] \\ &\quad\cdot\left(\frac{ae^{bT}-a+v}{v}\right)^{-\lambda/b}-\frac{k}{T}+\frac{ha}{b},\end{aligned} \quad (22)$$

under the inventory capacity constraints.

Before discussing the inventory constraints, we should compare S_1 in case 1 with S_2 in case 2.

$$S = \begin{cases} S_1 = \frac{1}{b}\ln\frac{ae^{bT}-a+v}{v}, & \tau > \frac{1}{b}\cdot\ln\frac{ae^{bT}-a+v}{v}; \\ S_2 = \frac{1}{b}\ln\frac{ae^{bT}+v-a-v\gamma e^{b\tau}}{v(1-\gamma)}, & 0 \le \tau \le \frac{1}{b}\cdot\ln\frac{ae^{bT}-a+v}{v}. \end{cases} \quad (23)$$

It is easy to prove $S_2 \ge S_1$, when they have the same T. For details see appendix, Proof 2.

In case 1, where $t=S_1$, the inventory reaches the maximum, which should be less than I_{max}, that is, $[e^{(T-S1)} - 1]/b \le I_{max}$. In case 2, if $\tau \in [0, (1/b)\ln((V-a)/V\gamma)]$, then where $t=S2$, the inventory reaches the maximum. And this time, the inventory should be less than Imax, that is, $[e^{(T-S_2)} - 1]/b \le I_{max}$. If $\tau \in [(1/b)\ln((V-a)/V\gamma), S_2]$, then where $t=\tau$, the inventory reaches its maximum. And the inventory should be less than I_{max}, that is, $(a-V)(e^{-b\tau} - 1)/b \le I_{max}$. Then, the inventory constraints are

Then, the inventory constraints are

$$\text{s.t.} \quad \frac{a\left[e^{b(T-S_1)}-1\right]}{b} \le I_{max}$$

$$\frac{a\left[e^{b(T-S_2)}-1\right]}{b} \le I_{max}$$

$$\frac{(a-v)\left(e^{-b\tau}-1\right)}{b} \le I_{max}$$

$$\Longrightarrow \max\left(\frac{a\left[e^{b(T-S_1)}-1\right]}{b}, \frac{a\left[e^{b(T-S_1)}-1\right]}{b}, \frac{(a-v)\left(e^{-b\tau}-1\right)}{b}\right) \le I_{max}. \quad (24)$$

We can prove that

$$\max\left(\frac{a\left[e^{b(T-S_1)}-1\right]}{b}, \frac{a\left[e^{b(T-S_1)}-1\right]}{b}, \frac{(a-v)\left(e^{-b\tau}-1\right)}{b}\right) = \frac{a\left[e^{b(T-S_1)}-1\right]}{b}. \quad ... (25)$$

For details see appendix, Proof 3.

So the inventory constraints convert into

$$\frac{a\left[e^{b(T-S_1)}-1\right]}{b} \le I_{max}. \quad (26)$$

Substituting (7) into (26), we have

$$\frac{a\left[e^{b(T-S_1)}-1\right]}{b} = \frac{a(v-a)\left(e^{bT}-1\right)}{b\left(ae^{bT}-a+v\right)} \le I_{max}. \quad (27)$$

Constraint formula (27) can be changed as follows:

$$\frac{a(v-a)\left(e^{bT}-1\right)}{b\left(ae^{bT}-a+v\right)} \le I_{max} \Longrightarrow a\left(v-a-bI_{max}\right)e^{bT} \le (v-a)\left(a+bI_{max}\right). \quad (28)$$

Since $V > a$, the right side of (28) is a positive constant. And the left side can be divided into two cases.

(1) If $V - a - bI_{max} \le 0$, we have $e^{bT} > 0$. So the left side of (28) is negative and inequality (28) is always true. Then the inventory capacity constraints have no practical significance.

(2) If $V - a - bI_{max} > 0$, obviously the left side of (28) $(T) = (V - a - bI_{max})e^{bT}$ is a monotonic increasing function of T. To satisfy the constraints, there should be

$$0 < T \le T_{max} = \frac{1}{b}\ln\frac{(v-a)\left(a+bI_{max}\right)}{a\left(v-a-bI_{max}\right)}. \quad (29)$$

The inventory capacity constraints are successfully converted into the scope limitation of the decision variable values, which becomes simpler and more intuitive. In summary, the expected annual profits maximization model changes into

$$\begin{aligned}
\max_T \; & AP[T] \\
&= \frac{1}{T}v\left[\left(p-c-\frac{h}{b}\right)+\gamma\left(\sigma-p+\frac{h}{b}\right)\right] \\
&\quad\cdot\int_0^{(1/b)\ln((ae^{bT}-a+v)/v)} \lambda e^{-\lambda\tau} \\
&\quad\cdot\frac{1}{b}\ln\left(\left(ae^{bT}+v-a-v\gamma e^{b\tau}\right)\cdot\left(v(1-\gamma)\right)^{-1}\right)d\tau \\
&\quad+\frac{1}{T}\left\{\gamma v\left(\sigma-p+\frac{h}{b}\right)\right\} \\
&\quad\cdot\left[\left(\frac{1}{\lambda}+\frac{1}{b}\ln\frac{ae^{bT}-a+v}{v}\right)\cdot\left(\frac{ae^{bT}-a+v}{v}\right)^{-\lambda/b}-\frac{1}{\lambda}\right] \\
&\quad+\frac{1}{T}\left[\left(p-c-\frac{h}{b}\right)\frac{v}{b}\ln\frac{ae^{bT}-a+v}{v}\right] \\
&\quad\cdot\left(\frac{ae^{bT}-a+v}{v}\right)^{-\lambda/b}-\frac{k}{T}+\frac{ha}{b} \\
\text{s.t.}\quad & T_{lb}<T\le\min\left[\frac{1}{b}\ln\frac{(v-a)(a+bI_{\max})}{a(v-a-bI_{\max})},T_{ub}\right] \\
&\qquad\text{if } v-a-bI_{\max}>0 \\
& T_{lb}<T<T_{ub},\quad \text{if } v-a-bI_{\max}\le 0.
\end{aligned} \tag{30}$$

Model (30) can be regarded as a unary nonlinear programming model. Thus the problem can be transformed into unary nonlinear programming problem.

Formula (22) shows that the profit function contains the integral of transcendental functions, and there is no analytic expression. Thus it is difficult to use derivatives and gradient information to prove the concavity of the profit function, and the difficulty and complexity of proving it are far beyond the traditional inventory model. However, is a continuous function. According to the maximum and minimum theorem of continuous function nature theorem on a closed interval, we can determine that the profit function has a maximum within the specified interval. Based on this, this paper designs the following optimization algorithm.

COMPUTATIONAL ALGORITHM

There are a variety of commonplace methods to solve nonlinear programming models, for instance, Golden Section, Fibonacci, Interpolation, Bisection, Steepest Descent, Newton, quasi-Newton, and Conjugate Gradient and so on. Those traditional algorithms are simple with fast convergence. However, they have relatively high requirements on initial value and functional structure. For more general computational application, scholars have proposed a variety of intelligent algorithms, such as Genetic Algorithms, Simulated Annealing Algorithm, Ant Colony Algorithm, and Chaotic Mixing Algorithm with their own advantages and broad applications, respectively. To overcome the defects of a single algorithm, integrating several algorithms in solving one problem has been prevailing in optimization field through mutually advantageous complementation among them.

Artificial fish swarm algorithm (AFSA) is one kind of integrated algorithms as above mentioned. Li [33] is among the early batch of studies to propose AFSA, which is an intelligent optimization algorithm based on fish behavior in nature. Peng [34] also demonstrates this approach in detail and compares its advantage to other algorithms. To learn more about the growing path of AFSA, Neshat et al. [35] is a good reference with excellent comprehensive literature review. Imitating and realizing three kinds of individual fish's behavior, that is, feeding and rear-ends, clusters underlie the working basis of the algorithm optimality seeking with the associated system parameters design including fish scale size N, the step, the congestion factor delta, and the perceived distance visual, Li et al. [36] and Li et al. [37]. Although the initial AFSA has lots of obvious virtues, there still exists improving space to abate its some weakness for application, which has been pointed out in detail and also improved to apply in diverse settings, say, Wang et al. [38], Li et al. [37], Huang and Zhou [39], Chen et al. [40], Cheng et al. [41], Peng [42], and Tsai and Lin [43].

Drawing on the fact that the concavity of the profit function cannot be clearly identified in this paper, we select the artificial fish school algorithm to solve our problem while we make a substantially specific problem-featured improvement for this algorithm inspired by existing literature aforementioned. Here we develop an improved artificial fish school algorithm by combining the adapting parameters strategy and tabu search heuristics simultaneously to enhance the local search ability and convergence rate of initial ASFA. However, our focus in this study is mainly on formulating, modeling, and analysing the focal managerial problem rather than the algorithm improvement. This proposed algorithm is just served to solve the nonlinear problem and will not be compared with other algorithms. We need to emphasize the fact that comprehensively integrating more extensive computational methods and heuristics together finally distinguishes our algorithm design from that in existing researches.

The Parameter Adapting Strategy

Strategy of Changing Step Length. A fixed step length is usually adopted in most of literature for searching optimizer in an artificial algorithm. The step length will affect the converging rate, speed, and accuracy of the searching process. The longer step length will incur higher converging rate and searching speed, but lower accuracy. On the other hand, a small step size will help local search with a slower global converging rate with higher accuracy. The varying step length will be much better in balancing the converging rate and searching accuracy. In this paper, the step is set as (31), a progressively decreasing function as addressed in Huang and Zhou [39] and Chen et al. [40]:

$$\text{step}_{\text{gen}+1} = \left(\frac{1}{2} + \frac{1}{\text{gen}}\right) \times \text{step}_{\text{gen}}. \tag{31}$$

Strategy of Changing the Congestion Factor Delta. In order to speed up the searching rate, avoid falling into the plight of local optimum and premature and suppress the congestion factor. In the iterative process, we adopt the congestion attenuation function set in (32) as presented in detail in Zheng [44] to progressively reduce congestion factor to enhance local search convergence rate:

$$\text{delta}_{\text{gen}+1} = \varepsilon \times \text{delta}_{\text{gen}}, \quad \varepsilon \in (0, 1). \tag{32}$$

Strategy of Changing the Perceived Distance Visual. According to Chen et al. [40], we set a perceived distance motion function (33) to make the perceived distance decreases with the iterative number increasing so as to improve the local intensity and solution accuracy:

$$\text{visual}_{\text{gen}+1} = \left(1 - \frac{\text{gen}}{\text{maxgen}}\right) \times \text{visual}_{\text{gen}}. \tag{33}$$

Tabu Search Heuristics

To solve the continuous problem in the present study with infinite number of neighbors, using the tabu method of discrete problems cannot make the algorithm effectively searching the global optimal solution. Thus we need to employ the neighborhood tabu for solving. Li et al. [37] elaborate the neighborhood tabu algorithm and propose a two-dimensional neighborhood concept for tabu algorithm. Inspired by their research and considering the specific situation in our current problem, we give out a onedimensional neighborhood concept applied for our problem solving. Divide current feasible solution interval $[T_{lb}, T_{ub}]$ into K fields falling within a circle with center $(T_{ub} + T_{lb})/2$ and the radius r as shown in Figure 4.

In the gen-th interaction, the current position of ith fish is T_i, and another position is T_j. T_j can be the optimal point in the view of the ith fish and also can be the center point or another point. The optimization researching time is denoted as *tryno*; the optimization coefficient θ is a randomly generated number between (0, 1) by systems.

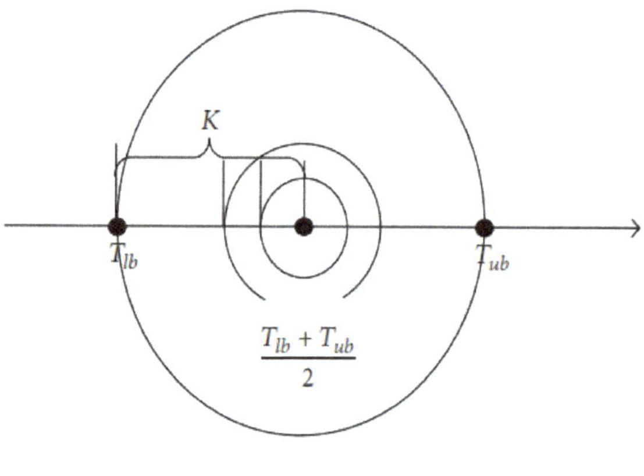

Figure 4: One-dimensional neighborhood concept.

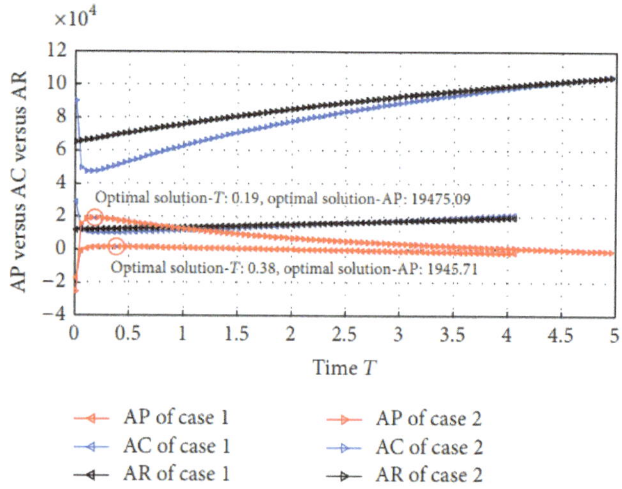

Figure 5: The total annual costs, the total annual revenue, and the total annual profits changes in the two examples.

The next position of ith fish is determined by the following formula:

$$T_{i|\text{next}} = T_i + \theta \times (T_j - T_i). \tag{34}$$

The process of tabu search is as follows.

Step 1. T_i is the current state of artificial fish, n is the counter, and n=1.

Step 2. We get a new point $T_{i|\text{next}}$ from (34) and then calculate $AP_{i|\text{next}}$, and n=n+1.

Step 3. Compare the size of AP_i with $AP_{i|\text{next}}$. If $AP_i < AP_{i|\text{next}}$, turn to Step 4; if not, go to Step 5.

Step 4. Estimate $T_{i|\text{next}}$ belonging to which k neighborhood and determine whether the kth neighborhood exists in the tabu list. If it does not exist, the optimization is successful, and then insert the kth neighborhood into the tabu list. If the tabu list is full, update the list according to the FIFO principle and then end the optimization process. However, if it does exist, go to Step 5.

Step 5. Estimate if n is greater than the maximum attempts time *tryno*; if so, the optimization process ends in failure. Otherwise, go to Step 2.

Behavior of a Single Fish

Applying the aforementioned established neighborhood tabu search process to our problem, we get the individual behavior of a single fish including foraging, clustering, rear-end, and searching for historical best fish. Here we introduce these four behaviors.

(1) The position of a single fish in current fish swarm is T_i. The fish randomly chooses a position T_j in their sensing range, $T_j = T_i + rand \times$ visualgen, and then calculate the tabu optimization. If we get the optimization successfully, we take $Ti|$next as the next mobile location of the current fish. If not, randomly move to another step, that is, $Ti|\text{next} = T_i + rand \times$ stepgen.

(2) Clustering. A single fish position in current fish swarm is T_i. In this fish's vision, the number of other fish is n_f and the optimal position of the fish is T_j. Compare

the size of AP_j/n_f with delta × AP_i. When AP_j/n_f > delta × AP_i, calculate the tabu optimization. If we get the optimization successfully, denote $Ti|\text{next}$ as the next mobile location of the current fish. If not, turn to foraging behavior.

(3) Rear-End. A single fish position in current fish swarm is T_i. In this fish's vision, the number of other fish is n_f and the geometric center position of the fish is T_j. Compare the size of AP_j/n_f with delta × AP_i. When AP_j/n_f > delta × AP_i, calculate the tabu optimization. If we get the optimization successfully, denote $T_i|_{\text{next}}$ as the next mobile location of the current fish. If not, turn to foraging behavior.

(4) A single fish position in current fish swarm is T_i. Learning from bulletin board, we have the historical best fish's position T_j and then calculate the tabu optimization. If we get the optimization successfully, denote $T_{i|\text{next}}$ as the next mobile location of the current fish. If not, turn to foraging behavior.

The Overall Improved AFSA Integrating Adapting Parameters and Tabu Search

Step 1. Set the population size N, the moving step $step$, the congestion factor delta, the perceived distance $Visual$, the maximum number of attempts $trynumber$, and the length of tabu list $Ntabu$. Make discrete solution space into K neighborhood and limit the maximum number of iterations $maxgen$, making gen = 1.

Step 2 Initialize the fish swarm (T_1, T_2, T_3,...,) in solution space. Calculate the objective function value of current fish swarm (AP_1, AP_2, AP_3, . . . , AP_N), the optimal fish T_j in the current fish swarm, and the historical best fish Tbest. Record the information of historical best fish on the bulletin board.

Step 3. Each fish in the fish swarm gets three best $T_{i|\text{next}}$ through the three best behaviors: clustering, rear-end, and searching for historical best fish. Select a larger target value of $T_{i|\text{next}}$ and update their position. After comparing with the historical best fish, update the information on bulletin board.

Step 4. If gen ≥ maxgen, then stop the algorithm. If not, make stepgen+1 = ((1/2)+(1/gen))×step$_{\text{gen}}$, delta$_{\text{gen+1}}$ = ε×delta$_{\text{gen}}$, and visual$_{\text{gen+1}}$ = (1 − (gen/max$_{\text{gen}}$)) × visual$_{\text{gen}}$. If gen = $_{\text{gen + 1}}$, then turn to Step 3.

NUMERICAL STUDIES AND ANALYSIS

Numerical Examples

To make numerical examples, we set the range of decision variable $T \in (0, 5]$.

Example 1. The inventory is limited; the values of each parameter are as follows:

$$[a, b, h, k, p, c, \sigma, \lambda, \gamma, v, I_{\max}] = [300, 0.3, 8, 200, 40, 30, 2, 0.1, 12, 2000, 1500]. \quad (35)$$

In this example, $V - a - bI_{\max} > 0$, so the limit of inventory maximum capacity is working. According to (29), we calculate the range of maximum value of the decision variable T and compare the result with 5. Then select a smaller one to be the maximum value of the decision variable T. Using the improved artificial fish swarm algorithm, we calculate the optimal solution of this example [T, AP] = [0.38, 1945.71]. The results and changing trend are shown in Figure 5.

Example 2. The inventory cannot reach the maximum; the values of each parameter are as follows:

$$[a, b, h, k, p, c, \sigma, \lambda, \gamma, v, I_{\max}] = [1000, 0.5, 15, 500, 65, 45, 4, 0.3, 6, 3000, 4500]. \quad (36)$$

In this example, $V - a - bI_{\max} \le 0$, so the limit of inventory capacity constraints do not need to be considered. The range of decision variable T is (0, 5]. Within this range, we calculate the optimal solution of this example by iterative optimization, which is [T, AP] = [0.19, 19475.09].

Sensitivity Analysis

The sensitivity analysis on Example 1 follows.

(1) The Impact of Inventory-Dependence Coefficient b on the Optimal Solution. Assign b with different value and observe the changing tendency of the profit function and the optimization.

As can be seen from Figure 6, with the increase of b, the optimal production cycle T^* and the maximum annual profits AP also gradually increase. An increasing b implies that the impact of current inventory on the demand is getting greater. The higher the current inventory level is, the greater the demand becomes. Within the inventory capacity, decision maker will try to increase the inventory in order to reach higher sales income. So the maximum annual profit AP gradually increases. But to run out a higher level of inventory, it will require a longer cycle. Therefore the optimal production cycle T^* will show an increasing trend.

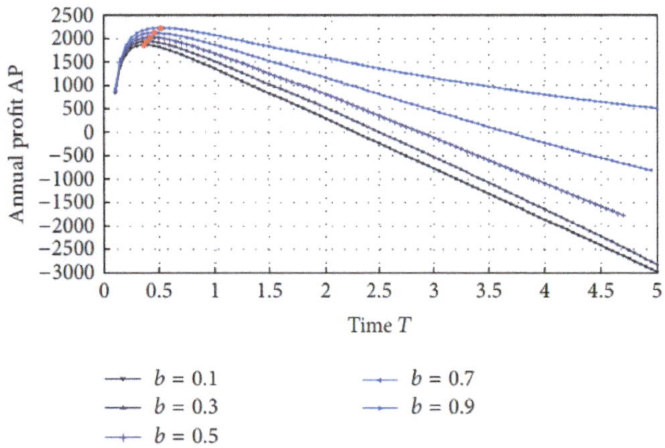

Figure 6: The sensitivity analysis of b.

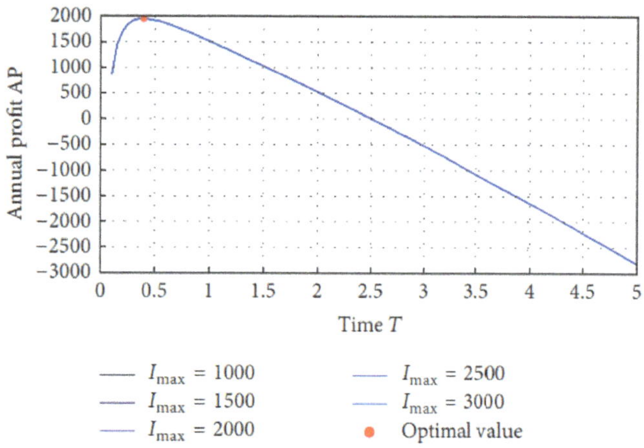

Figure 7: The sensitivity analysis of i_{max}.

With the change of b, range of decision variable T constantly changes. When b is relatively small and $b < (V - a)/I_{max}$, the maximum value of T is the min one of $(1/b)\ln((V- a)(a+bI_{max})/a(V- a-bI_{max}))$ and 5. $(1/b)\ln((V- a)(a + bI_{max})/a(V - a - bI_{max}))$ is increasing as b increases. When $(1/b)\ln((V - a)(a + bI_{max})/a(V - a - bI_{max})) \leq 5$ is satisfied, the maximum value of T also increases, which is reflected on the length changes of the corresponding profit curve. For example, the profit curve is longer with $b = 0.7$ than $b = 0.5$. When b increases to $b \geq (V - a)/I_{max}$ and the maximum value of T is the preset 5, it is not difficult to find out that the upper limit of T becomes 5 with $b = 0.9$.

(2) The Impact of Inventory Holding Space Bound I_{max} on the Optimal Solution. In Figure 7, we plot the profit-time curves varying as the inventory holding capacity changes. The results show that those curves are almost overlapped, which implies the high capacity has few influence on the optimal solution. It can also be seen from Figure 8 that when Imax changes, the optimal solution will remain unchanged, based on the premise that the change of I_{max} still ensures the optimal solution T^* within the optimization search scope of T; otherwise, the optimal solution $T*$ and the maximum annual profit AP^* are both likely to change.

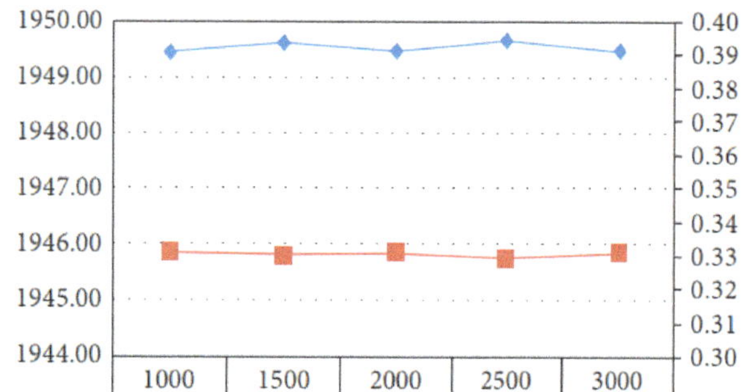

	1000	1500	2000	2500	3000
AP*	1945.84	1945.82	1945.84	1945.76	1945.84
T*	0.39	0.39	0.39	0.39	0.39

Figure 8: The changing trend of T^* and AP^* when Imax changes.

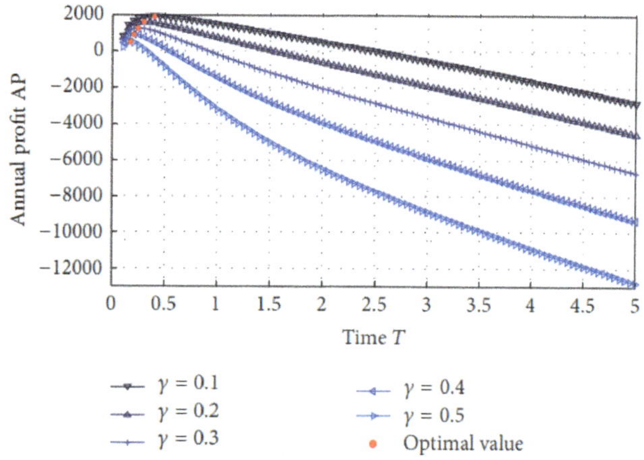

Figure 9: The sensitivity analysis of γ.

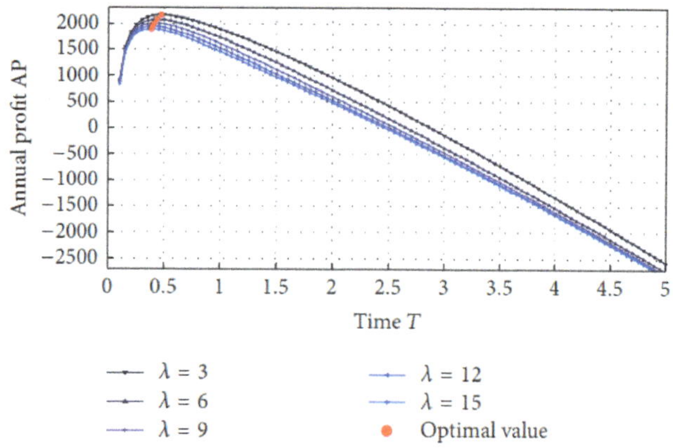

Figure 10: The sensitivity analysis of λ.

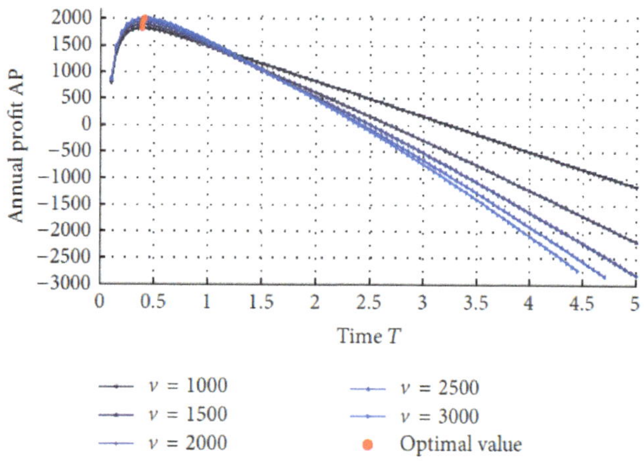

Figure 11: The sensitivity analysis of V..

(3) The Impact of Defective Rate γ on the Optimal Solution. When there is defective production, it is essentially a loss. With the same production cycle T, the annual profit will decrease. When defective rate increases, decision makers will reduce the probability of uncontrolled state. The probability of downtime is exponentially distributed; the shorter the production time is, the smaller the probability of downtime will

be. So when γ increases, the optimal production cycle T decreases.

(4) The Impact of the Parameter of Exponential Distribution λ on the Optimal Solution. λ is the parameter of the exponential distribution, which the system downtime τ obeys. $1/\lambda$ is the expected value of downtime. With an increasing λ, $1/\lambda$ gets smaller which means that the system fault occurs quicker. If we keep the fixed production cycle T in this condition, the uncontrolled production state will last longer. There will be more defects in the output and the total profit will decrease. Thus the decision maker will decrease the production cycle T to reduce the frequency of system failure and the defective output rate. However, even if λ increases continuously, T will not continue reducing. Because a decreasing T means the startup time of this machine in one year increases, the fixed costs increase. Affected by the fixed costs, T will stop decreasing and increase instead.

(5) The Impact of Productivity V on the Optimal Solution. An increasing V means the improvement of production efficiency and acceleration of inventory accumulation. So it can accumulate enough inventory in a shorter production time, and the average system inventory level will improve. As the demand is associated with the current inventory, the demand will also increase. When the value of T is small, the system stays at a relatively low inventory level. With an increasing demand, the increase in sales revenue is greater than the increase in inventory costs, so the annual profit will increase with V (Figure 12). When the value of T is larger, the inventory level of system is relatively higher. At this point, with the climbing demand, the increase in sales revenue is not enough to offset the increase in the inventory. So the annual profit will decrease when V increases.

When V increases, the maximum annual profit AP^* continues increasing, and the optimal production cycle T^* decreases firstly and then increases. For a fixed T, when the value of V is small, the system stays at a relatively low inventory level. With an increasing demand, the increase in sales revenue is limited and not enough to offset the increase in the inventory. So when V increases, the optimal production cycle T^* needs to be reduced in order to reduce the inventory costs.

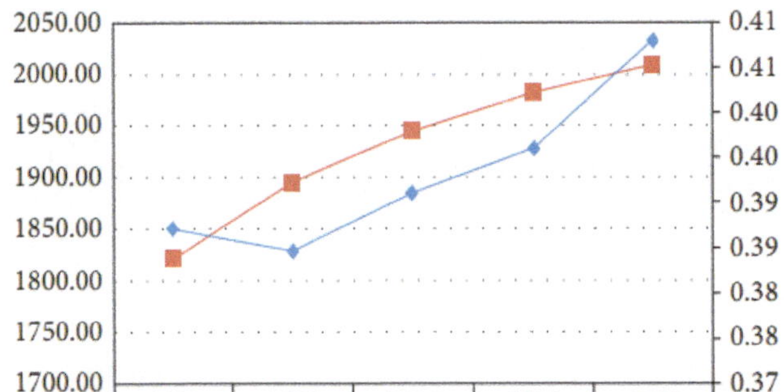

	1000	1500	2000	2500	3000
AP*	1822.29	1895.16	1945.84	1982.24	2009.59
T*	0.39	0.38	0.39	0.40	0.41

Figure 12: The changing trend of $T^*$and AP^* when V changes

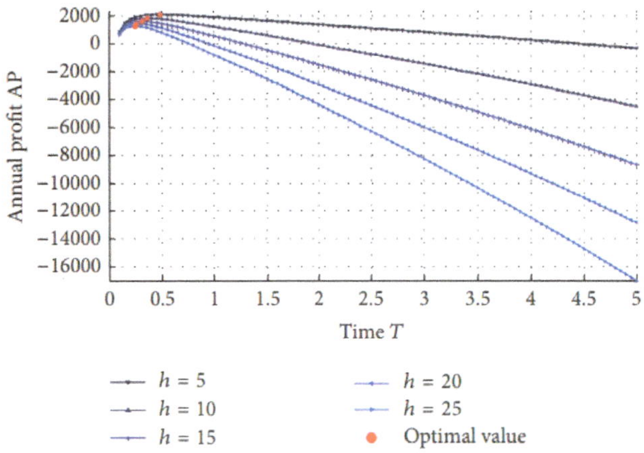

Figure 13: The sensitivity analysis of h.

When the value of V is larger, the inventory level of system is higher and the demand increases considerably, resulting in the fact that the increase in sales revenue is greater than the increase in inventory costs. So when V continues climbing, the optimal production cycle T^* will appear as an upward trend.

(6) The Impact of Inventory Holding Cost h on the Optimal Solution. When h increases, the inventory costs will aggrandize with it and the total annual profit will decrease. So decision makers will reduce the production cycle T to reduce the annual average inventory level in order to cut down the inventory costs. When h is larger, the annual profit curve will drop more rapidly with an increasing T.

CONCLUSION

This study has investigated production optimization problem arising from a manufacturing system with likely imperfect production and instantaneously inventory-dependent market demand. We have constructed constrained nonlinear programming models to depict this problem by incorporating the randomness of production imperfectness, inventory-demand dependence, and stock holding space bound. For seeking the optimal annual total profit, an improved artificial fish swarm algorithm combined with tabu search was devised to solve the model with numerical studies and sensitivity analysis. This research adds to our understanding of managing the manufacturing system with new characteristics as discussed in the present paper. It contributes to providing the method and strategy of determining optimum production cycle time to obtain best system performance. This study also shows us the necessity and power of integrate operations and sales. However, the research is limited in several ways. For instance, we only confine our study to a single manufacturer, which can be extended to a supply chain system. Furthermore, we consider the demand is only inventory-dependent with deterministic dependence sensitivity parameter. Other random factors can be added to the demand function in the future research.

APPENDIX

Proof 1. Consider

$$I_{2-2} = \frac{v\gamma e^{b\tau} - (v-a)}{b}e^{-bt} + \frac{v(1-\gamma)-a}{b}, \quad \tau \le t \le S_2. \tag{a1}$$

Seeking the first derivative $dI2-2/dt$ = (V $-a-$ V$\gamma e^{b\tau}$)$^{-bt}$, e^{-bt} is constantly greater than zero. When $\tau \in [0, (1/b)$ ln((V$-$ a)/Vγ)], V $-a-$ V$\gamma e^{b\tau} \ge 0$, and $dI_{2-2}/dt \ge 0$, $I_{2-2}(t)$ is an increasing function and the inventory increases in period of $\tau \le t \le S_2$; when $\tau \in ((1/b)\ln((V - a)/V\gamma), S_2]$,Proof 3. The inventory constraints are $-a-$ V$\gamma e^{b\tau} < 0$, and $dI_{2-2}/dt < 0$, $I_{2-2}(t)$ is a decreasing function and the inventory decreases in period of $\tau \le t \le S2$.

Proof 2. There is

$$S = \begin{cases} S_1 = \frac{1}{b}\ln\frac{ae^{bT}-a+v}{v}, & \tau > \frac{1}{b}\ln\frac{ae^{bT}-a+v}{v}; \\ S_2 & \\ = \frac{1}{b}\ln\frac{ae^{bT}+v-a-v\gamma e^{b\tau}}{v(1-\gamma)}, & 0 \le \tau \\ & \le \frac{1}{b}\ln\frac{ae^{bT}-a+v}{v}. \end{cases} \tag{a2}$$

Make $(\tau) = (1/b)\ln((ae^{bT} + V - a - V\gamma e^{b\tau})/V(1-\gamma))$; it is obvious that when τ increases, $f(\tau)$ decreases and $f(\tau)$ is the decreasing function of τ. When $\tau = (1/b)\ln((ae^{bT} - a + V)/V)$, then $f(\tau)$ gets the minimum value and $f\min = (1/b)\ln((ae^{bT} + V - a)/V) = S_1$, so $S_2 \ge S_1$.

Proof 3. The inventory constraints are

$$\text{s.t.} \quad \frac{a\left[e^{b(T-S_1)}-1\right]}{b} \le Q_{\max}$$
$$\frac{a\left[e^{b(T-S_2)}-1\right]}{b} \le Q_{\max}$$
$$\frac{(a-v)\left(e^{-b\tau}-1\right)}{b} \le Q_{\max}$$
$$\Longrightarrow \max\left(\frac{a\left[e^{b(T-S_1)}-1\right]}{b}, \frac{a\left[e^{b(T-S_1)}-1\right]}{b}, \frac{(a-v)\left(e^{-b\tau}-1\right)}{b}\right) \le Q_{\max}. \tag{a3}$$

From Proof 2, we can know that $S2 \ge S1$, and it is obvious that $[e^{(T-S1)} - 1]/b \ge a[e^{b(T-S_2)} - 1]/b$. There is $a[e^{b(T-S_1)} - 1]/b = a(V-a)(e^{bT}-1)/b(ae^{bT}-a+V)$, and make $\varphi(\tau) = (a-V)(e^{-b\tau} - 1)/b$ and $(1/b)\ln((V-a)/V\gamma) < \tau \le (1/b)\ln((ae^{bT}-a+V)/V)$. Then we can get $V > a$. And when τ increases, (τ) increases and (τ) is the increasing function of τ. When $\tau = (1/b)\ln((aebT - a + V)/V)$, then $\varphi(\tau)$ reaches the maximum value $\varphi\max = a(V - a)(ebT - 1)/b(aebT - a + V) = a[e^{b(T-S1)} - 1]/b$. From another layer of significance, there are no defective products in the production process when = $(1/b)\ln((ae^{bT} - a + V)/V)$. The production process carries the same as case 1 and the maximum inventory is $[e^{(T-S1)} - 1]/b$.

In summary, we can get

$$\max\left(\frac{a\left[e^{b(T-S_1)}-1\right]}{b}, \frac{a\left[e^{b(T-S_1)}-1\right]}{b}, \frac{(a-v)\left(e^{-b\tau}-1\right)}{b}\right)$$
$$\le Q_{\max} \Longrightarrow \frac{a\left[e^{b(T-S_1)}-1\right]}{b} \le Q_{\max}. \tag{a4}$$

The The maximum and minimum theorem: assuming that (x) is continuous on a closed interval $[a, b]$, it must have a minimum and maximum value on $[a, b]$. (An important property for the continuous function on a closed interval)

$$ae^{bT} - a + v = ve^{bS},$$
$$S = \frac{1}{b}\ln\frac{ae^{bT}-a+v}{v},$$
$$e^{bS} = \frac{ae^{bT}+v-a-v\gamma e^{b\tau}}{v(1-\gamma)},$$
$$S = \frac{1}{b}\ln\frac{ae^{bT}+v-a-v\gamma e^{b\tau}}{v(1-\gamma)},$$

$$\begin{aligned} E(S) &= \int_0^{(1/b)\ln((ae^{bT}-a+v)/v)} \lambda e^{-\lambda\tau} \times \frac{1}{b}\ln\frac{ae^{bT}+v-a-v\gamma e^{b\tau}}{v(1-\gamma)}d\tau \\ &+ \int_{(1/b)\ln((ae^{bT}-a+v)/v)}^{+\infty} \lambda e^{-\lambda\tau} \times \frac{1}{b}\ln\frac{ae^{bT}-a+v}{v}d\tau \\ &= \frac{1}{b}\left[\int_0^{(1/b)\ln((ae^{bT}-a+v)/v)} \lambda e^{-\lambda\tau} \times \ln\frac{ae^{bT}+v-a-v\gamma e^{b\tau}}{v(1-\gamma)}d\tau \right. \\ &\left. + \left(\frac{ae^{bT}-a+v}{v}\right)^{-\lambda/b} \times \ln\frac{ae^{bT}-a+v}{v}\right]. \end{aligned} \tag{a5}$$

Conflict of Interests

The authors declare that there is no conflict of interests regarding the publication of this paper.

ACKNOWLEDGMENTS

The authors would like to thank the editors and anonymous referees for their time, patience, and helpful suggestions devoted to the review of this paper. Their constructive comments and suggestions have concretely improved and enriched this paper much further. This work is partially supported by NSFC Grants (no. 71071107).

REFERENCES

1. M. J. Rosenblatt and H. L. Lee, "Economic production cycles with imperfect production processes," IIE Transactions, vol. 18, no. 1, pp. 48–55, 1986.
2. S. S. Sana, "A production-inventory model in an imperfect production process," European Journal of

Operational Research, vol. 200, no. 2, pp. 451–464, 2010. ·

3. R. Gupta and P. Vrat, "Inventory model for stock-dependent consumption rate," Opsearch, vol. 23, no. 1, pp. 19–24, 1986.
4. T. L. Urban, "Inventory models with inventory-level-dependent demand: a comprehensive review and unifying theory," European Journal of Operational Research, vol. 162, no. 3, pp. 792–804, 2005.
5. N. H. Shah and K. T. Shukla, "Optimal production schedule in declining market for an imperfect production system," International Journal of Machine Learning and Cybernetics, vol. 1, no. 1–4, pp. 89–99, 2010.
6. H. L. Lee and M. J. Rosenblatt, "Simultaneous determination of production cycle and inspection schedules in a production system," Management Science, vol. 33, no. 9, pp. 1125–1136, 1987.
7. H. L. Lee and M. J. Rosenblatt, "Production and maintenance planning model with restoration cost dependent on detection delay," IIE Transactions, vol. 21, no. 4, pp. 368–375, 1989.
8. M.-J. Liou, S.-T. Tseng, and T.-M. Lin, "Effects of inspection errors to the imperfect EMQ model," IIE Transactions, vol. 26, no. 2, pp. 42–51, 1994.
9. C. H. Kim and Y. Hong, "An optimal production run length in deteriorating production processes,"International Journal of Production Economics, vol. 58, no. 2, pp. 183–189, 1999.·
10. K.-J. Chung and K.-L. Hou, "An optimal production run time with imperfect production processes and allowable shortages," Computers and Operations Research, vol. 30, no. 4, pp. 483–490, 2003.
11. C.-Y. Guo, J.-S. Hu, and L. Wang, "Economic production quantity models for items with defective rate and shortages in fuzzy environment," Computer Integrated Manufacturing Systems, vol. 14, no. 11, pp. 2161–2166, 2008.
12. J.-S. Hu, Y.-P. Ji, C.-Y. Guo, and R.-Q. Xu, "Fuzzy random imperfect production system with shortages,"Control and Decision, vol. 26, no. 2, pp. 213–220, 2011.
13. R. I. Levin, C. P. McLaughlin, R. P. Lamone, and J. F. Kottas, Productions Operations Management: Contemporary Policy for Managing Operating Systems, McGraw-Hill, New York, NY, USA, 1972.
14. E. A. Silver and R. Peterson, Decision Systems for Inventory Management and Production Planning, John Wiley & Sons, New York, NY, USA, 2nd edition, 1982.
15. B. N. Mandal and S. Phaujdar, "An inventory model for deteriorating items and stock-dependent consumption rate," Journal of the Operational Research Society, vol. 40, pp. 483–488, 1989.
16. T. K. Datta and A. K. Pal, "Note on an inventory model with inventory-level-dependent demand rate,"Journal of the Operational Research Society, vol. 41, no. 10, pp. 971–975, 1990.
17. V. M. Dixit and N. H. Shah, "A production inventory model for exponentially increasing demand and production," Modeling, Measurement and Control, vol. 22, no. 3-4, pp. 15–24, 2001.
18. R. C. Baker and T. L. Urban, "Deterministic inventory system with an inventory level-dependent demand rate," Journal of Operational Research Society, vol. 39, no. 9, pp. 823–831, 1988.
19. H. Soni and N. H. Shah, "Optimal ordering policy for stock-dependent demand under progressive payment scheme," European Journal of Operational Research, vol. 184, no. 1, pp. 91–100, 2008.
20. J. Min, Y.-W. Zhou, and J. Zhao, "An inventory model for deteriorating items under stock-dependent demand and two-level trade credit," Applied Mathematical Modelling, vol. 34, no. 11, pp. 3273–3285, 2010.
21. N. H. Shah, A. R. Patel, and K.-R. Lou, "Optimal ordering and pricing policy for price sensitive stock-dependent demand under progressive payment scheme," International Journal of Industrial Engineering Computations, vol. 2, no. 3, pp. 523–532, 2011.
22. J.-T. Teng, I.-P. Krommyda, K. Skouri, and K.-R. Lou, "A comprehensive extension of optimal ordering policy for stock-dependent demand under progressive payment scheme," European Journal of Operational Research, vol. 215, no. 1, pp. 97–104, 2011. ·
23. Y.-W. Zhou, Y. Zhong, and J. Li, "An uncooperative order model for items with trade credit, inventory-dependent demand and limited displayed-shelf space," European Journal of Operational Research, vol. 223, no. 1, pp. 76–85, 2012.
24. M. Goh, "EOQ models with general demand and holding cost functions," European Journal of Operational Research, vol. 73, no. 1, pp. 50–54, 1994.
25. H.-M. Wee, "Joint pricing and replenishment policy for deteriorating inventory with declining market,"International Journal of Production Economics, vol. 40, no. 2-3, pp. 163–171, 1995.·
26. G. Padmanabhan and P. Vrat, "EOQ models for perishable items under stock dependent selling rate,"European Journal of Operational Research, vol. 86, no. 2, pp. 281–292, 1995.
27. J. Ray and K. S. Chaudhuri, "An EOQ model with stock-dependent demand, shortage, inflation and time discounting," International Journal of Production Economics, vol. 53, no. 2, pp. 171–180, 1997. ·
28. B. R. Sarker, S. Mukherjee, and C. V. Balan, "An order-level lot size inventory model with inventory-level dependent demand and deterioration," International Journal of Production Economics, vol. 48, no. 3, pp. 227–236, 1997.
29. B. C. Giri and K. S. Chaudhuri, "Deterministic models of perishable inventory with stock-dependent demand rate and nonlinear holding cost," European Journal of Operational Research, vol. 105, no. 3, pp. 467–474, 1998.
30. H. Yan and T. C. E. Cheng, "Optimal production stopping and restarting times for an EOQ model with deteriorating items," Journal of the Operational Research Society, vol. 49, no. 12, pp. 1288–1295, 1998. ·

31. M. Mandal and M. Maiti, "Inventory of damagable items with variable replenishment rate, stock-dependent demand and some units in hand," Applied Mathematical Modelling, vol. 23, no. 10, pp. 799–807, 1999.
32. K.-J. Chung and S.-F. Tsai, "Inventory systems for deteriorating items with shortages and a linear trend in demand-taking account of time value," Computers & Operations Research, vol. 28, no. 9, pp. 915–934, 2001.
33. X. L. Li, A new intelligent optimization-artificial fish swarm algorithm [Ph.D. thesis], Zhejiang University, Zhejiang, China, 2003.
34. H. Peng, Research on the production strategy and payment policy for the items with inventory-dependent demand [M.S. thesis], College of Management and Economics, Tianjin University, Tianjin, China, 2013.
35. M. Neshat, G. Sepidnam, M. Sargolzaei, and A. N. Toosi, "Artificial fish swarm algorithm: a survey of the state-of-the-art, hybridization, combinatorial and indicative applications," Artificial Intelligence Review, vol. 42, no. 4, pp. 965–997, 2012.
36. X. Li, Z. Shao, and J. Qian, "An optimizing method based on autonomous animats: fish-swarm algorithm," Systems Engineering—Theory & Practice, vol. 22, no. 11, pp. 32–38, 2002.
37. L. Li, S.-C. Chi, and G. Lin, "Tabu fish swarm algorithm and its application to slope stability analysis,"Engineering Mechanics, vol. 23, no. 3, pp. 6–10, 2006.
38. C.-R. Wang, C.-L. Zhou, and J.-W. Ma, "An improved artificial fish-swarm algorithm and its application in feed-forward neural networks," in Proceedings of the International Conference on Machine Learning and Cybernetics (ICMLC ‹05), pp. 2890–2894, Guangzhou, China, August 2005.
39. H. Huang and Y. Zhou, "A global optimization complex functions method based on improved artificial fish swarm algorithm," Journal-Guangxi Normal University (Natural Science Edition), vol. 26, no. 1, pp. 194–197, 2008.
40. G. Z. Chen, J. Q. Wang, C. J. Li, and X. Lu, "An improved artificial fish swarm algorithm and its applications," Systems Engineering, vol. 27, no. 12, pp. 105–110, 2009.
41. Y. Cheng, M. Jiang, and D. Yuan, "Novel clustering algorithms based on improved artificial fish swarm algorithm," in Proceedings of the 6th International Conference on Fuzzy Systems and Knowledge Discovery (FSKD ‹09), vol. 3, pp. 141–145, IEEE, Tianjin, China, August 2009.
42. Y. Peng, "An improved artificial fish swarm algorithm for optimal operation of cascade reservoirs,"Journal of Computers, vol. 6, no. 4, pp. 740–746, 2011. ·
43. H.-C. Tsai and Y.-H. Lin, "Modification of the fish swarm algorithm with particle swarm optimization formulation and communication behavior," Applied Soft Computing Journal, vol. 11, no. 8, pp. 5367–5374, 2011.
44. X. Zheng, The improvement and application for the artificial fish swarm algorithm [M.S. thesis], Shanghai Maritime University, Pudong, China, 2006.

CHAPTER 14 A CORE MODEL FOR PARTS SUPPLIERS SELECTING METHOD IN MANUFACTURING SUPPLY CHAIN

Guorong Chen[1,2,3] Jiangyuan Zhao[2] and Juli Deng[1]

[1]School of Electronic and Information Engineering, Chongqing University of Science and Technology, Chongqing 401331, China

[2]CAIA, Dhurakij Pundit University, Bangkok 10210, Thailand

[3]Faculty of Engineering, University of Ottawa, Ottawa, ON, Canada K1N 6M9

ABSTRACT

Service-oriented manufacturing is the new development of manufacturing systems, and manufacturing supply chain service is also an important part of the service-oriented manufacturing systems; hence, the optimal selection of parts suppliers becomes one of key problems in the supply chain system. Complex network theories made a rapid progress in recent years, but the classical models such as BA model and WS model can not resolve the widespread problems of manufacturing supply chain, such as the repeated attachment of edge and fixed number of vertices, but edges increased with preferential connectivity, and flexible edges' probability. A core model is proposed to resolve the problem in the paper: it maps the parts supply relationship as a repeatable core; a vertex's probability distribution function integrating the edge's rate and vertex's degree is put forward; some simulations, such as the growth of core, the degree distribution characteristics, and the impacting of parameter, are carried out in our experiments, and the case study is set also. The paper proposed a novel model to analyze the manufacturing supply chain system from the insights of complex network.

INTRODUCTION

Manufacturing services spawned the manufacturing system's range, which needs the dominant firms collaborating with their parts suppliers beyond geographical boundaries, but how to select the most optimal parts suppliers in the supply chain becomes the new difficulty. In order to resolve the problem of optimal selecting part suppliers in manufacturing supply chain, many scholars devote their researches in the topics from those fields.

Citation: Guorong Chen, Jiangyuan Zhao, and Juli Deng, "A Core Model for Parts Suppliers Selecting Method in Manufacturing Supply Chain," Mathematical Problems in Engineering, vol. 2015, Article ID 530147, 11 pages, 2015. doi:10.1155/2015/530147.

The first field is the cooperation models in the manufacturing supply chain. Holland proposed the mode of cooperation and collaboration of supply chain management [1]; Sarmah et al. went into the coordination models between the business entities to gain lower supply chain cost [2]. Kelle and Akbulut discussed the software tools of ERP in supply chain for information sharing, cooperation, and cost optimization [3]; Nagarajan and Sošić discussed the game-theoretic analysis method in cooperation supply chains [4]; Fiala discussed the information sharing between suppliers, manufacturers, distributors, retailers, and customers [5]; Savaskan et al. discussed the closed loop supply chain models with product remanufacturing [6]; Ma and Chen researched on the elements and characters in integrated logistics' supply chain [7]; Xu and

Ma researched on the trust mechanism between enterprises in supply chain and suggested that the cultivation of mutual trust of enterprises is the core of supply chain management [8]; and Love et al. researched on the interorganizational relations in construction industry [9].

The second field is the dynamic analysis, evaluation methods, and the business forecasting in the supply chain. Liang et al. put forward nonlinear programming problems to solve the DEA (Data envelopment analysis) models for supply chain efficiency evaluation [10]; Leng and Zhu reviewed the side-payment contracts in two-person nonzero-sum supply chain games and discussed its application [11]; Zhu et al. researched on the confirmation of a measurement model for green supply chain management practices implementation [12]; Cachon and Zipkin discussed the competitive and cooperative inventory policies in a two-stage supply chain [13]; Sarkis proposed a strategic decision framework for green supply chain management [14]; Huang et al. discussed an analysis of manufacturer-retailer supply chain coordination in cooperative advertising [15]; Ferguson discussed the "commitment time frame problem" between an end product manufacturer and a parts supplier [16]; Surana et al. researched the context of supply-chain networks [17]; Gupta and Maranas discussed the managing demand uncertainty method in supply chain planning and proposed a tool for evaluating and actively managing the exposure of enterprises assets (such as inventory levels and profit margins) to market uncertainties [18]; Kim et al. go through an empirical investigation in IT alignment between supply chain partners to enhance customer value creation [19]; and Connelly et al. discussed the perspectives of six prominent organizational theories, such as apply real options theory, internationalization theory, organizational economics, resource dependence theory, social network theory, and institutional theory, and how to effect the development of supply chain [20].

The third field is the complex network analysis approach and its applications. Stanley Milgram pointed out the famous theories of six degrees of separation; he observed that the distant of human in the world is about six degrees; Watts built a special website to check the theories [21]; and game of Kevin Bacon checked the theories as Bacon degree [22], and some scholars in Oakland University go through the theories as Erdos degree [23]; all of those experiments proved the theories are valid. Erdős and Rényi study the social network based on the perspective of graph theory for the first time in the world and published a famous paper on the evolution of random graphs [24]; Watts and Strogatz researched on the modeling of small-world networks, characteristics, and mechanism [25, 26]; Barabási and Albert studied some kinds of complex networks and put forward the statistical mechanism of small world networks, scale-free networks, complex networks, and evolutionary theory [27, 28]; Gregoriades A and Sutcliffe studied the reliability of complex systems based on load forecasting and planning methods [29]. Shi et al. studied the calculation model and vertex degree distribution of complex networks, such as clustering coefficient estimates [30, 31]; Wang research network synchronization and control technology [32]; Li and Chen studied the control principles and methods of complex networks [33]. Guoning and Zheren discussed the components product family relations, described the relationship between the network diagram of parts products using the weighted directed tree network method, calculated the statistical properties of the relationship between the network components, studied the revolution process of the parts products relationship network diagram, and proposed a prediction approach of the amount of parts [34]. The study is completely broken through the traditional way of manufacturing research information system and provides a guideline for our research.

From the reviews above, we can see that the complex network theory and its applications have made a rapid progress in recent years, and it can integrate the system engineering and topology and became a new approach to be used to analyze the part suppliers' optimal selecting problems of manufacturing supply chain.

But there have been rare reports of complex network theories applied in manufacturing supply chain. Based on the preliminary studies [35], this paper will focus on the part supplier's optimal selecting problems and presents a core model for the part suppliers optimal selecting problems in manufacturing supply chain. It will describe the application problem in Section 2; analyze the shortcomings of two classical complex network models named BA model and WS model in Section 3; put forward a core model, discuss the designs and description of the model, and present the probability formula of selecting part suppliers in detail in Section 4; and simulate the model and analyze the characteristics in Section 5.

THE PROBLEM DESCRIPTION

The supply relationship of parts in manufacturing system intertwined together and formed a complex relationship network. The dominant firm wants some parts for assembling its production. If there are many parts suppliers meeting the given quality and times requirements, the cost is the key factors to affect their decision that those parts suppliers should to be selected. Supposing the parts consisted of some parts, the parts are made up of subparts and so on; the supply relation forms a supply relationship network, so the manufacturing supply chain has the sketch as Figure 1.

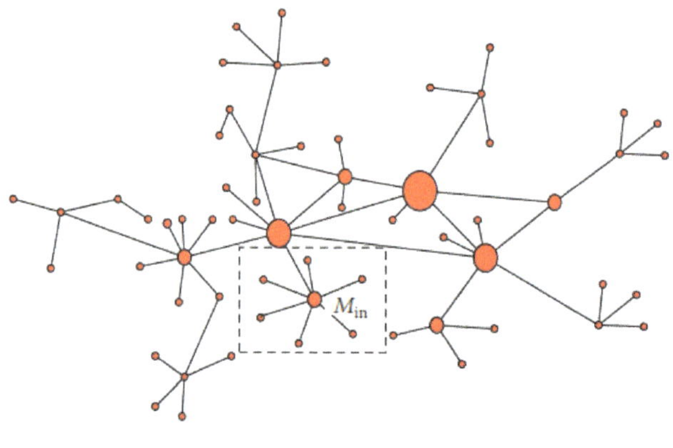

Figure 1: The supply relationship network.

It is not easy to resolve the suppliers selecting problem of the whole network. In the paper, we select one vertex that has many suppliers to analyze; other vertices' characteristics are easy to arrive using the same method.

THE RELATIONSHIP MODEL RESEARCHES AND THE NEW CHARACTERISTICS OF MANUFACTURING SUPPLY CHAIN

Toward the manufacturing supply chain, the similar researches mainly were two famous models, WS model [25] and BA model [27].

The WS Model [25]

Watts and Strogatz reported the WS model firstly in their "Collective dynamics of 'small-world' networks" in 1998 (by Nature). Newman and Watts improved it as NW model in 1999 [36]. The WS model starts from a ring lattice with n vertices and k edges per vertex then rewire each edge at random with probability p. This model shows a graph between regularity (p=0) and disorder (p=1), for intermediate values of , and the graph is a small-world network that is highly clustered like a regular graph, yet with small characteristic path length, like a random graph; it is shown in Figure 2. The characteristic can be used in large scale network, with some characteristic as that the number of vertices is fixed; number of edges is fixed, rewriting edges random.

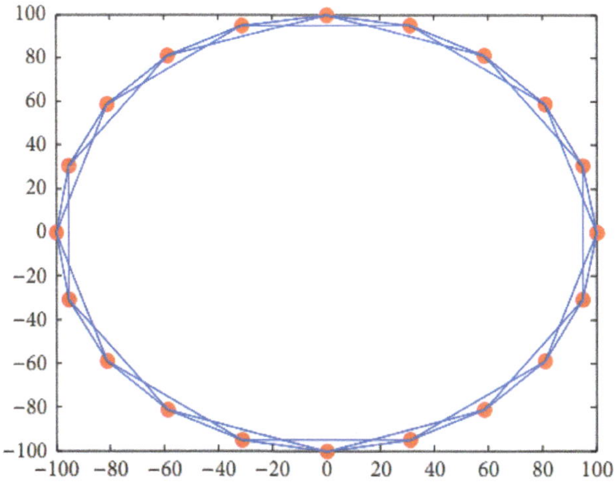

(a) $p = 0$

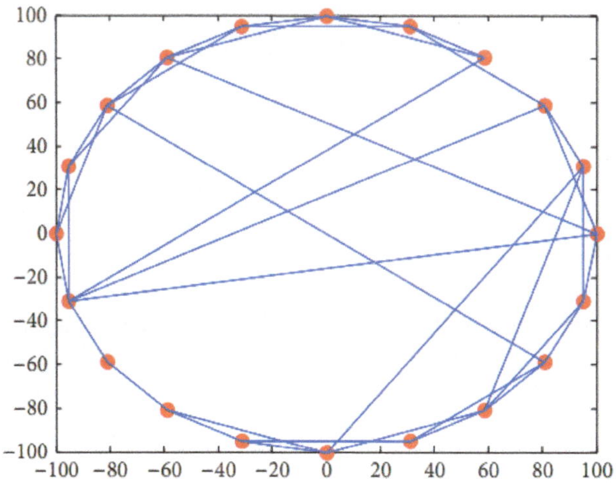

(b) $p = 0.2$

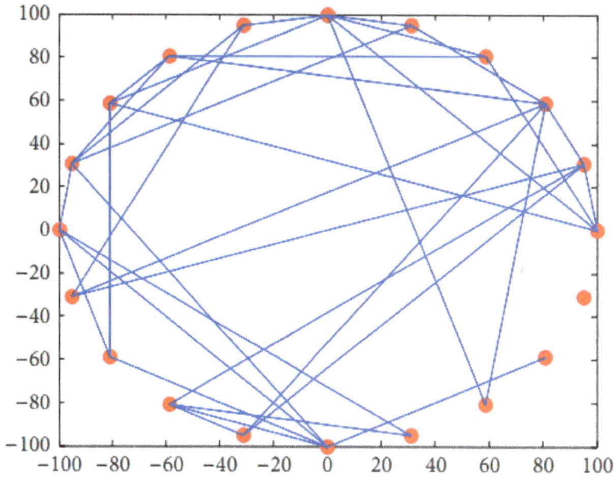

(c) $p = 1$

Figure 2: The WS model.

The BA Model

The BA model is proposed by Barabási and Albert in their "Emergence of scaling in random networks" in 1999 (by Science) [27]. The BA model should meet two conditions: at first, the graph is opened and new vertices should continue to be added to the system; secondly, the attachment of the edges has the characteristic of preferential connectivity. So, the degree of vertices with rich-get-richer phenomenon can be detected easily in BA networks, and the growth process is shown as in Figure 3.

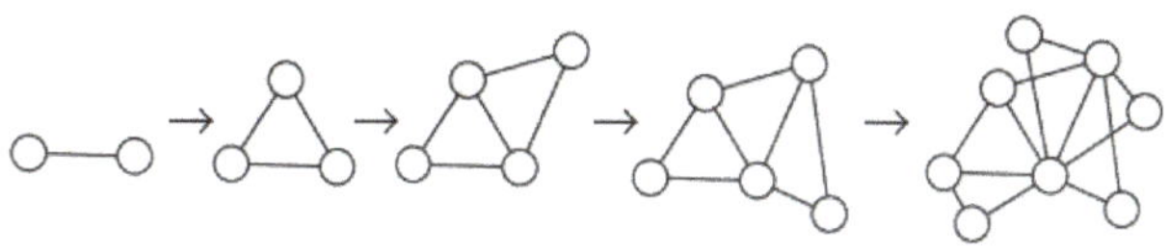

Figure 3: The growth of BA model.

The Performances of Manufacturing Supply Chain Network

Now, let us trace the development of the manufacturing supply chain network from its origin. In Figure 1, at first, the vertex M_{in} is new one and has no parts supplier, which means the core diagram only has candidate vertices with degree 1; then, some parts need to be produced from the suppliers, which means some edges should be added in the network. Which are the next vertices to produce the parts? It may be decided by random at the first time, the following selections maybe select other vertices different from the first vertices; perhaps the new selection is more efficient than the first one. Then, a most efficiency route arrived; the follow-up selections will tend to the same route to produce the parts; the selection will strengthen the ability of the notes in the chains, the abilities improvement of the vertices will firm the tendency to form fixed manufacturing supply chains. From the description above, we can see that the manufacturing supply chain network has some characteristics different from the BA model and WS model (see Figures 2 and 3).

First, both models suppose that the edges attached any two vertices without any repetition, but, in the manufacturing supply chain system, the edges maybe reduplicate, and it usually is repeated many times with the development of the manufacturing supply chain.

Second, in WS model, the attachment probability of new edge is fixed and comes randomly; in BA model, the attachment probability of new edge is be added along with vertices and with preferential connectivity. But those cannot meet the real supply chain network's request, and the new edges add into the network with preferential connectivity but do not need to add new vertices, so the number of vertices is fixed but the edges will keep increasing with preferential connectivity.

Third, the WS model and the BA model suggested that the edges probability of attachment is constant, but, in supply chain network, because of the limitation of production ability and the distance between them, the probability usually changes periodically. So, it needs a new model to meet the needs of manufacturing supply chains.

THE CORE MODEL

The Core Model of Manufacturing Supply Chain

In order to resolve the problems mentioned in WS model and BA model, our model maps the supply relationship network from Figure 1 to the core diagram in Figure 4; in Figure 4(a), it constituted by some vertices and some edges that the core has one input vertex and has n selective supply vertices. The supply vertices may be defined as part suppliers; the edges may be defined as the costs between the input vertex/dominant firm M_{in} and the special supply vertex. In the selecting process, the director of edges directs the collected to the supply vertices. If it needs to decide the parts supplier of vertex M_k, the core iterated on the vertex M_k, as shown in Figure 4(b).

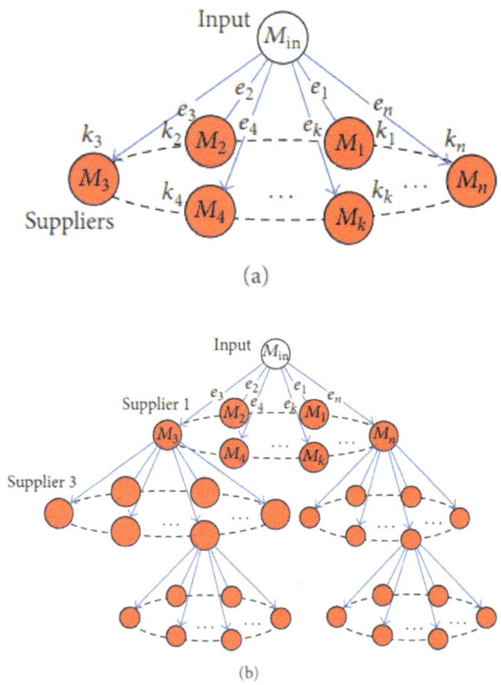

Figure 4: The core model.

Now, towards Figure 4(a), if k_i is used to express the degrees set of vertex i, obviously, larger k_i means more experience to produce the candidate parts. The edge E_i is the supply relationships between the vertex E_i to the supply vertex M_i, and lower value of E_i shows that the transportation costs from vertex M_i to the supply vertex i are cheaper and means that vertex i tends to be selected as the candidate unit. The object of production scheduling is to optimize the manufacturing systems so as to produce the parts using the minimize costs in the given during time, and it may be described as the following formula:

$$O = \min\{O_1, O_2, O_3, \ldots, O_n\}, \quad (0 < i < n). \tag{1}$$

Therefore, O_i is the cost of vertex i.

The Vertex Probability Distribution in Manufacturing Supply Network

In order to describe the growing process of the manufacturing supply network, we supposed it satisfied the following 4 assumptions.

- The number of vertices in the network is fixed as N.

It does not allow the edges' start and end to join the same vertices.

- The edges of network may grow freely only with preferential connectivity.
- The probability of changes meets a special function discussed next.

Supposing the core has $m0$ edges initially, now m (m<N) new edges should be added into the core diagram every fixed time interval t. and the attachments should select the vertices with biggest probability. After time interval t, the edges in the network will be $m_0 + mt$ edges.

The first factor to be considered is the degree of vertex; it means the truth, cooperation times, or selection history.

Obviously, the selection process is a binomial distribution function, as in the following formula:

$$p\{X = t\} = k_i. \tag{2}$$

The normalized result is the following formula:

$$p_i = \frac{k_i}{\sum_{j=1}^{N} k_j}. \tag{3}$$

However, the degree can not involve all the factors. It is only an experiment index; a bigger degree means the supplier has more experiment of operation, or it has more relation resources or skills. The edges rate should to be taken into consideration, and it is a cost index; it is to be decided by such as transportation costs of the parts, ready resources, and other costs. So, the relationship between the vertices degree and the edges rates should be taken into consideration. The preresearch shows that it is a linear function if only it is taken into consideration one factor of the vertex degree or the edges rates individually. Toward the vertex, the vertex degree and the edges rates are a binomial distribution; the probability distribution meets the function as in formula

$$p\{X = i\} = k_i^{\tau} r_i^{1-\tau}, \quad (0 \le \tau \le 1). \tag{4}$$

The normalized results may be expressed as

$$p_i = \frac{k_i^{\tau} r_i^{1-\tau}}{\sum_{j=1}^{N} k_j^{\tau} r_j^{1-\tau}}, \quad (0 \le \tau \le 1), \tag{5}$$

where k_i is the degree of vertex i, it may be decided as the operation times, and it may be obtained by computing the indegree; hence, k_i may be computed as the following formula:

$$k_i = \sum_{j=1}^{t} k_{\text{in-}i}, \tag{6}$$

where r_i is the rate of the attachment, E_i means the costs attaching vertex i, and the affecting factor includes the diverting and other costs. Then, r_i is

$$r_i = \frac{1}{E_i}, \tag{7}$$

where τ is the adjustable parameter between k_i and r_i, and it make it easier to adjust the vertices' degree and the edge's rates.

Formula (5) gives us a quantitative method to compute the probability distribution of suppliers at a time point. Each supplier has different calculating result and the results will change along with time because k_i changes with the time. If the time is unlimited, the parts supply chain plan tends to the most effective route, and other suppliers have no chances to obtain the orders. So, the operation cost is

$$C_{\min} = \sum_{j=1}^{N} C_{\text{Min-}i}, \quad (t\Gamma \le T). \tag{8}$$

Therefore, $C_{\text{Min-}i}$ is the minimizing vertex's cost of vertex i.

On the contrary, if the time is very urgent, the time should exceed the shortest time request so as to finish all the parts of the batch. Maybe, the cost is the largest one:

$$C_{\max} = \sum_{j=1}^{N} C_i, \quad \left(t\Gamma = \frac{T}{n}\right). \tag{9}$$

Therefore, $\sum_{j=1}^{N} C_i$ means all of the suppliers should work regardless of its inefficiency. Usually, the optimal plan is between the two extreme cases, which means the operation tending to some effective suppliers and other suppliers may allocate fewer amounts of orders.

SIMULATION STUDY

Now, let us study the core model in simulation to check it. In our experience, the rate of edges (the reciprocal of transportation costs) came from random number between 0 and 1; each vertex has initial degree of 1.

The Growth Process of Core Model

The growth of core is studied first in our experiment, and the results are shown in Figures 5 and 6.

Where the adjustable parameter of degree and edge is $\tau = 0.1$, the rates of edge come from random number; the vertex's number is 20, and the number following the vertex name in Figure 5 is the vertex' degree.

From Figure 5, we can see that, with new edges added into the core model and the degree of core growing with the time, some vertices tend to grow, but other vertices' degree keeps constant. M6, M8, and M12 have the greatest degrees while degree = 1000, that means those suppliers may keep lowest costs, and they are the best selections. On the contrary, the degrees of M1, M2, M4, M5, M7, M9, and M20 are 1 or 2, which means those suppliers have seldom chances to cooperate with the center vertex M_{in}.

The Degree Distribution of the Core Model

The degree distribution of growing process is shown in Figure 6: it shows that the key suppliers have a small number, but those vertices supplied a large number of parts; on the contrary, a large number of other vertices only supplied few parts (Figures 6(a) and 6(b)). The number of suppliers and the numbers of vertices in a core show a power law, vertex with rich degree becomes richer (Figures 6(c) and 6(d)), and the reality is almost like it. If the supplier has no chance to attend the cooperation in the earlier time, it needs to have a much lower cost to attend the core later; the cooperation of core is decided by lower cost and more trust.

Figure 7 shows the cooperation tendency chart in the core: the suppliers in higher degree (Figure 7(a)) or hot color (Figure 7(b)) mean the supplier has closer cooperation relationship with the core center vertex. Figure 5 shows the cooperation tendency chart in the core: the suppliers in higher degree (Figure 5(a)) or hot color (Figure 5(b)) mean the supplier has closer cooperation relationship with the core center vertex.

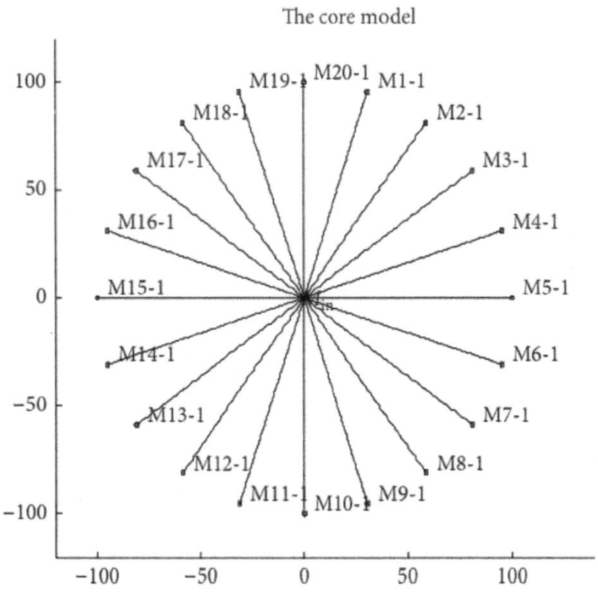

(a) Initial status

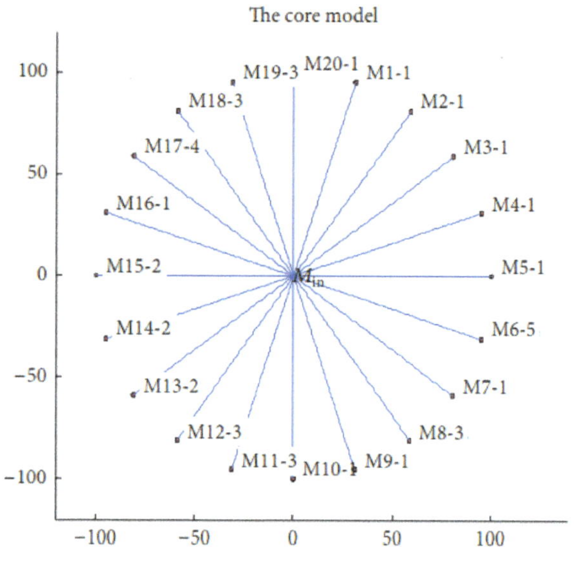

(b) Degree = 30

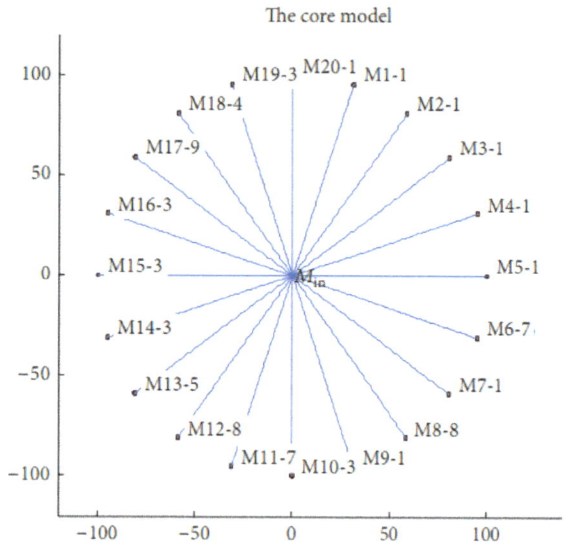

(c) Degree = 40

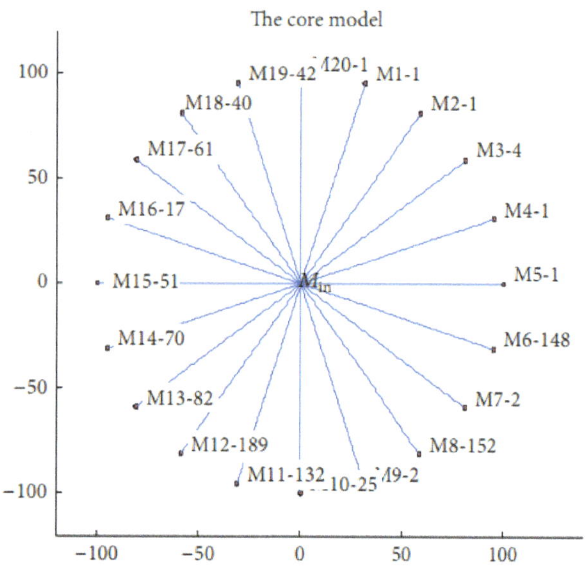

(d) Degree = 1000

Figure 5: The growth of the core.

The Impact of Parameters

The Impact of Edge's Rate

Formula (5) shows that the edge's rate is one of the key factors for the core. The edge's rate comes from random number in our experiments above, and it affected the core chart which tends to smooth. What will happen if the edge's rate keeps constant? The experiment results show in Figure 8.

It shows that the degree distribution is a single peak function when the rates of edge are constant; it is obviously different from the rates of edge which are not constant. It may be explained from formula (5), while the rates of edge are constant means that the probability of vertex is only decided by the degree; once a vertex has a bigger degree than other vertices, the vertex's probability will be the biggest one in the core; other vertices have no chance to have the new edge.

The Impact of Adjustable Parameter

The adjustable parameter t adjusts the importance of vertex's rate and the degree; it can impact the selection of new edge's attachment. Figure 9 shows the action of adjustable parameter$t$from t=0 to t=1. While t=0, the degree does not act on any action according to formula (5); the degree of the core is a random function, and the difference of vertex degree is wave in a narrow range (the wave motion is between 30 and 45 while t=0, shown in Figure 9(a)); while t=1, the edge's rate does not take action, and only one vertex degree keeps increasing. With the value of t increasing, the wave motion of vertex's degree becomes more and more wider (the wave motion is amplified between 20 and 50 while $t = 0.01$, shown in Figure 9(b); between 1 and 200 while $t = 0.1$ in Figure 9(c); between 1 and 350 while $t = 0.2$ in Figure 9(d); between 1 and 1000 while $t = 0.5$ in Figures 9(e) and 9(f)). It also shows that degrees of vertex tend to difference, and the amount of suppliers attending to supply chain becomes more and more smaller; only one supplier supplies the parts while $t = 0.5$.

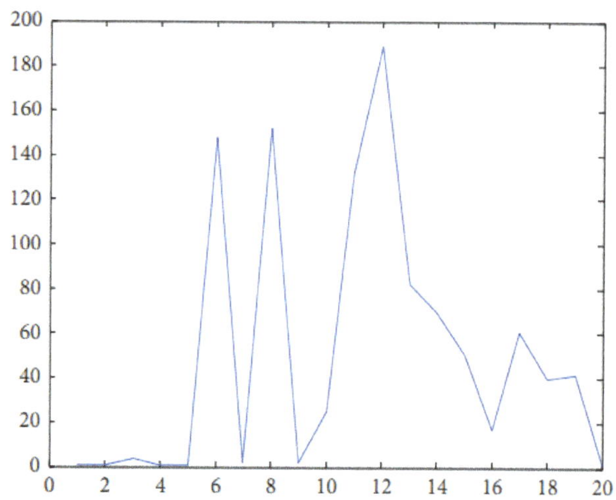

(a)

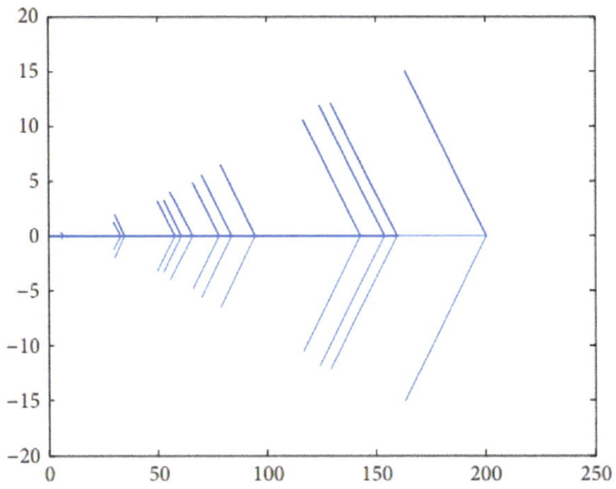

(b)

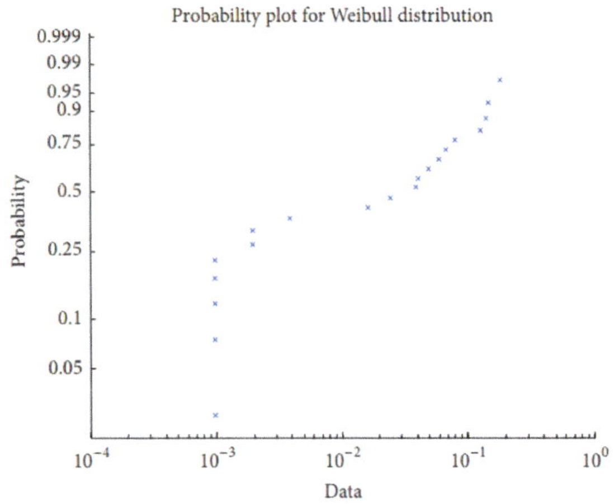

(c)

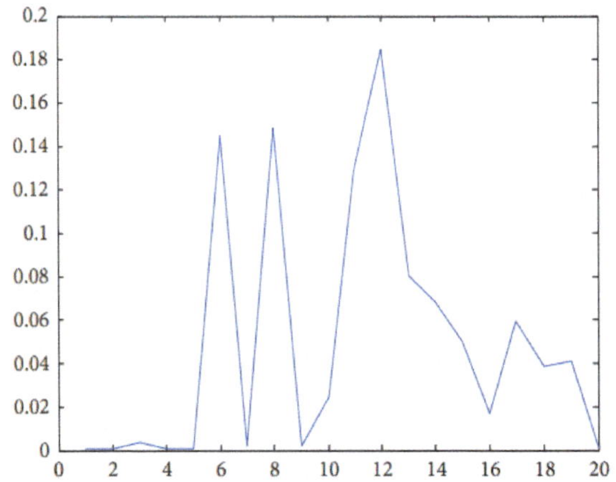

(d)

Figure 6: The degree distribution of the core while time = 1000.

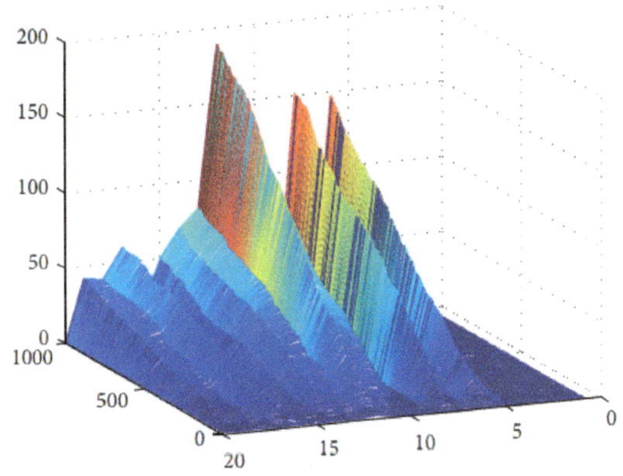

(a)

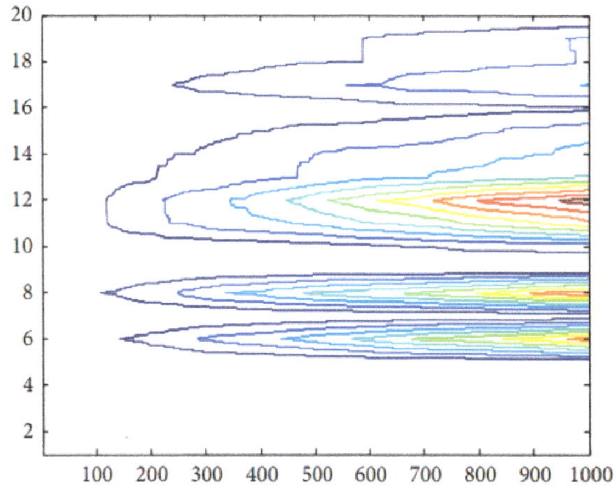

(b)

Figure 7: The cooperation relationship and tendency.

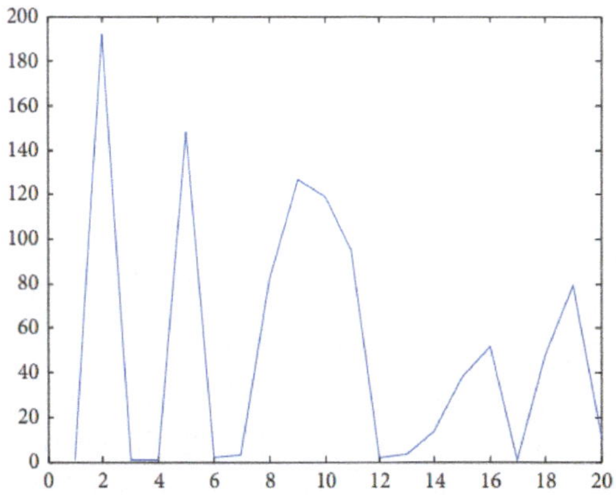

(a) The rates of edge are constant

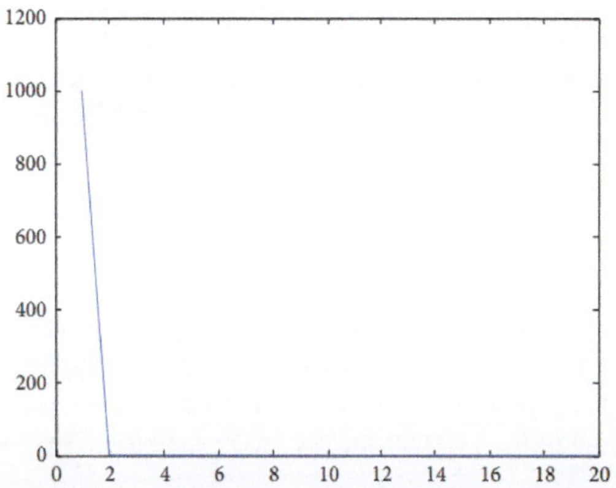

(b) The rates of edge are not constant

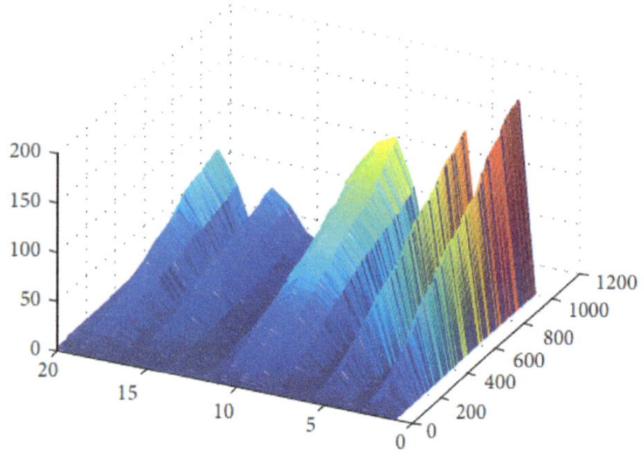

(c) The rates of edge are constant (3D)

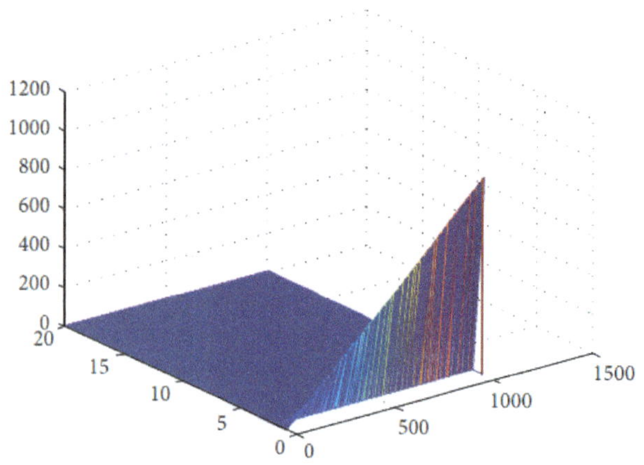

(d) The rates of edge are not constant (3D)

Figure 8: The impact of edge's rate.

The Impact of Time/Experience

Figure 10 shows the growth process and tendency of the core with the time; the supply relationship is very chaotic at first, but it will tend to become different with the development of core; the supply relationship of the core tends to differentiation and stablility after a long time.

The Results of the Simulations

The simulations show that the supply chain is a power law network, and the selecting amount tends differently with the time. A little amount of vertices supply the most amount of parts; on the contrary, most of the vertices have little chances to attend the supply chain, and the supply relationship is the effect of Matthew; on the other hand, if a supplier improved their product and reduced the cost of parts, they have the chance to attend the supply chain network. So, the supply chain network is dynamic and clustered.

Formula (5) shows the directions of how to improve the chance of attending the supply chain: the first way is that the parts supplier keeps a closer relationship to the core firms in the network, which means that there is a larger k_i; another way is to keep a lower cost, which means that the supplier should reduce the cost of the parts, such as sale of lower parts price or producing the product locally (to reduce the transportation cost).

The simulations also show the difference between the model proposed in the paper and the traditional model: our model disclosed the running principle of the supply chain and gives the ways to how to improve the chance to obtain the orders, so it is a novel and practical method to analyze the supply chain.

CASE ANALYSES

The IT Manufacturing Supply Chains Development at Chongqing China

The IT manufacturing industry of Chongqing China was very weak at 2000, but it had a rapid expansion while Hewlett-Packard (Chongqing) Co., Ltd, and Acer (Chongqing) Co., Ltd, began to produce the notebook computer in recent years and the notebook computer accounts for one-fifth of global production [37, 38].

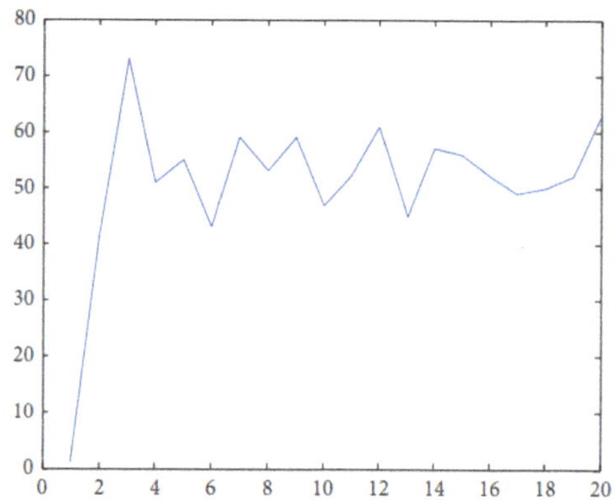

(a) $\tau = 0$

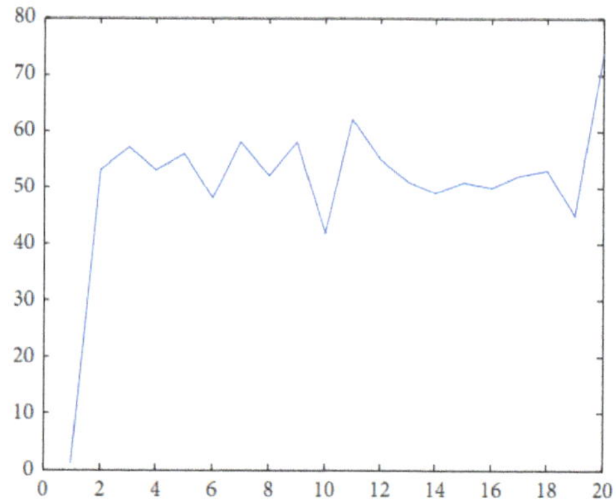

(b) $\tau = 0.01$

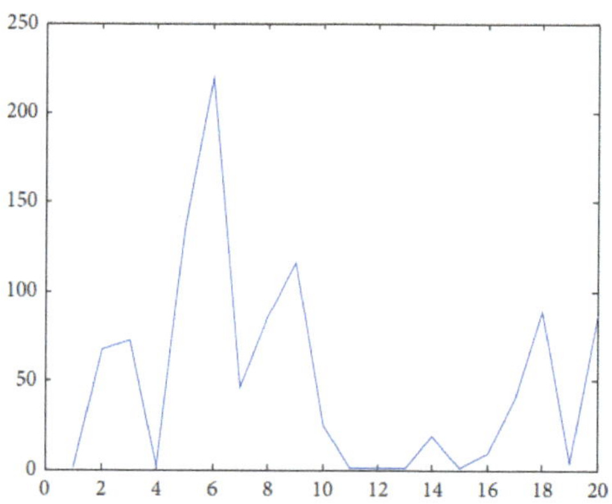

(c) $\tau = 0.1$

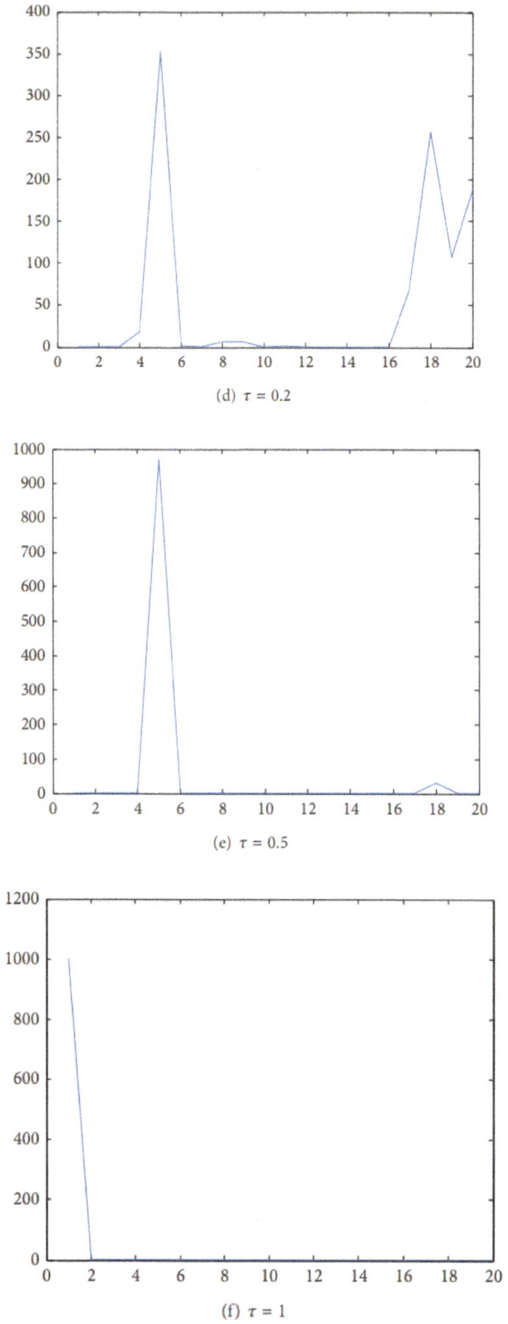

Figure 9: The impact of adjustable parameter.

Why did it expand so quickly? It clustered 5 large computer enterprises, 6 large foundries, and 817 parts manufacturers in Xiyong Chongqing till February 11, 2014, and the industry gathering circle is about 1 hour drive. One very interesting phenomenon is that these companies are new plants following the large enterprises such as HP, Acer, and Cisco. The foundries and parts manufacturers built new production plant in Chongqing in order to decrease the transportation costs to the lowest level, so they have the lowest, calculated from formula (5); the lowest gained the biggest probability distribution, and they will obtain the greatest opportunities for cooperating with the large computer enterprises, and the industry aggregation phenomenon proved their action in practice.

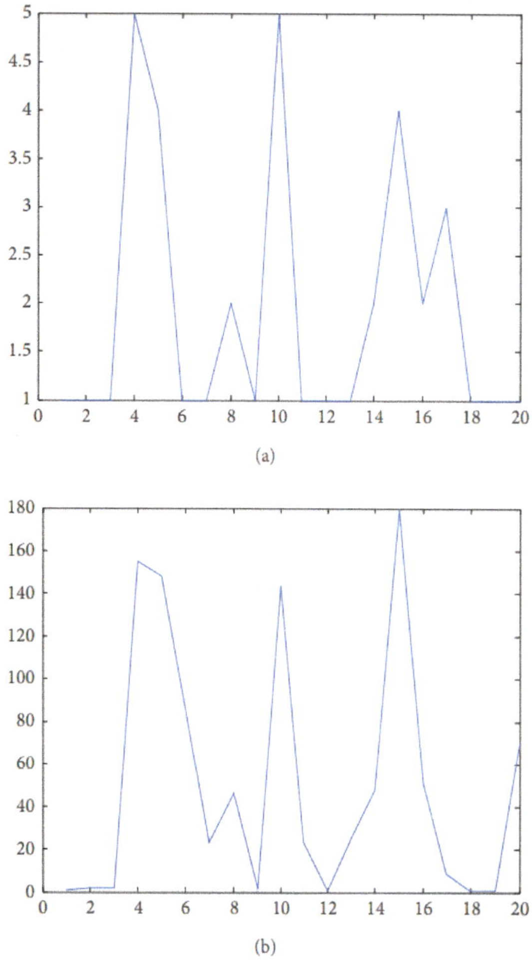

Figure 10: The growth process and tendency of the core.

Jiangmen Hi-Tech Development Zone

Jiangmen Hi-tech Development Zone began at 1992 [39]; it is one of the Hitech development zones of China. But the park was one of the worst Hi-tech development zones in China before 2010. It introduces itself as being able to serve for many industries, such as LED, advanced and other industries, but only a few enterprises grow up in the zone. Why is that? Because there almost are no large enterprises, and the products produced here need to be sold to the enterprises far away from there, which means the product costs have higher than other part suppliers; in other words, it has least probability distribution. In order to develop better, Jiangmen Hi-tech Development Zone replaned the industrial layout and direction and set up three professional parks, such as auto parts professional park; FAW Volkswagen and other auto manufacturers built their factories in the park; it speeded up all the auto parts development, so the professional park developed very fast in recent years; Jiangmen Hi-tech Development Zone is an important professional park in south China now.

SUMMARY

Aiming at the three difficulties of classical complex network theories to resolve the optimal selection of parts suppliers, a core model is proposed in the paper after reviewing famous BA model and WS model; it maps the parts supply relationships of manufacturing supply chain as a repeatable core. A vertex's probability distribution formula is put forward; it mainly involves the edge's rate and vertex's degree. Some simulation studies, such as the growth of core, the degree distribution characteristics, and the impacting of parameter experiment, were carried out. At last, two cases were set out to prove the correctness in the paper. The core model can be used to resolve some difficulties—with repeating edge attachment, fixed vertices number but increasing edges with preferential connectivity, and flexible edges probability.

It did not consider the interactions of other cores and the consistency of costs in our experiments; the calculating results may be a little deviated, and we plan to resolve those problems in next steps.

ACKNOWLEDGMENTS

This research is supported by the Natural Sciences Fund of China (Grant no. 51275547) and the Natural Science Fund Projects in Chongqing (Grant no. cstc2012gg-yyjs40019, CSTCjjA1481).

REFERENCES

1. C. P. Holland, "Cooperative supply chain management: the impact of interorganizational information systems," The Journal of Strategic Information Systems, vol. 4, no. 2, pp. 117–133, 1995.·
2. S. P. Sarmah, D. Acharya, and S. K. Goyal, "Buyer vendor coordination models in supply chain management," European Journal of Operational Research, vol. 175, no. 1, pp. 1–15, 2006.
3. P. Kelle and A. Akbulut, "The role of ERP tools in supply chain information sharing, cooperation, and cost optimization," International Journal of Production Economics, vol. 93-94, pp. 41–52, 2005.
4. M. Nagarajan and G. Sošić, "Game-theoretic analysis of cooperation among supply chain agents: review and extensions," European Journal of Operational Research, vol. 187, no. 3, pp. 719–745, 2008.
5. P. Fiala, "Information sharing in supply chains," Omega, vol. 33, no. 5, pp. 419–423, 2005.
6. R. C. Savaskan, S. Bhattacharya, and L. N. van Wassenhove, "Closed-loop supply chain models with product remanufacturing," Management Science, vol. 50, no. 2, pp. 239–252, 2004.·
7. S.-H. Ma and X.-Y. Chen, "The study for elements and characters of integrated logistics capability,"Chinese Journal of Management, vol. 1, no. 1, pp. 107–111, 2004.
8. S.-J. Xu and S.-H. Ma, "A study on trust crisis between inter-firm in supply chain in china," Computer Integrated Manufacturing System, vol. 8, no. 1, pp. 51–53, 2002.
9. P. E. Love, Z. Irani, E. Cheng, and H. Li, "A model for supporting interorganizational relations in the supply chain," Engineering Construction and Architectural Management, vol. 9, no. 1, pp. 2–15, 2002.
10. L. Liang, F. Yang, W. D. Cook, and J. Zhu, "DEA models for supply chain efficiency evaluation," Annals of Operations Research, vol. 145, pp. 35–49, 2006.
11. M. Leng and A. Zhu, "Side-payment contracts in two-person nonzero-sum supply chain games: review, discussion and applications," European Journal of Operational Research, vol. 196, no. 2, pp. 600–618, 2009.
12. Q. Zhu, J. Sarkis, and K.-H. Lai, "Confirmation of a measurement model for green supply chain management practices implementation," International Journal of Production Economics, vol. 111, no. 2, pp. 261–273, 2008.
13. G. P. Cachon and P. H. Zipkin, "Competitive and cooperative inventory policies in a two-stage supply chain," Management Science, vol. 45, no. 7, pp. 936–953, 1999.·
14. J. Sarkis, "A strategic decision framework for green supply chain management," Journal of Cleaner Production, vol. 11, no. 4, pp. 397–409, 2003.
15. Z. Huang, S. X. Li, and V. Mahajan, "An analysis of manufacturer-retailer supply chain coordination in cooperative advertising," Decision Sciences, vol. 33, no. 3, pp. 469–494, 2002.
16. M. E. Ferguson, "When to commit in a serial supply chain with forecast updating," Naval Research Logistics, vol. 50, no. 8, pp. 917–936, 2003.
17. A. Surana, S. Kumara, M. Greaves, and U. N. Raghavan, "Supply-chain networks: a complex adaptive systems perspective," International Journal of Production Research, vol. 43, no. 20, pp. 4235–4265, 2005. ·
18. A. Gupta and C. D. Maranas, "Managing demand uncertainty in supply chain planning," Computers and Chemical Engineering, vol. 27, no. 8-9, pp. 1219–1227, 2003. ·
19. D. Kim, S. T. Cavusgil, and E. Cavusgil, "Does IT alignment between supply chain partners enhance customer value creation? An empirical investigation," Industrial Marketing Management, vol. 42, no. 6, pp. 880–889, 2013.
20. B. L. Connelly, D. J. Ketchen, and G. T. M. Hult, "Global supply chain management: toward a theoretically driven research agenda," Global Strategy Journal, vol. 3, no. 3, pp. 227–243, 2013.
21. http://smallword.columbia.edu.
22. http://www.columbia.edu/cu/news/01/12/duncanWatts.html.
23. http://www.oakland.edu/enp/erdpaths.
24. P. Erdős and A. Rényi, "On the evolution of random graphs," Magyar Tudományos Akadémia Matematikai Kutató Intézetének Közleményei, vol. 5, pp. 17–61, 1960.
25. D. J. Watts and S. H. Strogatz, "Collective dynamics of "small-world" networks," Nature, vol. 393, no. 6684, pp. 440–442, 1998.
26. D. J. Watts, "The "new' science of networks," Annual Review of Sociology, vol. 30, pp. 243–270, 2004.
27. A.-L. Barabási and R. Albert, "Emergence of scaling in random networks," American Association for the Advancement of Science. Science, vol. 286, no. 5439, pp. 509–512, 1999.
28. R. Albert and A.-L. Barabási, "Statistical mechanics of

complex networks," Reviews of Modern Physics, vol. 74, no. 1, pp. 47–97, 2002.

29. A. Gregoriades and A. Sutcliffe, "Workload prediction for improved design and reliability of complex systems," Reliability Engineering and System Safety, vol. 93, no. 4, pp. 530–549, 2008.·

30. D. Shi, Q. Chen, and L. Liu, "Markov chain-based numerical method for degree distributions of growing networks," Physical Review E, vol. 71, no. 3, Article ID 036140, 9 pages, 2005.

31. D. Shi, L. Liu, S. X. Zhu, and H. Zhou, "Degree distributions of evolving networks," EPL (Europhysics Letters), vol. 76, no. 4, pp. 731–737, 2006.

32. X. F. Wang, "Complex networks: topology, dynamics and synchronization," International Journal of Bifurcation and Chaos, vol. 12, no. 5, pp. 885–916, 2002.·

33. Z. Li and G. Chen, "Global synchronization and asymptotic stability of complex dynamical networks,"IEEE Transactions on Circuits and Systems II: Express Briefs, vol. 53, no. 1, pp. 28–33, 2006.

34. L. F. Y. Q. Q. Guoning and H. Zheren, "Universality analysis method of parts for product family based on complex network," Chinese Journal of Mechanical Engineering, vol. 11, article 014, 2005.

35. C. Guorong, D. Juli, S. Jinliang, and Y. Ping, "Relationship network analysis of manufacturing system limited by business lines," TELKOMNIKA Indonesian Journal of Electrical Engineering, vol. 10, no. 6, pp. 1503–1509, 2012.

36. M. E. J. Newman and D. J. Watts, "Renormalization group analysis of the small-world network model,"Physics Letters A, vol. 263, no. 4–6, pp. 341–346, 1999.

37. http://wjj.cq.gov.cn/cqgy/gygk/cqgyfzndbg/40451.htm.

38. http://news.jrcq.cn/news/2014-02/11/1818.html.

39. http://www.jip.gov.cn/default.html.

CHAPTER 15

AN ADAPTIVE MAINTENANCE MODEL ORIENTED TO PROCESS ENVIRONMENT OF THE MANUFACTURING SYSTEMS

Xun Gong[1,2] Yixiong Feng[1] Hao Zheng[1] and Jianrong Tan[1]

[1]State Key Lab of Fluid Power Transmission and Control, Zhejiang University, Hangzhou 310027, China

[2]Robotics and Microsystems Center, Soochow University, Suzhou 215006, China

ABSTRACT

We explored an adaptive maintenance model of the process environment to diagnose progressive faults in manufacturing systems. Progressive faults are usually caused by deterioration of the operating environment or aging and show stochastic properties. Many researchers have reported how to detect faults on the machine body in manufacturing systems. However, little research has been conducted on the process environment which causes progressive faults. To tackle this problem, we explored an adaptive maintenance model to detect progressive faults and repair the process environment on the E-repair location. When a difference of the environmental factor state is detected, it will combine the transcription factor and the state enzyme to locate fault source. Then the comprehensive maintenance program is derived to repair the operating environment while eliminating progressive faults. For the purpose of validation, this model was implemented on the process environment of the air separation plant. And the simulation experiments validated the feasibility and effectiveness of this method.

INTRODUCTION

Some large complex manufacturing systems are often operated under high pressures, at high temperatures, with fast material flows and complex manufacturing mechanism. In production, facility malfunction, environmental fluctuation, or feed stream instability can introduce a variety of process disturbances, which would aggravate the load of the equipment, accelerate wear and tear on the components, and increase the consumption of electricity or power. Severe combined disturbance propagations in a plant can be destructive. Obviously, such security threatening situations should be detected early, the potential impact on production should be precisely monitored, and operational solutions should be derived quickly.

Failure process in practical engineering applications mainly includes fault diagnosis and maintenance. Publishing of Beard's doctoral dissertation in 1971 marked the birth of fault diagnosis technology [1]. Since then, the fault diagnosis technology has become a research focus and scholars have conducted extensive and in-depth researchin two groups: the model-based methods: some intelligent classification algorithms, such as artificial neural networks (ANNs) and support vector machines (SVM), have been successfully used for fault diagnosis of mechanical systems [2–5]. In machine

Citation: Xun Gong, Yixiong Feng, Hao Zheng, and Jianrong Tan, "An Adaptive Maintenance Model Oriented to Process Environment of the Manufacturing Systems," Mathematical Problems in Engineering, vol. 2014, Article ID 537452, 10 pages, 2014. doi:10.1155/2014/537452.

condition monitoring and fault diagnosis, some researchers have used this as a tool for classification of faults [6, 7]; the data-driven methods: this group of methods monitored and collected the input and output signals of the manufacturing process [8–11]. It extracted fault features from a large number of practical samples, described the relationships between faults and symptoms, and then constructed deep knowledge of expert systems [12, 13].

The concept of preventive maintenance (PM) involves the performance of maintenance activities prior to the failure of equipment [14, 15]. One of the main objectives of PM is to reduce the failure rate or failure frequency of the equipment. This strategy contributes to minimizing failure costs and machine downtime (production loss) and increasing product quality [16]. Reliability-centered maintenance emphasizes on equipment reliability and the consequences of equipment failure as the main basis for maintenance strategy [17, 18]; fault limited strategy decides whether to maintain the system or not by the failure rate and reliability as indicators [19]; condition-based maintenance strategies are monitoring the system [20, 21]; engineering systems maintenance strategy mainly considers the economic relations among the system devices [22].

The studies of environmental factors that affect product performance have focused on environmental simulation test before the operational process [23]. These methods exposed the defects of product components by the reliability enhancement testing. The adaptability of product was improved according to the scheduled test environment. The existing fault diagnosis and maintenance techniques are mostly oriented to device components or the system itself but not deep enough to the environmental factors stress which causes the failure [24].

From the year of failure statistics of the US airborne electronic equipment [25], it can be found that the fault caused by the temperature accounted for 22.2%; by the vibration accounted for 11.38%; by moisture accounted for 10%; by the dust accounted for 4.16%; by the salt spray accounted for 1.94%; by the impact accounted for 1.11%; by other causes accounted for 47.3%. From the above statistics it can be seen that the 52% of the total fault of the equipment system failures is caused by environmental stress factors of temperature, vibration, humidity, and pollution. At the same time the environmental stress factors also affect the validity of the detection data, thereby affecting the accuracy of fault diagnosis and blocking the maintenance work.

Motivated by those problems, we want to propose a maintenance method of the process environment to bridge the gap. The paper is presented as follows. Section 1 reviews briefly the development of the diagnosis and the maintenance. Section 2 presents the prerequisites of the method by analyzing the fundamental of the environmental stress response. Section 3 proposes the adaptive maintenance model to repair the abnormal process environment to be normal. Section 4 analyzes the temperature sensitivity of the air separation process (the precooling system/purification system/booster expansion turbine/refrigeration system). Section 5 presents the experiments on the air separation plant to test the feasibility and effectiveness of the adaptive maintenance model. Section 6 highlights findings of the paper and suggests potential research directions.

PREREQUISITE

Environmental stress response is defined as follows: during the P-F interval in the operating environment, it monitors the environmental factors of the equipment/system (the time from the potential failure to functional failure of the equipment called P-F interval). Once the early warning is in the potential failure state, the equipment/system is diagnosed timely to find out the disturbance source of the environmental factors. The operating environment is maintained in real time on the environmental repair location to avoid the duration of the potential failure state and the functional failure. It is shown in Figure 1.

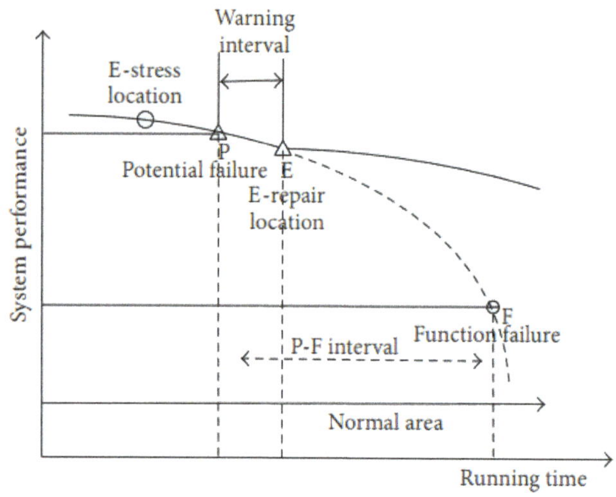

Figure 1: System performance of environmental stress response.

By the above description of the fundamental of the environmental stress response and Figure 1, it can be seen that the presence, occurrence, diagnosis, and maintenance of the operating environment potential failure have their own prerequisites. The adaptive maintenance model of the equipment/system operating environment in this paper requires some prerequisites like the following:(1)a certain degree of fault sensitivity to environmental stress;(2)determining an obvious potential failure state P;(3)less than P-F interval time length of fault warning and carrying out the environmental stress response at the environmental restoration point E;(4)the minimum P-F interval which must be long enough to arrange prevention and the environmental stress response in the potential failure process, but not the functional failure process.

The failure cumulative effect of the system is caused by environmental stress and it declines the system performance seriously. Various preventive techniques are used to diagnose potential failure timely to avoid the occurrence of functional failure. The environmental stress response on the environmental repair location repairs the operating failure environment to normal, to extend the lifetime of the system.

METHODOLOGY

Diagnostic Description

The mechanical device is constituted by the function, behavior, structure, carriers, and other design elements. In the mechanical system design theory, there are reciprocating mapping relationships among function domain, behavior domain, and carrier domain [26]. During operation, the device performance degrades because of fluctuations in E-factors. As shown in Figure 2, in order to diagnose the potential fault of the operating environment of the equipment in an abnormal operating environment, environment domain, monitoring domain, and state domain are increased beside equipment design elements domains.

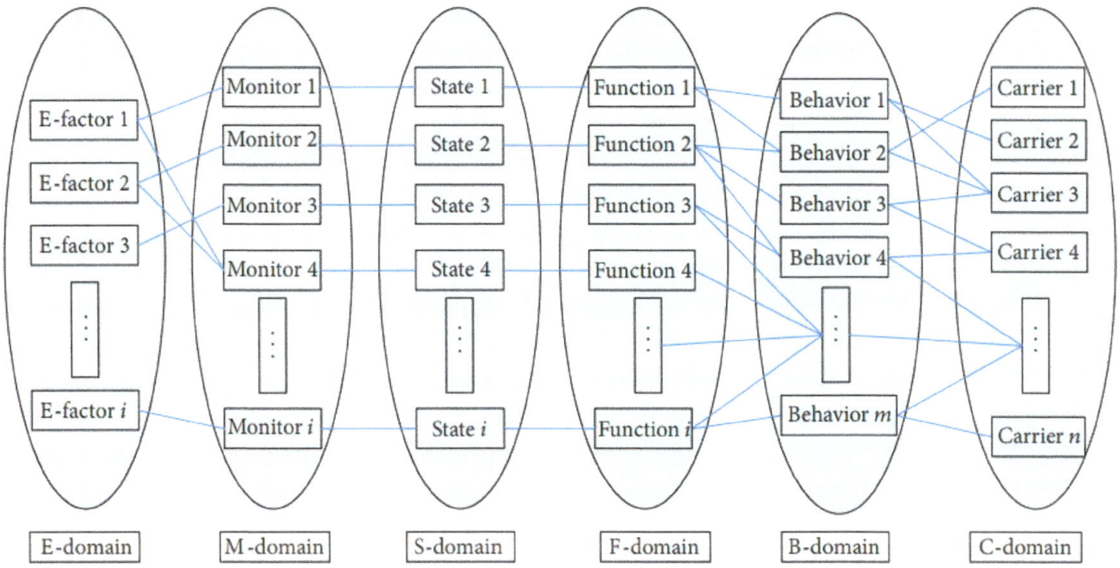

Figure 2: Domain structure mapping model.

Environmental domain is the set of E-factors in the equipment operating condition. The real-time state of the operating environment is gathered by monitoring the E-factors timely. The state data is traced back to the source of fluctuations in E-factors by diagnosing and analyzing.

The temperature is one of the E-factors which the equipment must face. Its fluctuations affect the performance of the system. For example, high temperature could cause thermal aging, structure changing, or physical expansion, while low temperature could cause material physical contraction, and temperature changes could cause the expansion and contraction, the institution stress, and so on.

Humidity is another E-factor. The high humidity stress can cause moisture accumulation and electrochemical reaction resulting in potential failures, while the low humidity stress can cause materials dried, grain, or other reactions.

The E-factors are various and random in reality, and there are mutual interactions among them. There are other E-factors leading to the system failure such as vibration/pressure/salt spray.

Adaptive Maintenance Model

The environment of the factory in which the equipment locates is extremely complex. There are a variety of disturbance sources causing the random fluctuations of the operating environment. The equipment is susceptible to progressive fault state under the E-factors stress during operation. If the abnormal operating environment is not diagnosed and maintained timely, the fault behavior is transferred and diffused among the mechanical components, and the progressive fault state also accumulates quickly. Eventually the function failure of the device would occur. Because of this, an adaptive maintenance model is proposed to diagnose and maintain progressive fault which is caused by environmental stress. It is shown in Figure 3. The adaptive maintenance model is composed of two parts. It is shown that the fluctuations of the manufacturing performance are caused by E-factors in part 1. And the monitoring and maintaining process are constructed in part 2.

The random fluctuations of the operating environment are considered during the system design process. E-factors are adjusted through the adaptive maintenance model based on the progressive fault state of the system. The method keeps the system in the normal operating environment.

Firstly, the system is running in the normal environment. According to the statistics, the stochastic dynamic environmental stresses are divided into several kinds such as temperature stress, humidity stress, vibration stress, and others.

Secondly, the stresses act on various components of the system causing a variety of physical and chemical reactions and affect the performance of the whole system to the progressive fault. If the progressive fault acts constantly on the system accumulating the fault state, that would lead to the function failure of the system. In the function failure process, the traditional fault diagnosis methods can be used for diagnosis and maintenance.

Thirdly, signal monitoring and recognition technology would be applied to the state of the system environment in the progressive fault process.

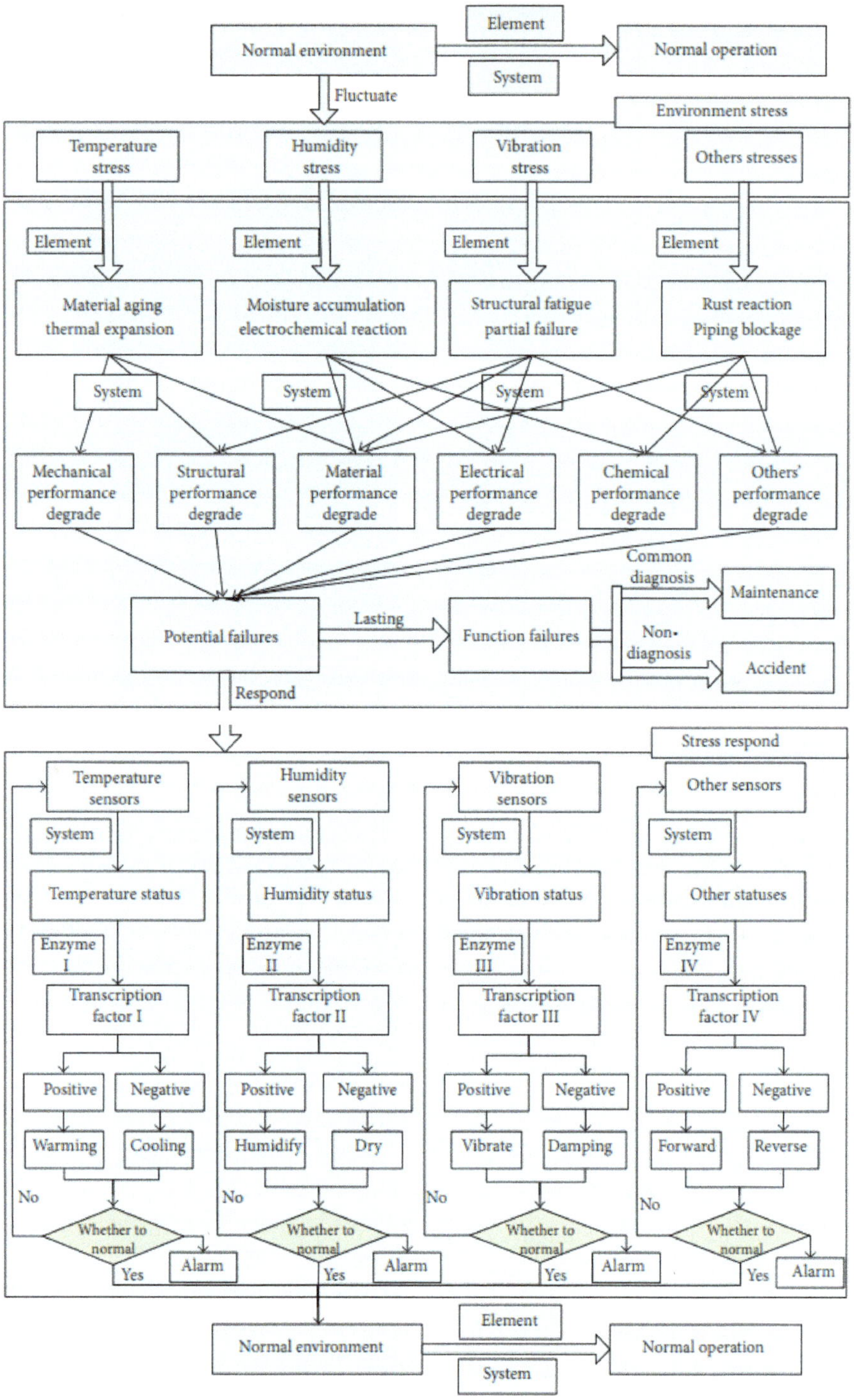

Figure 3: The adaptive maintenance model.

Fourthly, transcription factors and the array of state enzyme are constructed by the real-time operation of the system environment state and comapped to the expert system of the fault diagnosis established by the artificial intelligence methods. Environmental repair programs to the corresponding progressive fault states are obtained in the expert system.

Fifthly, the model uses transcription factors to activate or deactivate the components of the environmental repair programs and the state enzyme to adjust the trend and extent of the components. Then the closed-loop detection conditioning system is formed to repair the dynamic E-factors. And the progressive fault is removed by responding to the online environmental stress timely to maintain the normal operating environment of the system.

The specific steps are described as follows.

(1) Determine the system environment stress monitoring sites, L:

$$L = \{l_i \mid i = 1, 2, \ldots, n\}. \tag{1}$$

(2) The environment data of the system is obtained in the normal operation from the historical data statistics, T^0:

$$T^0 = \{t_i^0 \mid i = 1, 2, \ldots, n\}. \tag{2}$$

(3) The range of the environment factors of the system is obtained during the normal operation through the analysis of the system performance, T':

$$T' = \{|t'_i| \mid i = 1, 2, \ldots, n\}. \quad (3)$$

(4) The state enzyme array is constructed based on the structure and the control parameters of the components of the environmental repair programs, E.

The state enzyme (e_{ij}) stands for the fluctuation extent of the failure/performance parameters of the same component of the system in different environments:

$$E = \{e_{ij} = f(\Delta T_{ij}) \mid i = 1, 2, \ldots, n;\ j = 1, 2, \ldots, m\}. \quad (4)$$

(5) The system state array is constructed by real-time monitoring of the E-factors of the operation system, T:

$$T = \{t_{ij} \mid i = 1, 2, \ldots, n;\ j = 1, 2, \ldots, m\}. \quad (5)$$

(6) The environmental stress transcription factor array is calculated by monitoring the E-factor state data, F:

$$F = \{f_{ij} \mid i = 1, 2, \ldots, n;\ j = 1, 2, \ldots, m\},$$

$$f_{ij} = \begin{cases} 1 & |t'_i| < |t_{ij} - t_i^0|, t_{ij} < t_i^0 \\ 0 & |t'_i| > |t_{ij} - t_i^0| \\ -1 & |t'_i| < |t_{ij} - t_i^0|, t_{ij} > t_i^0 \end{cases} \quad (i = 1, 2, \ldots, n). \quad (6)$$

Transcription factor is one of the concepts of genetics. The transcription factor array can be used in the field of fault diagnosis to determine the locations of system components which could be affected by the environmental stress and in the progressive fault state. The positive or negative of the matrix values shows out the trend of the progressive fault.

(7) By the transcription factor array F and the state enzyme array E comapping to the expert system of the fault diagnosis, obtain the comprehensive maintenance program for the environmental stress response M. While the transcription factor value (f_{ij}) is 1, it shows that the monitoring site (l_i) is under the environmental stress and the response is positive, and while the transcription factor value (f_{ij}) is 0, it shows that the monitoring site (l_i) is not under the environmental stress and it needs no response, and while the transcription factor value (f_{ij}) is −1, it shows that the monitoring site (l_i) is under the environmental stress and the response is negative.

The repair effects of the environmental restoration program (M) is determined by comparing the environment state data of the operating system (such as T_i and T_{i+1}). If it is found out that the data and the trend do not match or exceed the regulatory range of the comprehensive maintenance (M), then the alarm would be worn. Once in this kind of situation, the components of the system should be diagnosed or the system should be upgraded to prevent functional failure happening.

Table 1: Monitoring data of the booster expansion turbine

Contents	Status			
	1	2	3	4
Outlet temperature (°C)	−157.7	−161.7	−165.7	−167.7
Oxygen content in oxygen (%)	0.99953	0.99944	0.99933	0.99927
Nitrogen content in nitrogen (%)	0.9998634	0.9998636	0.9998638	0.9998639
The ratio of oxygen extraction (%)	0.6731	0.6840	0.6951	0.7007

SENSITIVITY ANALYSIS

The air separation plant has the typical characteristics such as electrohydraulic system coupling, the complex spatial structure, and the high failure risk. The operation reliability and the product quality of the air separation plant are sensitive to environmental stresses such as temperature, humidity, vibration, and pressure. The temperature is one of the most important environmental stresses which the air separation plant must face. Its fluctuations affect the performance of the system.

The air separation process is mainly composed of a refrigerating system and rectification system, as shown in Figure 4. The simulation model includes the compressed air system, the precooling system, the purification system, the heat exchange system, the refrigeration system, and the distillation system. And there are parts of the air separation experimental setups shown in Figure 5 (the heat exchange system).

The precooling system of the air separation plant cools down by the circulating water. The temperature of the cooling water will decline with the drop of the atmospheric humidity and temperature and would enhance the cooling effect on the compressed air heat load; on the contrary, the temperature will rise with the rise of the atmospheric humidity and temperature and would decline the cooling effect.

The reduction of the cooling effect leads to the high temperature of the air before entering the purification system. And the high temperature air would affect the normal operation of the molecular sieve purification into the potential failure condition. Then the system production load is decreased, the productivity declines, and the power consumption increases.

85% to 90% of the cooling capacity is produced by the turbine expander of the full low pressure air separation plant. The purified air passes through the turbine expander cooling to form the raw solution. The raw solution is separated into gas products by the distillation column.

The air separation processes are simulated and the outlet temperatures of the booster expansion turbine are adjusted in the simulation. Collect the changes of the purity of the oxygen product and the nitrogen product, the oxygen extraction rate, and other data. Parts of the collected data are shown in Table 1 and Figure 6.

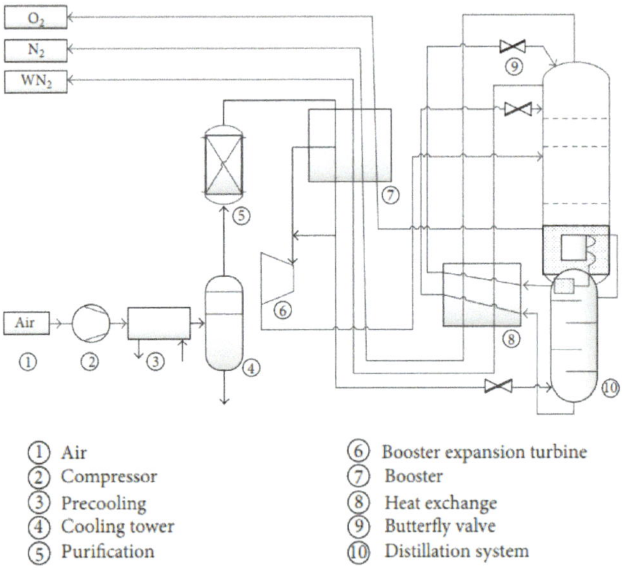

Figure 4: The simulation model of air separation process.

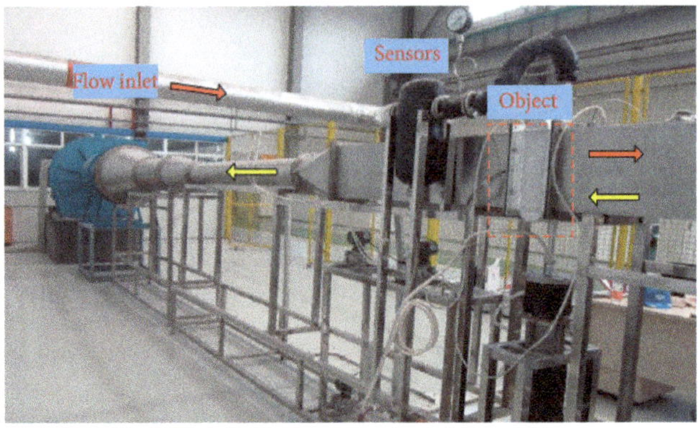

Figure 5: Parts of the air separation equipment.

It is shown directly in Figure 6. When the temperature of the booster turbo expander declines, the expansive air is blown into the feeding plate of the upper rectifying tower and its superheat drops down. That causes the vaporization of the reflux liquid in the rectifying section of the upper tower decrease. The liquid-gas ratio in the rectifying section is higher than at the original operational temperature. The oxygen fraction condenses fully from the vapor phases to the liquid phase because of the increase of the reflux liquid flow. The oxygen content of the vapor phase drops while the nitrogen content rises.

EXPERIMENTS

A series of experiments were carried out to validate the proposed approach. The air separation process is simulated based on the data collected in the field on a certain type (7500/15000) of air separation plant in the normal operating environment. Construct the normal operating state (S_0) of the simulated air separation according to the requirements of the air separation such as the convergence and the thermal coupling of the tower systems. And then monitor the Efactors such as inlet-outlet temperatures, pressures, and flows of the key equipments such as air compressors, precooling systems, purification systems, and the booster expansion turbine.

With reference to the system historical monitoring data, the process is simulated by adjusting E-factors on the air separation system. The simulation results are shown in Table 2.

Table 2: Statistics of the air separation process compared simulation

Parameters	S	The operation in the stress state but without the environmental stress response				The operation in the stress state and with the environmental stress response			
		T_1	T_2	T_3	T_4	T_5	T_6	T_7	T_8
Air compressor									
cp_in	T	41.3	41.3	41.3	41.3	41.3	41.3	41.3	41.3
cp_out	T	72.48	75.62	78.14	79.97	72.48	75.62	78.14	79.97
Precooling system									
c_in	T	72.57	75.47	77.98	80.51	72.57	75.47	77.98	80.51
c_out	T	18.88	19.56	20.17	20.85	18.88	17.32	15.41	13.82
Purification system									
p_in	T	18.96	18.81	19.19	19.91	18.96	17.29	15.27	13.68
p_out	T	29.87	31.56	34.24	37.11	29.87	25.75	21.29	17.14
Air booster									
pb_in	T	29.87	31.44	34.12	37.03	29.89	25.77	21.35	17.28
pb_out	T	96.2	98.4	102.3	105.7	96.2	82.7	75.3	65.8
Booster expansion turbine									
e_in	T	−106.68	−104.25	−102.67	−100.18	−106.68	−108.53	−110.24	−111.08
e_out	T	−158.2	−157.5	−155.7	−153.5	−158.2	−158.9	−159.8	−161.4
Distillation tower									
Flow	O	7395	7392	7388	7382	7395	7406	7419	7430
	N	14547	14542	14539	14537	14547	14561	14572	14581
	WM	11826	11829	11831	11837	11826	11821	11814	11809
Purity	O	0.999287	0.999311	0.999338	0.999354	0.999287	0.999012	0.998772	0.998616
	N	0.918654	0.918649	0.918642	0.918636	0.918654	0.918692	0.918706	0.918712

For the operations, select the temperature stress which is one of the E-factors as the object from Table 2. Detect mainly the inlet and outlet temperatures of the air compressor, the precooling system, the purification system, and the booster expansion turbine and monitor the components of the gas products from the rectification tower. Then construct the array (L) of the system monitoring locations:

$$L = \{\text{cp\_in}, \text{cp\_out}, \text{c\_in}, \text{c\_out}, \text{p\_in}, \text{p\_out}, \text{pb\_in}, \text{pb\_out}, \text{e\_in}, \text{e\_out}\}. \quad (7)$$

Construct the fluctuation range array (T') to maintain the healthy state based on the environment parameters of the operation system.

Consider T' = {40 60 60 4 4 6 6 30 10 3}. Construct the system state array (T) based on the simulated data in Table 2.

The monitoring data ($T_1 \sim T_4$) in Table 2 are obtained by the timing simulation of the air separation process when the system is in the stress state (S_8) but without the environmental stress response. It shows that the potential failure of the system accumulates constantly.

The monitoring data ($T_5 \sim T_8$) in Table 2 are obtained by the timing simulation of the air separation process when the system is in the stress state (S_8) and the environmental stress response.

Then $T_{10,4}$ and $T'_{10,4}$ can be obtained from Table 2:

$$T_{10,4} = \begin{vmatrix} 41.30 & 41.30 & 41.31 & 41.31 \\ 72.48 & 75.62 & 78.14 & 79.97 \\ 72.57 & 75.47 & 77.98 & 80.51 \\ 18.88 & 19.56 & 20.17 & 20.85 \\ 18.96 & 18.81 & 19.19 & 19.91 \\ 29.87 & 31.56 & 34.24 & 37.11 \\ 29.87 & 31.44 & 34.12 & 37.03 \\ 96.19 & 98.41 & 102.30 & 105.70 \\ -106.68 & -104.25 & -102.67 & -100.18 \\ -158.2 & -157.5 & -155.7 & -153.5 \end{vmatrix},$$

$$T'_{10,4} = \begin{vmatrix} 41.30 & 41.30 & 41.31 & 41.31 \\ 72.48 & 75.62 & 78.14 & 79.97 \\ 72.57 & 75.47 & 77.98 & 80.51 \\ 18.88 & 17.32 & 15.41 & 13.82 \\ 18.96 & 17.29 & 15.27 & 13.68 \\ 29.87 & 25.75 & 21.29 & 17.14 \\ 29.89 & 25.77 & 21.35 & 17.28 \\ 96.19 & 82.71 & 75.34 & 65.80 \\ -106.68 & -108.53 & -110.24 & -111.08 \\ -158.18 & -158.91 & -159.83 & -161.42 \end{vmatrix}. \quad (8)$$

Locations of system components which are in the potential failure can be detected in the positive stresses by the transcription factor array (F_8). Start a negative regulation scheme for the temperature stresses to cool down or cooling the ambient system environment, while the state enzyme array (E_8) is obtained by comapping and calculating.

Make the most obvious state of the environmental stress (S_8) as the state of the potential failure which is under the temperature stress in computing of the environmental stress response. According to formula 8, the transcription factor array (F_8) can be obtained as follows:

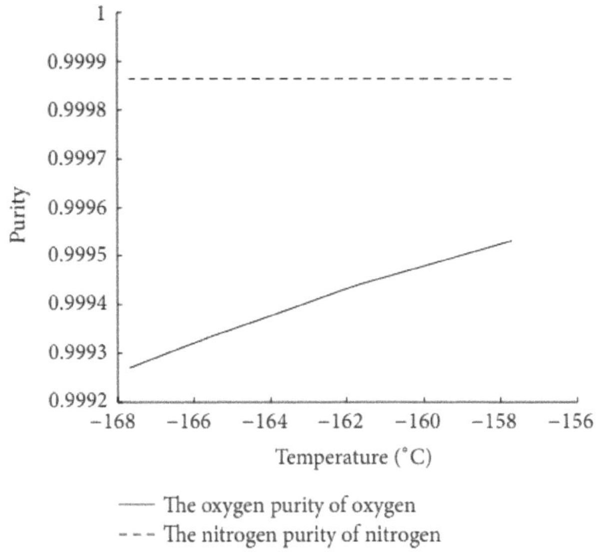

Figure 6: Diagrams of outlet temperatures of the booster expansion turbine and oxygen and nitrogen purity.

$$F_8 = \begin{cases} 26.8 < 40 & & \longrightarrow 0 \\ 36 < 60 & & \longrightarrow 0 \\ 36 < 60 & & \longrightarrow 0 \\ 5.98 > 4 & 12.2 < 18.18 & \longrightarrow -1 \\ 6.06 > 4 & 12.2 < 18.26 & \longrightarrow -1 \\ 14.07 > 6 & 15.8 < 29.87 & \longrightarrow -1 \\ 14.07 > 6 & 15.8 < 29.87 & \longrightarrow -1 \\ 37.12 > 30 & 59.08 < 96.2 & \longrightarrow -1 \\ 6.93 < 10 & & \longrightarrow 0 \\ 6.9 > 3 & -165.1 < -158.2 & \longrightarrow -1 \end{cases}$$

$$F_8 = \{0, 0, 0, -1, -1, -1, -1, -1, 0, -1\},$$

$$\begin{aligned} E_8 &= \{e_{ij} = f(\Delta T_{ij}) \mid i = 1, 2, \ldots, n;\ j = 1, 2, \ldots, m\} \\ &= \{0, 0, 0, 1.49, 1.52, 2.35, 2.35, 1.24, 0, 2.3\} \end{aligned} \quad (9)$$

Cool down the components which are under the potential failure and the ambient environment by operating the comprehensive maintenance program (M_8) to adjust the Efactors.

The comparisons of the air separation process simulation between those without (solid line) and those with (dotted line) the environmental stress response are shown in Figures 7 and 8. The outlet temperatures of the precooling system, the purification system, and the air compressor are shown in Figure 7. The inlet-outlet temperatures of the booster expansion turbine are shown in Figure 8.

It can be seen by the comparisons that it is effective to take the environmental stress response on the operational phase of the air separation system as shown in Table 3. The reducing of the temperature means low power consumption of the manufacturing system and the increasing of production means higher benefits. Also the normal operating environment would extend the service life of the system.

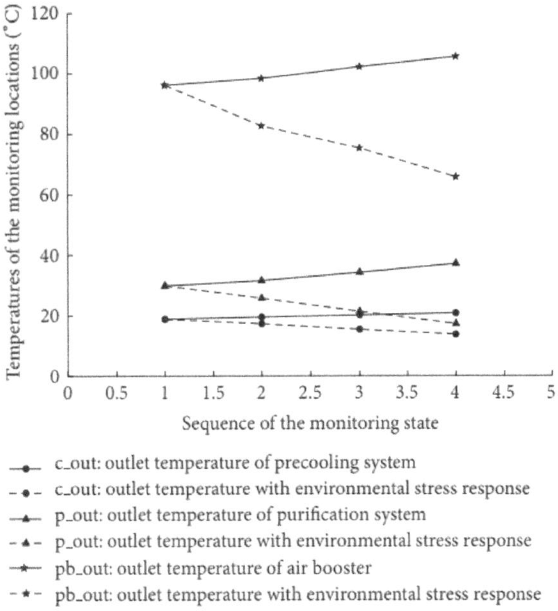

Figure 7: Contrast diagram of the running temperatures of the components.

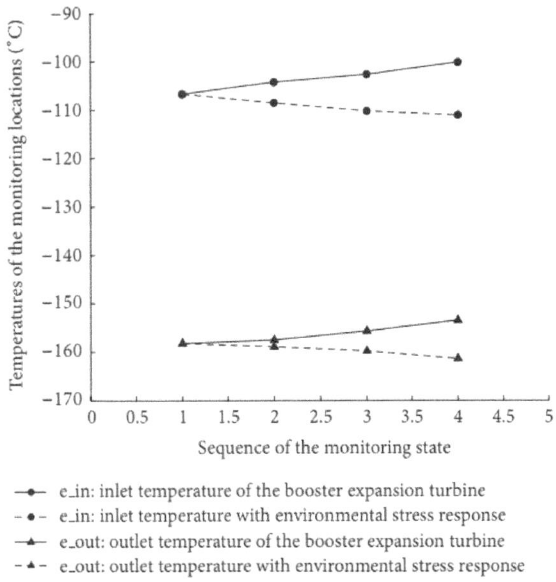

Figure 8: Contrast diagram of the inlet and outlet temperatures of the booster expansion turbine.

Table 3: Analysis of the effect of the adaptive maintenance model

Locations	Status			
	T_4 (°C)	T_8 (°C)	ΔT (°C)	Δ%
c_out	20.85	13.82	−7.03	−33.72
p_out	37.11	17.14	−19.97	−53.81
pb_out	105.7	65.8	−39.9	−37.75
e_in	−100.18	−111.08	−10.9	−10.88
e_out	−153.5	−161.4	−7.9	−5.15
F_O	7382	7430	48	+0.65
F_N	14537	14581	44	+0.30

CONCLUSION

We presented an adaptive maintenance model to repair the process environment which caused progressive faults in the air separation plant system. This maintenance approach includes the following. (1) The diagnostic model monitors the environmental states of the plants and also compares the inputs/outputs and presettings to detect faults. (2) The mapping structure is constructed with the I/O environmental states and behaviors of carriers, while the state enzyme and the transcription factor array are calculated through the expert system. (3) The comprehensive maintenance program is obtained by the comapping of the state enzyme and the transcription factor array for the environmental stress response.

For the future research, we suggest to optimize the deployment of the sensors for the model. Through preselection of sensor locations, it may improve the detection of the system with optimal cost and sensor configuration.

ACKNOWLEDGMENTS

This work was supported by the National Natural Science Foundation of China (No. 51322506 and 51175456), Zhejiang Provincial Natural Science Foundation of China (No. LR14E050003), the Fundamental Research Funds for the Central Universities, Innovation Foundation of the State Key Laboratory of Fluid Power Transmission and Control, and Zhejiang University K.P. Chao's High Technology Development Foundation. Sincere appreciation is extended to the reviewers of this paper for their helpful comments.

REFERENCES

1. R. V. Beard, Failure Accommodation in Linear Systems through Self-Reorganization, MIT, Cambridge, Mass, USA, 1971.
2. M. Demetgul, "Fault diagnosis on production systems with support vector machine and decision trees algorithms," The International Journal of Advanced Manufacturing Technology, vol. 67, no. 9–12, pp. 2183–2194, 2013.
3. J. Yang, Y. Zhang, and Y. Zhu, "Intelligent fault diagnosis of rolling element bearing based on SVMs and fractal dimension," Mechanical Systems and Signal Processing, vol. 21, no. 5, pp. 2012–2024, 2007.
4. S. F. Yuan and F. L. Chu, "Support vector machines-based fault diagnosis for turbo-pump rotor,"Mechanical Systems and Signal Processing, vol. 20, no. 4, pp. 939–952, 2006.
5. A. Widodo and B. Yang, "Support vector machine in machine condition monitoring and fault diagnosis,"Mechanical Systems and Signal Processing, vol. 21, no. 6, pp. 2560–2574, 2007.
6. A. Widodo, B. Yang, and T. Han, "Combination of independent component analysis and support vector machines for intelligent faults diagnosis of induction motors," Expert Systems with Applications, vol. 32, no. 2, pp. 299–312, 2007.
7. Y. Yang, D. Yu, and J. Cheng, "A fault diagnosis approach for roller bearing based on IMF envelope spectrum and SVM," Measurement: Journal of the International

Measurement Confederation, vol. 40, no. 9-10, pp. 943–950, 2007.

8. P. Stepanic, I. V. Latinovic, and Z. Djurovic, "A new approach to detection of defects in rolling element bearings based on statistical pattern recognition," International Journal of Advanced Manufacturing Technology, vol. 45, no. 1-2, pp. 91–100, 2009.
9. F. Pan, S. R. Qin, and L. Bo, "Development of diagnosis system for rolling bearings faults based on virtual instrument technology," Journal of Physics: Conference Series, vol. 48, article 467, 2006. ·
10. C. Angeli, "An online expert system for fault diagnosis in hydraulic systems," Expert Systems, vol. 16, no. 2, pp. 115–120, 1999.
11. C. Angeli and A. Chatzinikolaou, "On-line fault detection techniques for technical systems: a survey,"International Journal of Computer Science & Applications, vol. 1, no. 1, pp. 12–30, 2004.
12. B. Samanta and K. R. Al-Balushi, "Artificial neural network based fault diagnostics of rolling element bearings using time-domain features," Mechanical Systems and Signal Processing, vol. 17, no. 2, pp. 317–328, 2003.
13. T. Lindh, On the condition monitoring of induction machines [Ph.D. thesis], Lappeenranta University of Technology, 2003.
14. I. B. Gertsbakh, Models of Preventive Maintenance, North-Holland Publishing, Oxford, UK, 1977.
15. H. Löfsten, "Management of industrial maintenance—economic evaluation of maintenance policies,"International Journal of Operations and Production Management, vol. 19, no. 7, pp. 716–737, 1999.
16. J. S. Usher, A. H. Kamal, and W. H. Syed, "Cost optimal preventive maintenance and replacement scheduling," IIE Transactions, vol. 30, no. 12, pp. 1121–1128, 1998.
17. L. Pintelon and G. Waeyenbergh, "A practical approach to maintenance modelling," in Flexible Automation and Intelligent Manufacturing, J. Ashayeri, W. G. Sullivan, and M. M. Ahmad, Eds., pp. 1109–1119, Begell House, New York, NY, USA, 1999.
18. G. Waeyenbergh and L. Pintelon, "A framework for maintenance concept development," International Journal of Production Economics, vol. 77, no. 3, pp. 299–313, 2002.
19. H. Wang, "A survey of maintenance policies of deteriorating systems," European Journal of Operational Research, vol. 139, no. 3, pp. 469–489, 2002. ·
20. M. Wiseman, "Optimizing condition based maintenance," Plant Engineering and Maintenance, vol. 23, no. 6, pp. 57–71, 2001.
21. Y. Liao, G. Lang, and L. Qu, "Precession trend analysis and balancing strategy for rotors with multi-fault,"Journal of Mechanical Engineering, vol. 45, no. 8, pp. 45–51, 2009.
22. L. Dieulle, C. Bérenguer, A. Grall, and M. Roussignol, "Sequential condition-based maintenance scheduling for a deteriorating system," European Journal of Operational Research, vol. 150, no. 2, pp. 451–461, 2003. ·
23. M. Kearney, J. Marshall, and B. Newman, "Comparison of reliability enhancement tests for electronic equipment," in Proceedings of the Reliability and Maintainability Symposium, pp. 435–440, Singapore, January 2003.
24. R. Ahmad and S. Kamaruddin, "An overview of time-based and condition-based maintenance in industrial application," Computers and Industrial Engineering, vol. 63, no. 1, pp. 135–149, 2012.
25. R. D. Brillhart, D. L. Hunt, and H. Chimerine, "Multiple input excitation methods for aircraft ground vibration testing," Sound and Vibration, vol. 27, no. 1, pp. 77–85, 1993.
26. Y. Umeda, T. Takeda Tomiyama, et al., "Function behavior and structure," Application of Artificial Intelligence in Engineering, vol. 10, no. 4, pp. 177–193, 1990.

CHAPTER 16 MODELING AND VERIFICATION OF RECONFIGURABLE AND ENERGY-EFFICIENT MANUFACTURING SYSTEMS

Jiafeng Zhang,[1,2] Mohamed Khalgui,[3] Wassim Mohamed Boussahel,[2,4] Georg Frey,[2] ChiTin Hon,[5] Naiqi Wu,[5] and Zhiwu Li[5,6]

[1]School of Electro-Mechanical Engineering, Xidian University, Xi'an 710071, China

[2]Chair of Automation and Energy Systems, Saarland University, 66123 Saarbrücken, Germany

[3]University of Carthage, 1054 Carthage, Tunisia

[4]Zentrum für Mechatronik und Automatisierungstechnik, 66121 Saarbrücken, Germany

[5]Institute of Systems Engineering, Macau University of Science and Technology, Taipa, Macau

[6]Faculty of Engineering, King Abdulaziz University, Jeddah 21589, Saudi Arabia

ABSTRACT

This paper deals with the formal modeling and verification of reconfigurable and energy-efficient manufacturing systems (REMSs) that are considered as reconfigurable discrete event control systems. A REMS not only allows global reconfigurations for switching the system from one configuration to another, but also allows local reconfigurations on components for saving energy when the system is in a particular configuration. In addition, the unreconfigured components of such a system should continue running during any reconfiguration. As a result, during a system reconfiguration, the system may have several possible paths and may fail to meet control requirements if concurrent reconfiguration events and normal events are not controlled. To guarantee the safety and correctness of such complex systems, formal verification is of great importance during a system design stage. This paper extends the formalism reconfigurable timed net condition/event systems (R-TNCESs) in order to model all possible dynamic behavior in such systems. After that, the designed system based on extended R-TNCESs is verified with the help of a software tool SESA for functional, temporal, and energy-efficient properties. This paper is illustrated by an automatic assembly system.

INTRODUCTION

A reconfigurable manufacturing system (RMS) is designed at the outset for rapid change in structure, as well as in hardware and software components, in order to quickly adjust production capacity and functionality within a part family in response to sudden changes in market or in regulatory requirements [1]. A RMS should be designed with several configurations (behavior modes) to, respectively, meet different production requirements in various conditions. There are two types of reconfigurations: static and dynamic reconfigurations. Generally, a static reconfiguration is applied offline to modify a RMS extensively such as adjusting architecture of physical systems and removing obsoleted machines, whereas a dynamic system reconfiguration, to switch a RMS from one configuration to another at runtime, is applied with the aim of fault-tolerance or actively changing system behavior modes [2, 3]. This paper focuses on dynamic RMSs.

Traditionally, manufacturing is an energy-intensive process, using motors, steam, and compressed air systems to transform raw materials into durable goods and consumer products [4–6]. Recent research shows that switching machines of a manufacturing system into their energy-efficient modes

Citation: Jiafeng Zhang, Mohamed Khalgui, Wassim Mohamed Boussahel, et al., "Modeling and Verification of Reconfigurable and Energy-Efficient Manufacturing Systems," Discrete Dynamics in Nature and Society, vol. 2015, Article ID 813476, 14 pages, 2015. doi:10.1155/2015/813476.

when they are idle during production can make considerable contribution to the reduction of energy demand and thus can reduce carbon footprint as well as operating costs [7–15]. This paper takes the advantage of dynamic reconfigurations of machines of a RMS between their working modes and energy-efficient modes as a way of reducing system energy consumption. A RMS with such energy-efficient operations is called a reconfigurable and energy-efficient manufacturing system (REMS).

REMSs can be abstracted as reconfigurable discrete event systems (DESs) when only their logic behavior properties are investigated. In this paper, a reconfiguration is called a local reconfiguration, if it is applied for switching a machine of a REMS between its working mode and energy-efficient mode. A reconfiguration is named a global reconfiguration if it is applied for switching a REMS between different configurations.

A REMS should be able to reconfigure itself smoothly due to changed inner/outer environments at runtime. Meanwhile, normal unreconfigured events should go on occurring whenever they meet their occurrence preconditions. However, uncontrolled concurrence of reconfiguration events and normal events may cause faults such as deadlocks and overflow [16–19]. Therefore, the formal verification is of great importance during design stages.

Petri nets [20, 21] have found an extensive application to discrete event systems [22, 23], including automated flexible manufacturing systems [24–28] and reconfigurable systems [29]. Reconfigurable timed net condition/event systems (R-TNCESs) [30, 31] are reconfigurable extensions of timed net condition/event systems (TNCESs) [32, 33]. TNCESs [34, 35] have a visual graph expression, a clear modular structure, and an exact mathematical definition inherited from Petri nets [36–40]. In addition, they have a strong analysis software tool: SESA (http://homepages.engineering.auckland.ac.nz/vyatkin/tools/modelchekers.html) [41]. System behavior properties, such as state/event trajectories and temporal requirements, can be specified by Computation Tree Logic (CTL), extended CTL (eCTL), and timed CTL (TCTL) [42–44] before being checked by SESA automatically. If a property is satisfied by the system, the model checker will return "true". Otherwise, a counterexample will be returned. Therefore, TNCESs have been widely applied in verification and validation of industry control systems especially for manufacturing systems [45–47]. The verification of a R-TNCES can be performed with the assistance of SESA [30, 31].

However, R-TNCESs cannot fully meet our requirements for a REMS. In a R-TNCES, reconfiguration functions model system reconfiguration events and transitions model normal events. However, the concurrence of reconfiguration functions and transitions is forbidden in a R-TNCES, which is in fact inconsistent with system requirements of REMSs. As a result, formal verification of such complex systems cannot be performed.

Motivated by the fact aforementioned, this paper extends R-TNCESs. First, the reconfiguration functions of R-TNCESs are assigned with action ranges and concurrent decision functions. After that, they are divided into two types according to their action ranges: major and minor reconfiguration functions. The major ones are used to model global reconfiguration events, whereas the minor ones are applied to model local reconfiguration events. Accordingly, the dynamics of R-TNCESs is updated for these extensions such that the concurrence of reconfiguration events and normal events can be conditionally allowed to guarantee the system correctness. Afterwards, an implementation method for an extended R-TNCES is developed. Finally, the software tool SESA is applied to check system functional, temporal, and energy-efficient properties. An automatic assembly system is used to illustrate this work.

The paper is organized as follows. The system specification of REMSs and the applied automatic assembly system are depicted in Section 2. The drawbacks of R-TNCESs on analyzing REMSs and the proposed extended R-TNCESs are described in Section 3. The formal verification of a REMS based on extended R-TNCESs is illustrated in Section 4. Finally, Section 5 concludes this paper and briefly presents further studies.

RECONFIGURABLE AND ENERGY-EFFICIENT MANUFACTURING SYSTEMS

This paper treats a reconfigurable and energy-efficient manufacturing system (REMS) as a reconfigurable discrete event control system. This section presents system specification and interesting system dynamics before it illustrates them with an automatic assembly system.

System Specification

A REMS is designed with a set of configurations to meet various requirements in different execution environments. A configuration CON is defined as

$$\text{Con} = (\text{Com}, \text{Str}, \text{Dat}), \qquad (1)$$

where Com is a set of all activated components in Con, Str defines the structure, that is, the connection relationship and the communication protocol among components of Com, and Dat denotes the set of all global variables and parameters of Con.

A REMS is denoted by

$$\text{Sys} = \left(\sum, R_c\right), \qquad (2)$$

where $\sum$ is the set of n configurations and $R_c : \sum \rightarrow \sum$ is the reconfigurable controller dealing with system reconfigurations.

There are two types of system reconfigurations in a REMS: global and local reconfigurations. The former ones are applied for switching system configurations. The latter ones are applied for switching an activated component between its working mode and energy-efficient mode when the system is in a particular configuration.

A REMS starts running as described in one of these configurations. After that, it should be able to change into other configurations smoothly due to the detection of component faults or other well-defined conditions. In addition, in each configuration, local reconfigurations can be applied to components such that the components can reconfigure themselves into their energy-efficient modes to save energy when they are idle and turn back to their working modes when the system needs them.

Dynamics of a REMS can be described by the evolution of system states. The evolution is caused by the occurrences of events events. A REMS includes three types of normal events, local reconfiguration events, and global reconfiguration events. (1)If a normal event occurs, the system changes its state within its current configuration.(2)If a local reconfiguration event occurs, a component of current configuration switches into its energy-efficient mode or switches back into its working mode. (3)If a global reconfiguration event occurs, the system switches into another configuration.

Meanwhile, during a global or local reconfiguration, if normal events meet their occurring conditions and they are not modified by the occurring reconfiguration events, they should go on occurring. However, this kind of concurrence brings safety threat to the system, since they may cause unboundedness, deadlocks, and even other functional or temporal failings.

Running Example

An automatic assembly system, denoted by AAS, is applied to illustrate works presented in this paper. AAS includes three workstations ($W1$, $W2$, and $W3$) and four robots (Rb1, Rb2, Rb3, and Rb4). It is assumed that robots are high energy consumption machines. The respective time consumption of $W1$, $W2$, and $W3$ to finish a machining task is 40 time units, 30 time unites, and 50 time unites. The time consumption of both Rb1 and Rb2 to finish a task is 20 time units. The time consumption of both Rb3 and Rb4 to finish a task is 25 time units. The default working process diagram of AAS is shown in Figure 1.

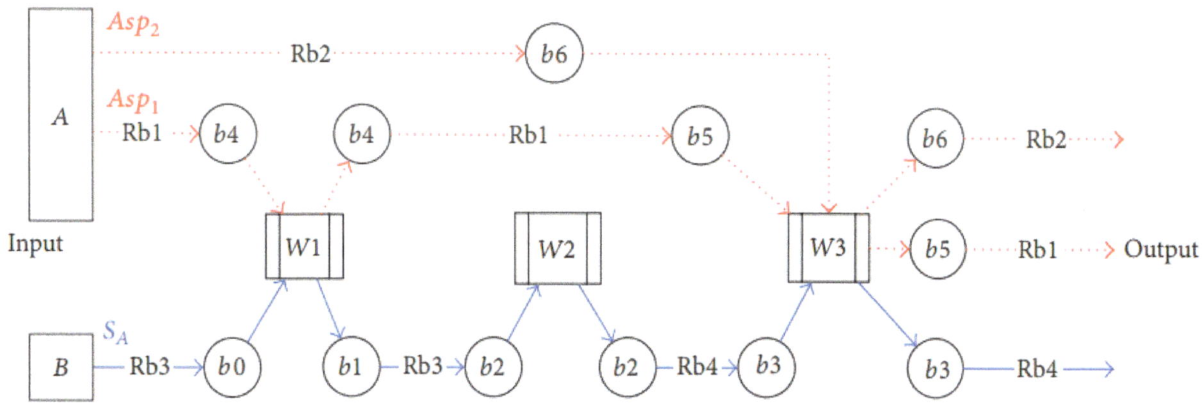

Figure 1: Default working process diagram of AAS.

The main function of AAS is to assemble machine parts into a subassembly of a vehicle, to be marked by S_A. Robots Rb1 and Rb2 move machine parts from the input into AAS, transfer machine parts between workstations, and remove trashy machine parts to the output. Dotted arrows in Figure 1 are used to denote the movements of machine parts during an assembly process. On the other hand, S_A is shifted along $W1$, $W2$, and $W3$ by robots Rb3 and Rb4. Solid arrows in Figure 1 are used to denote the movement of S_A. To make it clear, $b0$, $b1$,..., and $b6$ are used to denote positions where machine parts or S_A should be during an assembly process. The main assembly process is briefly described by the following three steps.

(1) The to-be-worked subassembly S_A is shifted from input B to $b0$ by Rb3. A machine part Asp_1 is delivered to $b4$ from the input A. After that, Asp_1 and S_A are preprocessed on $W1$. Then, the preprocessed S_A is moved to $b1$ automatically. The preprocessed $Asp1$ is moved to $b4$ automatically before being moved to position $b5$ by Rb1.

(2) S_A is transported to $b2$ from $b1$ by Rb3.Then, a second preprocess for S_A is done by $W2$. After that, S_A is shifted to $b3$ from $b2$ by Rb4.

(3) A machine part Asp_2 is delivered to $b6$ by Rb2. Then, $W3$ starts the assembly after S_A is in $b3$, preprocessed Asp_1 is in $b5$, and Asp_2 is in $b6$. After the assembly, the machined S_A is moved out by Rb4. Two other trashy machine parts are removed out of AAS by Rb1 and Rb2, respectively.

It is assumed that four behavior modes are designed for AAS. Their work processes are illustrated as follows.

(i): *Mode* 1: *Mode* 1 is the default mode as depicted in Figure 1, where all the robots are applied.

(ii): *Mode* 2: *Mode* 2 is a responding mode when Rb2 breaks down, where Rb1 should update itself to perform the function of Rb2.n of.

(iii): *Mode* 3: *Mode* 3 is applied when Rb4 breaks down during the execution of *Mode* 1, where the work of Rb4 has to be done by Rb3.

(iv): *Mode* 4: *Mode* 4 is applied when both Rb1 and Rb2 break down, where only Rb1 and Rb3 are applied. In this case, Rb1 should cover the function of Rb2 as in *Mode* 2 and Rb3 should cover the function of Rb4 as in *Mode* 3.

In each behavior mode, the applied robots should be able to reconfigure themselves into their energy-efficient modes when they are idle and reconfigure themselves back into their working modes when they have new tasks. A local reconfiguration for switching a robot from its working mode to its energy-efficient mode consumes one time unit. Likewise, a local reconfiguration for switching a robot from its energy-efficient mode back to its working mode consumes one time unit, as well.

To avoid the halt of a continuous production line, possible dynamic reconfigurations applied for switching AAS between these behavior modes are shown in Figure 2. The solid arrows denote global reconfigurations and dotted ones denote local reconfigurations.

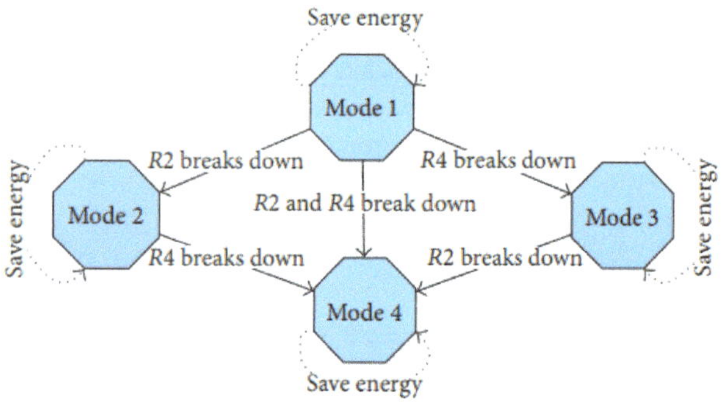

Figure 2: Possible reconfigurations in AAS.

It is assumed that a robot consumes one energy unit per time unit when it works in its working mode. However, it only consumes 30% energy units per time unit when it works in its energy-efficient mode. Note that the numerical value "30%" is an assumption by the authors to facilitate the quantitative analysis on energy-efficient operations. It does not come from any literature on industry systems.

Obviously, the possible reconfiguration events of AAS can occur simultaneously with many normal events in it. For example, when RB1 is being modified by a global reconfiguration or being switched into its energy-efficient mode, only its own work needs to stop for a while and the workstations and other running robots should do their jobs unaffectedly.

EXTENDED R-TNCES

Reconfigurable timed net condition/event systems (RTNCESs) [30, 31] are extensions of timed net condition/event systems (TNCESs) [34, 35]. Reconfiguration functions of R-TNCESs can be used to model global reconfiguration events of REMSs. However, they are not proper to model local reconfiguration events of REMSs directly. In addition, the concurrence of normal events and reconfiguration events is currently not allowed in R-TNCESs. Therefore, in order to perform correct formal verification of a REMS, this paper extends R-TNCESs. This section briefly recalls basic conceptions of R-TNCESs, analyzes the drawbacks of R-TNCEs on investigating REMSs, and represents the proposed extended R-TNCESs.

R-TNCES

Definition 1 (see [30]). A R-TNCES is a structure RN = (B, R), where B is a behavior module and R is a control module. The behavior module B is a union of n superposed TNCESs. For any $i \in \{0,...,n-1\}$, the TNCES Γi is denoted by $\Gamma_i = (N_{\Gamma i}, z_{0i})$ with $N_{\Gamma i} = (P_i, T_i, F_i, CN_i, EN_i, emi, DC_i)$. Then B can be represented as

$$\mathcal{B} = (\mathcal{P}, \mathcal{T}, \mathcal{F}, \mathcal{CN}, \mathcal{EN}, \mathcal{EM}, DC, \mathcal{Z}_0) \qquad (3)$$

$\mathrm{P} = P0 \cup P1 \cup \cdots \cup Pn-1$ (resp., $\mathrm{T} = T0 \cup T1 \cup \cdots \cup Tn-1$) is a superset of places (resp., transitions). $\mathrm{F} \subseteq (\mathrm{P}\times\mathrm{T})\cup(\mathrm{T}\times\mathrm{P})$ is a superset of flow arcs. $\mathrm{CN} \subseteq (\mathrm{P} \times \mathrm{T})$ (resp., $\mathrm{EN} \subseteq (\mathrm{T} \times \mathrm{T})$) is a superset of condition (resp., event) signals. $\mathrm{DC} = DC_0 \cup DC_1 \cup \cdots \cup DC_{n-1}$ is a set of time intervals to input flow arcs. EM: $\mathrm{T} \rightarrow \{\boxtimes, \boxdot\}$ maps an event processing mode (AND or OR) for each transition. Let $\mathrm{Z}_0 = (\mathrm{M}_0, \mathrm{D}_0)$, where $\mathrm{M}_0 : \mathrm{P} \rightarrow \{0, 1\}$ is the initial marking. $\mathrm{D}_0 : \mathrm{P} \rightarrow \{0\}$ is the initial clock position. $\Omega = \{N_{\Gamma 0}, N_{\Gamma 1}, ..., N_{\Gamma n-1}\}$ is a set of all TNCES structures that can be represented by RN.

The control module R is a set of reconfiguration functions. A reconfiguration function r is a structure r = (Cond, s, x). Cond $\rightarrow$ {true, false} is the precondition of r. $s{:}\Omega \rightarrow \Omega$ is the structure modification instruction. : $R(N_{\Gamma i}, z_{0i}) \rightarrow \mathrm{Z}_{0j}$ is the state correlation function, where Z_{0j} is a set of feasible initial states of Γ_j. $\star r = \Gamma_i = (N_\Gamma, z_{0i})$ (resp., $r \star = \Gamma_j = (N_{\Gamma j}, z_{0j})$) denotes the TNCES before (resp., after) the implementation of r.

Definition 2 (see [30]). A state Z of R-TNCES RN is a pair $[N_\Gamma, z]$, where $N_\Gamma \in \Omega$ identifies the activated TNCES with $N_\Gamma = (P, T, F, CN, EN, em, DC)$, and $z = (m, d)$ is a state of N_Γ with $m{:}P \rightarrow \{0, 1\}$ and $d{:}P \rightarrow \{0, 1, 2, . . .\}$.

In a R-TNCES, each TNCES in the behavior module models a configuration. For a R-TNCES RN, only one of the TNCESs of the behavior module B is activated at the beginning until a reconfiguration function is implemented. Other TNCESs with net structures defined in Ω can be activated only after implementing reconfiguration functions. At any time, only one of the TNCESs with net structures defined in Ω is activated.

If a reconfiguration function r=(Cond, s, x) meets its precondition, that is, Cond = True, it is enabled. A reconfiguration function can fire if it is enabled, that is, to implement it. The evolution of a R-TNCES depends on what events (reconfiguration functions or transitions) take place. Let Γ_i be the activated TNCES with $\Gamma_i = (N_\Gamma, z_{0i})$, where $N_{\Gamma i} = (P_i, T_i, F_i, CN_i, EN_i, em_i, DC_i)$. If a maximal step $u \in T_i$ fires, Γ_i evolves from its one inner state to another. However, if a reconfiguration function r fires, then Γi is transformed into Γj by changing its net structure and updating its state, where $\star r = \Gamma_i$, $r \star = \Gamma_j$, and $\Gamma_j = (N_{\Gamma j}, z_{0j})$.

Drawbacks of R-TNCESs

The TNCES models for the four behavior modes of AAS are denoted by $\Gamma_1 = (N_{\Gamma 1}, z1_0)$, $\Gamma 2 = (N_{\Gamma 2}, z2_0)$, $\Gamma_3 = (N_{\Gamma 3}, z3_0)$, and $\Gamma_4 = (N_{\Gamma 4}, z4_0)$, respectively. The set of all possible reconfiguration events of AAS is marked by R={r_1,s, r_2,s, r_3,s, r_4,s, r_{-1} 1,s, r_{-1} 2,s, r_{-1} 3,s, r_{-1} 4,s, $r_{1,2}$, $r_{1,3}$, $r_{1,4}$, $r_{2,4}$, $r_{3,4}$}. The reconfiguration event r_i, indicates a local reconfiguration that transforms robot Rbi into its energy-efficient mode and $r_{i,s}^{-1}$ is the reverse of $r_{i,s}$, that is, to transform robot Rbi from its energy-efficient mode into its working mode. The implementation of the events $r_{i,s}$ and $r_{i,s}^{-1}$ does not change the current behavior mode but can switch robot Rbi between its working mode and energyefficient mode according to its busy/idle status and waiting time. Finally, $r_{i,j}$ $(i \neq j)$ denotes a global reconfiguration event that transforms AAS from the configuration Mode i into Mode j.

The firing of a reconfiguration function of a R-TNCES changes the system configuration. As a consequence, if reconfiguration functions are applied to model local reconfiguration events for switching components between their working modes and energy-efficient modes directly, the number of system configurations should be enlarged. For example, configuration Mode 4 should be considered as four different configurations: (1) Both Rb1 and Rb3 are in their working modes, (2) Rb1 is in working mode and Rb3 is in energy-efficient mode, (3) Rb3 is in working mode and Rb1 is in energy-efficient mode, and (4) both Rb1 and Rb3 are in their energy-efficient modes. These four configurations are with the same structure. However, they should be verified separately. Obviously, this increases the verification cost and burdens the whole design process. Generally, transitions in a R-TNCES model normal events of a reconfigurable discrete event control system, whereas reconfiguration functions are used to model system reconfiguration events. However, the concurrence of reconfiguration functions and transitions is not allowed in R-TNCESs, which is in fact inconsistent with requirements of REMSs. To make it clearer, let us take the modules Rb3, Rb4, and *W*2 as an example. Their TNCES-based models in Mode 1 and Mode 4 are shown in Figures 3 and 4, respectively. The differences between them are marked by dotted lines.

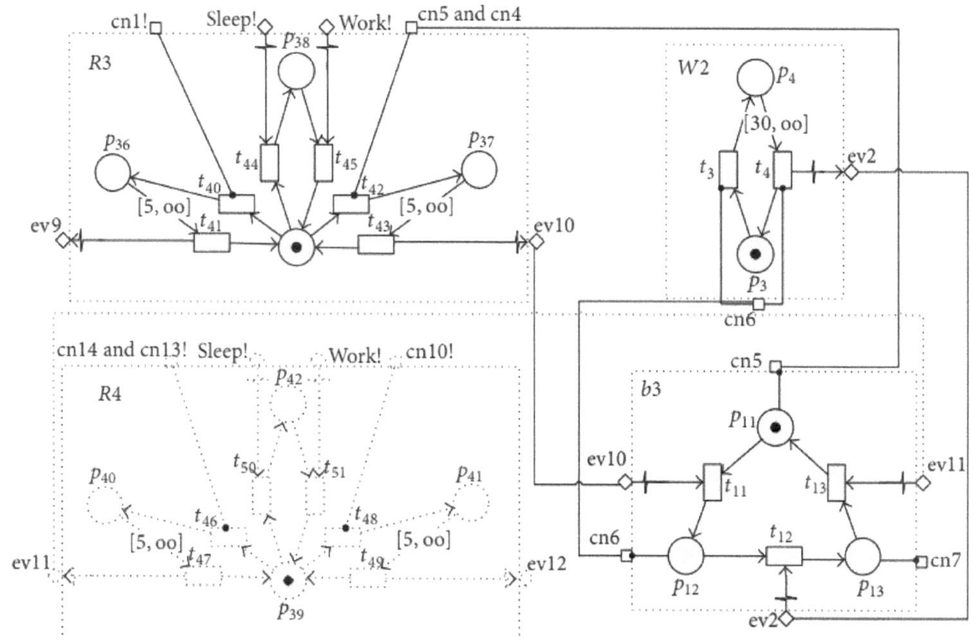

Figure 3: TNCES-based model of *R*3, *R*4, and *W*2 in Mode 1.

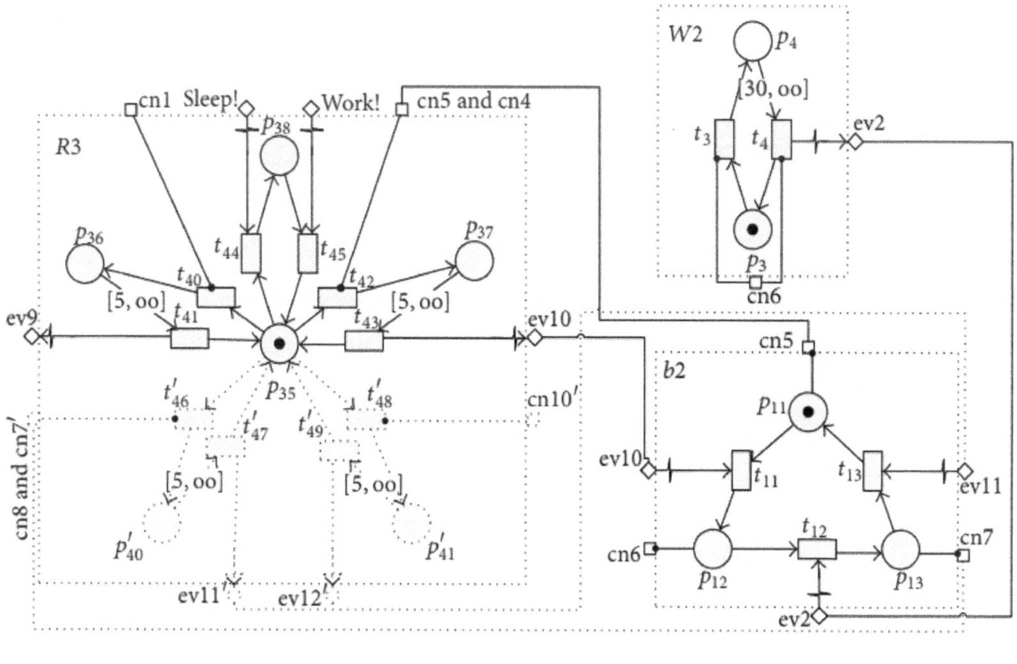

Figure 4: TNCES-based model of *R*3, *R*4, and *W*2 in Mode 4.

Example 3. Suppose that a reconfiguration function $r_{1,4}$ gets enabled at state $S3$ when AAS is in Mode 1. The physical meaning of $S3$ is that (1) Rb3 just finishes transporting S_A to $b2$ and (2) $W2$ is ready to process S_A. Assume that at this time a fault is detected in Rb4. Rb4 should be removed. Meanwhile, Rb3 must update itself soon in order to cover Rb4's task. According to the design requirements for AAS, $W2$ should go on working "naturally" at this time, that is, the enabled transition t_3 can fire at this state. However, the concurrence of reconfiguration functions and transitions is not allowed in R-TNCESs. Therefore, at state $S3$, only $r_{1,4}$ fires alone and AAS turns to the state $S1'$. Afterwards, t_3 fires, which leads to the next state $S2'$. However, if $r_{1,4}$ and t_3 fire together, AAS turns to the state $S2'$ directly without generating $S1'$. The state transition graph of this case is shown in Figure 5.

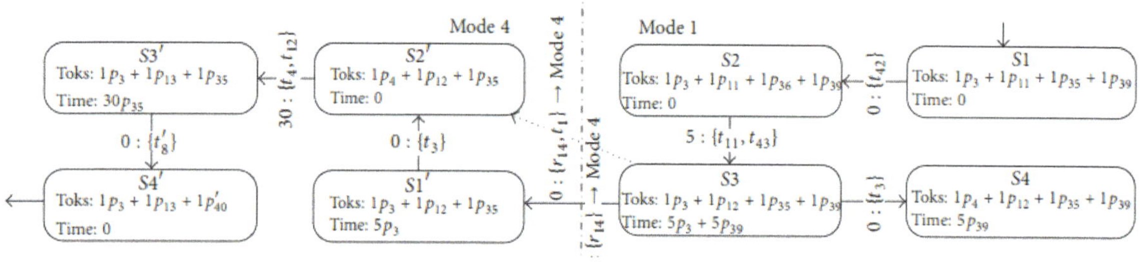

Figure 5: State transition graph of Example 3.

Example 4. Assume that two reconfiguration functions r_3,s and r_4,s get enabled simultaneously at state $S4$. The physical meaning of $S4$ is that (1) $W2$ just starts its work and (2) both Rb3 and Rb4 are idle. The firing of r_3 and $r_{4,s}$ only changes the states inside their modules but neither alters the system structure nor enables/disables any other transitions outside. That is to say, the firing of r_3,s and $r_{4,s}$ does not change the current system configuration. According to the design requirements for AAS, both Rb3 and Rb4 can reconfigure themselves into energy-efficient modes freely when they are idle for more than two time units. However, the concurrence of multiple reconfiguration functions is not allowed in RTNCESs. Therefore, at state $S4$, only r_4, or r_3,s fires alone. After that, the remaining one fires since it is still enabled. However, if r_3, and r_4,s fire together, AAS turns to the state $S7'$ directly. The state transition graph of this case is shown in Figure 6.

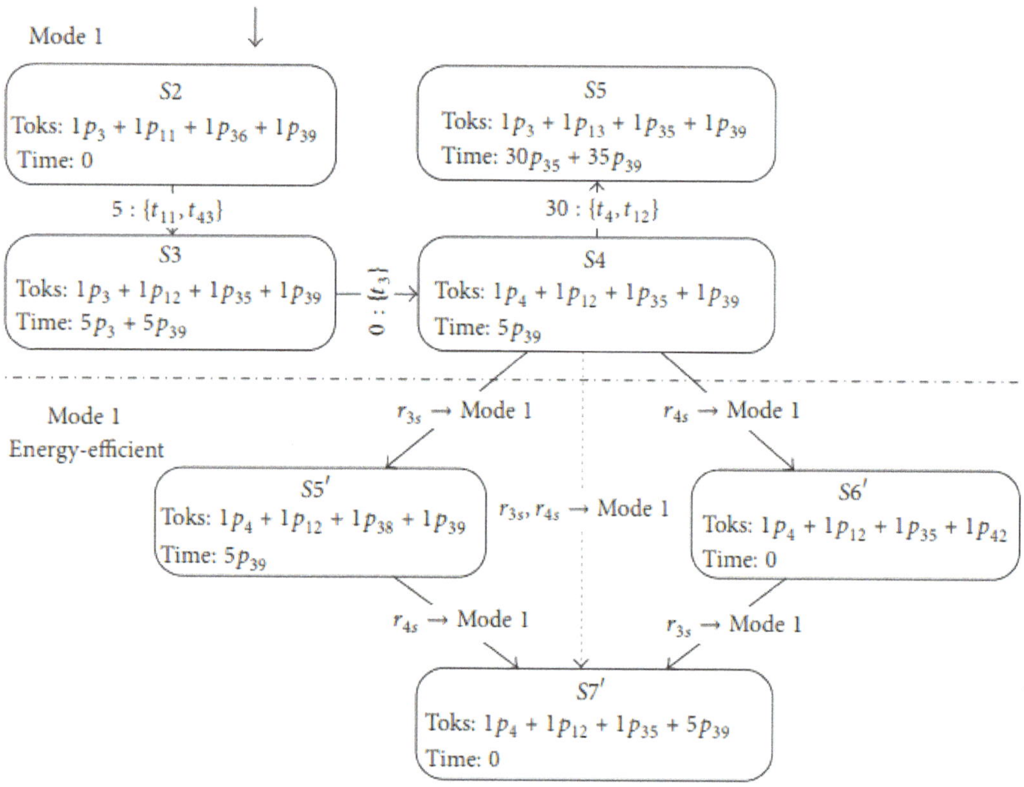

Figure 6: State transition graph of Example 4.

In conclusion, the original R-TNCESs are not sufficient to model a REMS. The reason can be explained from the following three aspects.(i)Reconfigurations at the component level only change component behavior modes between their working modes and energy-efficient modes rather than changing system configurations. If this kind of reconfigurations is modeled by reconfiguration functions directly, the number of system configurations should be enlarged, which increases the verification cost and burdens the whole design process.(ii)The concurrence of reconfiguration functions and transitions is not allowed in R-TNCESs. However, from the above examples, the concurrence of reconfiguration events and normal events is a common phenomenon in a REMS.(iii)Since the local reconfigurations for energy-efficient operations cannot be properly described, their corresponding dynamics and reasonable analysis cannot be performed.

To this end, this paper extends R-TNCESs to achieve two aims. First, all possible events including concurrent events that may occur in REMSs can be properly described. Second, the concurrence of reconfiguration functions and transitions should be controlled to ensure the system correctness.

Extended R-TNCESs

An extended R-TNCES has the same structure as the original R-TNCES. It is composed of a behavior module and a control module, denoted by eRN = {B, R}. The definition of system states is not changed, as shown in Definition 2 in Section 3.1. However, in the extended R-TNCES, reconfiguration functions are newly assigned with action ranges and concurrent decision functions. In addition, the firing rules of transitions and reconfiguration functions are updated such that they are conditionally allowed to fire concurrently.

Modified Reconfiguration Functions

In order to model the two types of reconfiguration events in a REMS directly, a concept, namely, action range, is developed for each reconfiguration function of a R-TNCES. In addition, a concurrent decision function is also assigned to a reconfiguration function to constrain concurrent transitions that may lead to undesired states such as deadlocks and overflow during a reconfiguration. For the sake of brevity, a reconfiguration function indicates a modified reconfiguration function in what follows.

Definition 5. A reconfiguration function r of an extended RTNCES eRN is a structure r=(Cond, s, x, Λ, Π). Cond → {true, false}is the precondition of $r.s$: Ω→Ω is the structure modification instruction. : $R(N_{\Gamma i}, z_{0i}) \rightarrow Z_{0j}$ is the state correlation function, where $Z0j$ is a set of feasible initial states of Γ_j. $\star r=\Gamma_i=(N_\Gamma, z_{0i})$ (resp., $r\star=\Gamma_j=(N_{\Gamma j}, z_{0j})$) denotes the TNCES before (resp., after) r fires.$\Lambda \in (N_{\Gamma i} \cup N_{\Gamma j})$ denotes the action range of r. $\Pi(r, Z) \rightarrow T$ is a concurrent decision function deciding a set of forbidden transitions that cannot fire together with r at state Z.

The reconfiguration functions of extended R-TNCESs are divided into two types: major and minor reconfiguration functions. For a reconfiguration function r=(Cond, s, x, Λ, Π) with $\star r=\Gamma_i=(N_{\Gamma i}, z_{0i})$ and $r\star=\Gamma_j=(N_{\Gamma j}, z_{0j})$, it is a major reconfiguration function if and only if $N_{\Gamma i}=N_{\Gamma j}$. Otherwise, it is a minor reconfiguration function. Let R_{ma} and R_{mi} denote the sets of major and minor reconfiguration functions of eRN, respectively.Then we have $R = R_{ma} \cup R_{mi}$ and $R_{ma} \cap R_{mi} = 0$.

The implementation (firing) of a major reconfiguration function changes the structure of the current activated TNCES, whereas the implementation (firing) of a minor reconfiguration function only adjusts partial states of the activated TNCES within its action range.

Similar to Petri nets, the "conflict" concept is proposed for two enabled reconfiguration functions. We have the following two cases.(1)For two reconfiguration functions within the same type, that is, both being minor or major reconfiguration functions, if their action ranges have intersections, they are conflicting.(2)For a minor reconfiguration function and a major reconfiguration function, if the action range of the minor reconfiguration function is not completely covered by that of the major reconfiguration function, they are conflicting.

If two reconfiguration functions are conflicting, they cannot be implemented simultaneously. The symbol r_1 □ r_2 denotes that reconfiguration functions r_1 and r_2 are not conflicting. Similar to the definition of steps in TNCES, a r-step in an extended R-TNCES is a maximal set of reconfiguration functions that can fire simultaneously at a particular state. A r-step should satisfy the following two conditions.

(1) For any two reconfiguration functions r_i and r_j ($r_i \neq r_j$) in a r-step γ, r_i and r_j are not conflicting; that is, $r_i \parallel r_j$.

(2) There does not exist any other maximal set of reconfiguration functions γ' such that $\gamma \subset \gamma'$

Accordingly, two r-steps γ_1 and γ_2 are conflicting, if $\exists r_1 \in \gamma_1$, $\exists r_2 \in \gamma_2$, $r_1 = r_2$, r_1 and r_2 are conflicting.

Dynamics of Extended R-TNCESs

Suppose that, at state Z = [$N\Gamma$, z], multiple reconfiguration functions get enabled, to be denoted by

$$R^* = \gamma_1 \cup \gamma_2 \cup \cdots \cup \gamma_g, \tag{4}$$

where γ_i ($i \in [1, g]$) is a maximal r-step at Z and, for all $i, j \in [1, g]$, $i = j$, γ_i and γ_j are conflicting. At the same state Z, the set of all enabled transitions is denoted by

$$T^* = u_1 \cup u_2 \cup \cdots \cup u_k, \tag{3}$$

where ui ($i \in [1, k]$) is a maximal step and, for all $i, j \in [1, k]$, $i = j$, u_i and u_j are conflicting. For more information on how these steps are computed, please see [34, 35].

As a consequence, different compositions of r-steps and steps can occur simultaneously at this state. Given an enabled reconfiguration function r, we use $D \cdot T$ (resp., $D \cdot P$) to denote the set of deleted transitions (resp., deleted places) and $A \cdot T$ (resp., $A \cdot P$) to denote the set of added transitions (resp., added places) by firing it, where $\star r=\Gamma_i=(N_{\Gamma i}, z_{0i})$, $r^* = \Gamma_j = (N_{\Gamma j}, z_{0j})$, $N_{\Gamma i} = (P_i, T_i, F_i, CN_i, EN_i, em_i, DC_i)$, and $N_{\Gamma j} = (P_j, T_j, F_j, CN_j, EN_j, em_j, DC_j)$. We have the following two cases.

(1) For a transition $t \in T_i$, if it is enabled simultaneously with a minor reconfiguration function r=(Cond, s, x, Λ, Π) at state Z = [N_Γ, z] and $t \notin \Lambda$, then t can fire simultaneously with r; that is, $t \notin \Pi(r, Z)$.

(2) For a transition $t \in T_i$, if it is enabled simultaneously with a major reconfiguration function r=(Cond, s, x, Λ,

Π) at state $Z = [N_\Gamma, z]$, then we have the following two subcases.

(A) A spontaneous transition t is forbidden to be concurrent with r at Z, if it meets one of the following conditions.

(i) If it is deleted by r, that is, $t \in D \cdot T$, it is forbidden by r; that is, $t \in \Pi(r, Z)$.

(ii)) If $t \notin D \cdot T$ and all its elements are not changed by firing r, then it is allowed to fire simultaneously with r. Formally, if ${}^{\bullet}t_i = {}^{\bullet}t_j, t_i^{\bullet} = t_j^{\bullet}, {}^{\sim}t_i = {}^{\sim}t_j, {}^{\sim}t_i = {}^{\sim}t_j = \emptyset$, and $em(t)_i = em(t)_j$, we have $t \notin \Pi(r, \mathcal{Z})$.

(iii) If $t \notin D \cdot T$, some of its elements are modified by r, which include its preset, postset, source places, and firing mode, and we have the following two cases.

(a)) The preset, source places, and firing mode of t decide whether t is enabled after the firing of r. Therefore, if its preset, source places, or firing mode is changed by r, it can fire simultaneously with r. Formally, if ${}^{\bullet}t_i \neq {}^{\bullet}t_j, {}^{\sim}t_i \neq {}^{\sim}t_j$, or $em(t)_i \neq em(t)_j$, then $t \notin \Pi(r, \mathcal{Z})$.

(b) The postset of$t$ does not change its enabling condition but influences the structure of the net. Therefore, it is forbidden by r. Formally, if $t^{\bullet}_i \neq t^{\bullet}_j$, we have $t \in \Pi(r, \mathcal{Z})$.

(B) A forced transition t is forbidden to be concurrent with r at Z, if it further meets one of the following conditions.

(i) Its firing mode is - and all of its forcing transitions are forbidden to be concurrent with r; that is, if $em(t) = em_j(t) =$ - and, for all $t' \in {}^{\sim}t, t' \notin \Pi(r, \mathcal{Z})$, then $t \in \Pi(r, \mathcal{Z})$.

(ii) Its firing mode is ⊠, and at least one of its forcing transitions is forbidden by r; that is, if $emi(t) = em_j(t) =$ ⊠ and $\exists t' \in {}^{\sim}t, t' \notin \Pi(r, \mathcal{Z})$, then $t \in \Pi(r, \mathcal{Z})$.

Since an extended R-TNCES allows the concurrence of multiple reconfiguration functions and transitions, the reachability graph of an extended R-TNCES is defined as follows.

Definition 6. The reachability graph of an extended R-TNCES eRN is a combination of several labeled directed graphs whose nodes are the states of eRN and whose arcs are of three kinds: steps, r-steps, and combinations of a step and a r-step. (i) The arc from state $[N_{\Gamma i}, zi]$ to state $[N_{\Gamma i}, z'i]$ is denoted by a step u represented by $[N_{\Gamma i}, z_i][u\rangle[N_{\Gamma i}, z'i]$, where $z'i \in R(N_{\Gamma i}, z_i)$. (ii) The arc from state $[N_\Gamma, z_i]$ to state $[N_{\Gamma j}, z_j]$ is labeled with a r-step γ represented by $[N_{\Gamma i}, z_i][\gamma\rangle[N_{\Gamma j}, z_j]$, if γ contains major reconfiguration functions with $N_{\Gamma i} = N_{\Gamma j}$. Otherwise, we have $N_{\Gamma i} = N_{\Gamma j}$ and $z_j \notin (N_{\Gamma i}, z_i)$.(iii) The arc from $[N_{\Gamma i}, z_i]$ to state $[N_{\Gamma j}, z_j]$ is labeled with a step and a r-step {R, u} represented by $[N_{\Gamma i}, z_i][\gamma, u\rangle[N_{\Gamma j}, z_j]$, if γ contains major reconfiguration functions with $N_{\Gamma i} = N_{\Gamma j}$. Otherwise, we have $N_{\Gamma i} = N_{\Gamma j}$.

Obviously, the graphical representation of an extended RTNCES model is the same as that of a R-TNCES model. However, system dynamics get enriched along with the changes of reconfiguration functions. If we use an extended R-TNCES to model AAS, the graphical TNCES models shown in Figures 3 and 4 are still correct. However, their reachability graphs get enriched during same reconfiguration. Example 7. A fragment of the reachability graph of the extended R-TNCES-based model of the example composed of Rb3, Rb4, and $W2$ is shown in Figure 7. AAS starts running in *Mode* 1. When it arrives at state $S3$, two minor reconfiguration functions r_3, and $r_{4,s}$ get enabled and fire simultaneously to reconfigure robots Rb3 and Rb4 into their energy-efficient modes. After 28 time units, they reconfigure back to working modes. Assume that $R4$ is detected to have a fault at state $S10$, the major reconfiguration function $r_{1,4}$ gets enabled. In the meantime,$_3$ gets enabled simultaneously with $r_{1,4}$. Therefore, $t3$ fires simultaneously with $r_{1,4,}$ which leads to the transformation of AAS into *Mode* 4.

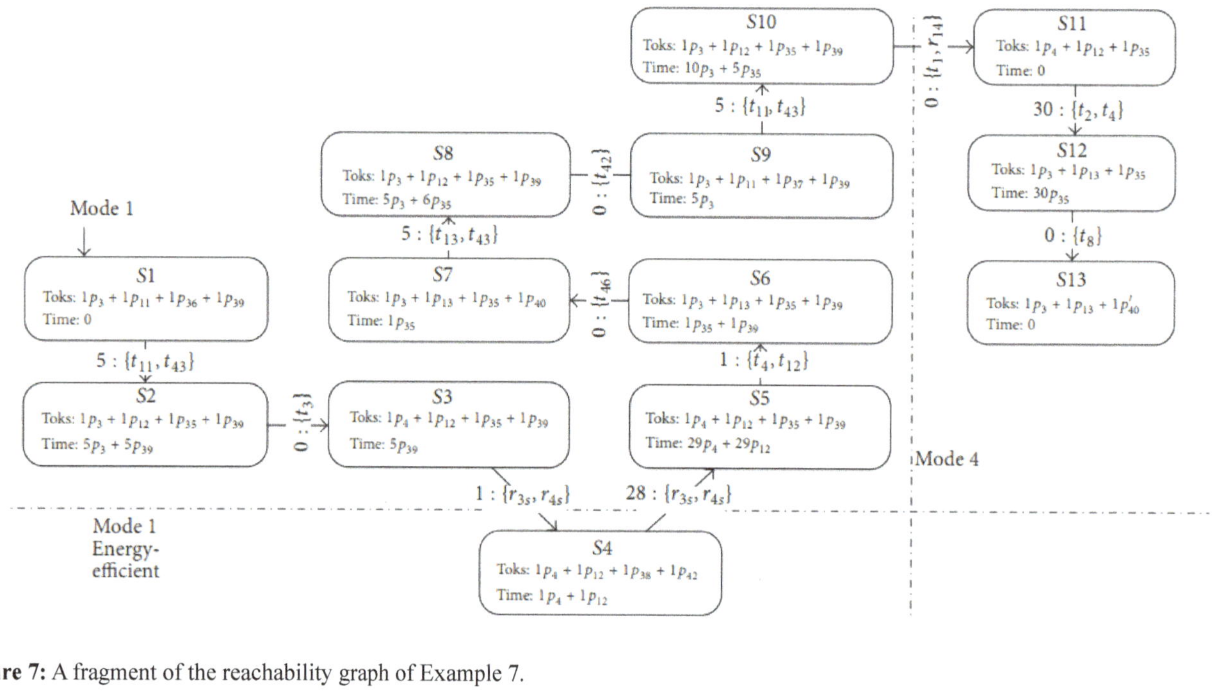

Figure 7: A fragment of the reachability graph of Example 7.

Verification of REMSs Based on Extended R-TNCESs

In order to perform correct formal verification of AAS, an extended R-TNCES-based model should be built for it. The extended R-TNCES based model of AAS is marked by $eRN_{AAS} = \{\mathscr{B}, \mathscr{R}\}$, $\mathscr{B} = \Gamma_1 \cup \Gamma_2 \cup \Gamma_3 \cup \Gamma_4$, $\mathscr{R} = \mathscr{R}_{ma} \cup \mathscr{R}_{mi}$, $\mathscr{R}_{ma} = \{r_{1,2}, r_{1,3}, r_{1,4}, r_{2,4}, r_{3,4}\}$, and $\mathscr{R}_{mi} = \{r_{1,s}, r_{2,s}, r_{3,s}, r_{4,s}, r^{-1}_{1,s}, r^{-1}_{2,s}, r^{-1}_{3,s}, r^{-1}_{4,s}\}$. We have $\Omega = \{N_{\Gamma_1}, N_{\Gamma_2}, N_{\Gamma_3}, N_{\Gamma_4}\}$. The four major reconfiguration functions are conflicting with each other. The minor reconfiguration functions $r_{i,}$ and $r^{-1}_{i,s}$ ($i \in [1, 4]$) are conflicting but others are not. The behavior module of eRN_{AAS} is shown in Figure 8, where elements drawn by dotted lines are possibly modified during the implementation of a major reconfiguration function. In order to apply automatic model checking to an extended RTNCES, a TNCES-based nested state machine is developed to implement its control module.

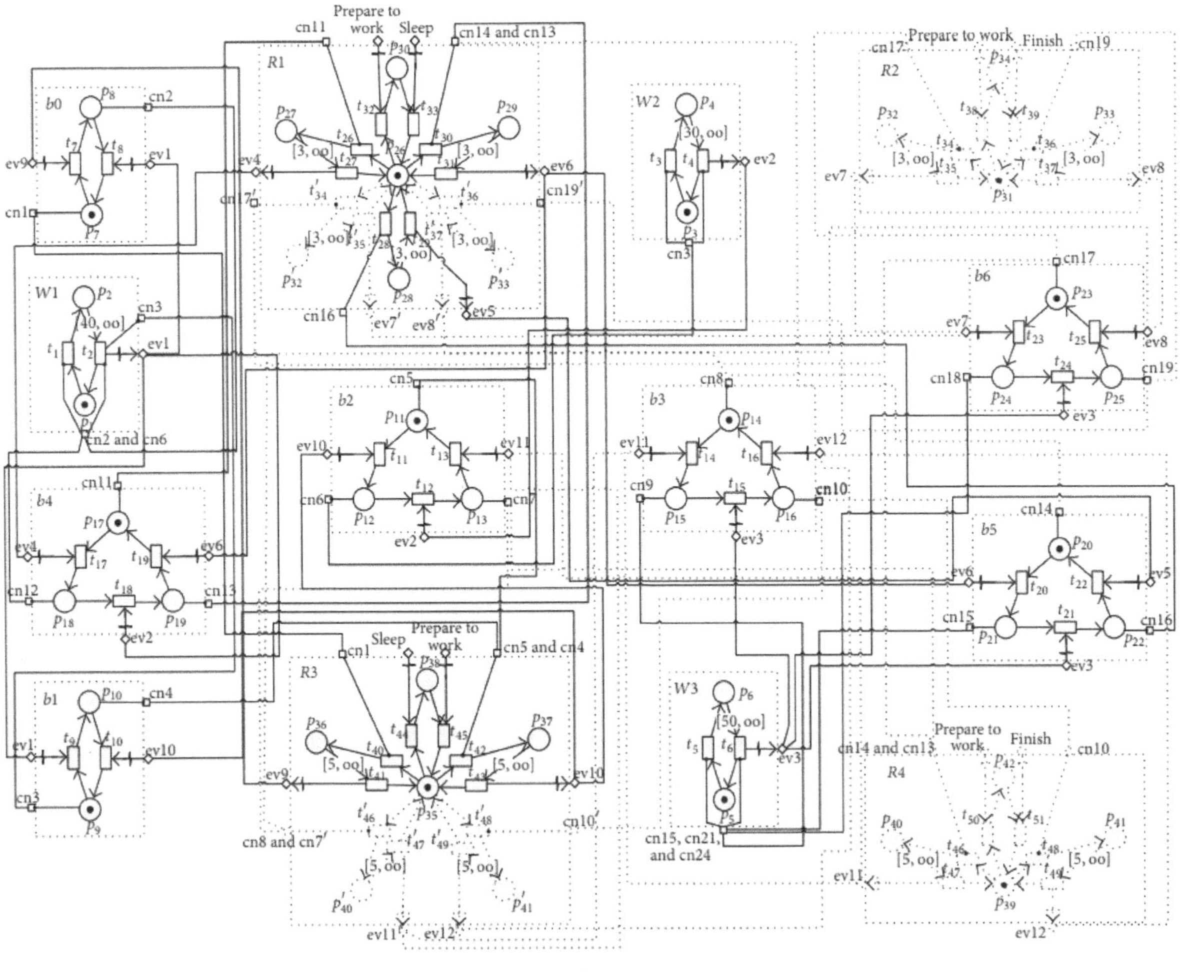

Figure 8: Behavior module of .

Implementation of Extended R-TNCESs

First of all, major reconfiguration functions are grouped according to their action ranges. A set of state machines specified by TNCESs, which are called Major changers, is defined. Each state machine corresponds to a group of major reconfiguration functions that share the same action range. In a particular Major changer, each transition corresponds to a major reconfiguration function. The transitions in a state machine cannot fire simultaneously, which means that these modeled major reconfiguration functions by one state machine are conflicting with each other. Firing a transition st in a Major changer implies that a major reconfiguration function is implemented. A Major changer is formalized as follows:

$$\text{Major\_changer} = (P, T, F, em, z_0), \tag{6}$$

where, for any $t \in T, |{}^{\bullet}t| = |t^{\bullet}| = 1, \sum M_0(P) = 1$, which means that only one place in P owns a token at the initial state, and $em : T \rightarrow \{\boxtimes\}$. The precondition Cond can be modeled by input event/condition signals from external to transitions in a Major changer.

In addition, an actuator denoted by Actuator is defined for each place sp in all Major changers, which is marked by Actuator = Act(sp). Each actuator is composed of a place ap and a transition at only, where · ${}^{\bullet}ap = ap^{\bullet} = \{at\}$, ${}^{\bullet}at = at^{\bullet} = \cdot = \{ap\}$, and $(ap) = 1$. When the place sp in a Major changer receives a token, the actuator Actuator = Act(sp) is activated. An Actuator is formalized as follows:

where $|P| = |T| = 1$, ${}^{\bullet}at = at^{\bullet} = \{ap\}$, ${}^{\bullet}ap = ap^{\bullet} = \{at\}$, $m_0(P) = 1$, and $em : T \rightarrow \{\boxtimes\}$.

Similar to major reconfiguration functions, minor reconfiguration functions are grouped according to their action ranges. A set of state machines specified by TNCESs, which are called Minor changers, is defined. Each state machine corresponds to a group of minor reconfiguration functions. If the action ranges of two minor reconfiguration functions are the same, they are modeled by transitions in a Minor changer.

If the action range of a group of minor reconfiguration functions, to be modelled by a Minor changer, is completely covered by that of a group of major reconfiguration functions, to be modeled by a Major changer, then this Minor changer is activated while this Major changer is activated.

A is formalized as follows:

$$\text{Minor\_changer} = (P, T, F, em, z_0), \tag{8}$$

where, for any $t \in T, |{}^{\bullet}t| = |t^{\bullet}| = 1, \sum M_0(P) = 1,$, which means that only one place in $T \rightarrow \{\boxtimes\}$. owns a token at the initial state, and . The precondition can be modeled by input event/condition signals from external to transitions in a .

Example 8. Figure 9 depicts the TNCES-based control module of eRNAAS. It has only one Major changer, since the four major reconfiguration functions share the same action range. It has four Minor changers, since the four robots have four distinguished action ranges. Places p_1, p_2, p_3, and p4 in Major changer correspond to Mode 1, Mode 2, Mode 3, and Mode 4, respectively.When t_3 fires, the major reconfiguration function $r_{1,4}$ is implemented. Robots Rb3 and Rb1 are applied in every mode of AAS. Therefore, minor reconfiguration functions that transform them between energyefficient modes and working modes are activated in every system behavior mode. Moreover, it is possible for them to fire simultaneously with other major reconfiguration functions.

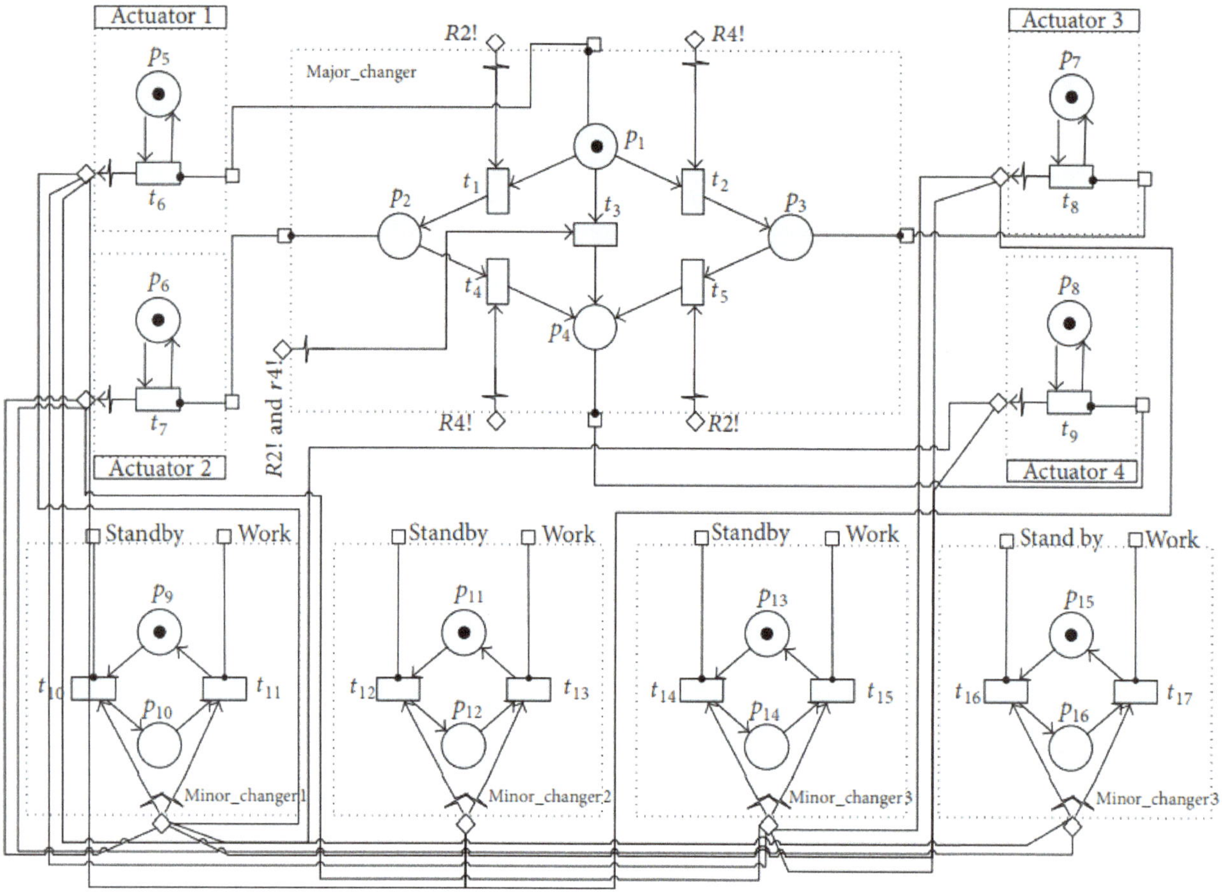

Figure 9: Control module of eRN_{AAS}.

Formal Verification of AAS

Since the time when a major reconfiguration function can get enabled and fire cannot be predicted, this paper applies an instruction insertion method to simulate AAS. In addition, eRNAAS evolves according to fired maximal steps and r-steps. Assume that AAS should finish 100 subassemblies. It starts with Mode 1. At time $t1$ when it finishes the 60th subassembly, it reconfigures into Mode 2 due to the fault detection of Rb2. Then, it goes on working in Mode 2. At time $t2$ when the 91st subassembly is being processed, it transforms into Mode 4 according to the fault detection of Rb4. During the whole process, minor reconfigurations, that is, transforming robots between their working modes and energy-efficient modes, are applied.

SESA is applied to compute the reachability graph of this whole process. A minimal path regarding time consumption from the initial state to the objective state is computed in each mode. In Mode 1, it generates 23044 states, taking 6990 time units to finish assembly of the first 60 subassemblies in the minimal path. In Mode 2, it generates 85259 states, costing 4127 time units to finish assembling the next 30 subassemblies in the minimal path. Finally, in Mode 4, it generates 195007 states, taking 1525 time units to finish assembling the last 10 subassemblies in the minimal path. Note that two states can be considered to be same if and only if they have the same token numbers and time status. Since each TNCES-based model of the behavior modes of AAS is a well-designed control system, they are proved to be qualified according to SESA, where eCTL based functional properties and TCTL based temporal properties are checked. In addition, the following eCTL formula is applied to the control module of eRNAAS:

$$Z_0 = EX\langle t4ANDt12\rangle X\langle p12 = 1\rangle . \quad (9)$$

This formula is proved to be false by SESA. Transition t_{12} corresponds to minor reconfiguration function $r_{2,}$. Therefore, it can fire only when AAS is in Mode 1 or Mode 2.The following formula is proved to be true:

$$Z_0 = EX\langle t2ANDt10\rangle X\langle p10 = 1\rangle . \quad (10)$$

It means that when robot RB4 breaks down, two reconfiguration functions $r_{1,3}$ and $r_{1,s}$ are possible to fire simultaneously.

The triggering conditions of minor reconfiguration functions can be computed previously. There are several possible state/event paths showing system behavior from the initial state to the objective state, at which 100 subassemblies are finished. We select a minimal path regarding time for each TNCES-based model of the three configurations, to be denoted by Path = $Z_1, Z_2, \ldots, Z_n$, where energy-efficient operations are not included. That is to say, all robots should stay in their working modes in this case although they should wait for a period of time before the next task comes. After that, based on the states on this path, the time when a minor reconfiguration function gets enabled and fires can be computed. For example, if an activated robot starts to wait at a particular state Z_i, at which the system time is τ_1, a search is performed along this minimal path at τ_1. If it is found that at Z_j the robot works again, at which the system time is τ_2, then the time delay $\Delta_\tau = \tau_2 - \tau_1$ between these two states is obtained. The round local reconfigurations for switching a robot between its working mode and energy-efficient mode take two time units. Therefore, if the time delay is larger than two, that is, $\Delta_\tau > 2$, a local reconfiguration can be applied to this robot. The system time for reconfiguring this robot from its working mode to its energy-efficient mode is τ_1. The system time for reconfiguring this robot from its energy efficient mode to its working mode is τ_{2-1}.

The time of robots on their energy-efficient modes in minimal paths is computed during the assembly of 100 subassemblies. They are shown in Table 1 together with the whole system uptime in each mode. Take Mode 1 as an example. Assume that Rb1 consumes one energy unit per time unit in its working time but only consumes 30% energy unit per time unit in its energy-efficient mode. In *Mode* 1, if there is no minor reconfiguration applied to Rb1 for saving energy, it will consume 6990 energy units. However, it only consumes $6990 - 3233 + 30\% \times 3233 = 4726.9$ energy units in Mode 1 if minor reconfigurations are applied when it is idle. In the same way, the energy saved by the robots during this simulation is shown in Table 2, where the third row shows the energy consumption of each robot if no minor reconfigurations are applied, the fourth row shows the energy consumption of each robot when minor reconfigurations are applied, and the last row shows the saved energy of each robot during this process.

Table 1: Time of robots on their energy-efficient modes

Configuration	Mode 1				Mode 2			Mode 4	
System uptime	6690				4127			1525	
Robot	Rb1	Rb2	Rb3	Rb4	Rb1	Rb3	Rb4	Rb1	Rb3
Time on energy-efficient mode	3233	4455	3818	2643	1004	2458	2428	523	435

Table 2: Energy consumption of robots

Configuration	Mode 1				Mode 2			Mode 4	
Robot	Rb1	Rb2	Rb3	Rb4	Rb1	Rb3	Rb4	Rb1	Rb3
Energy 1	6690	6690	6690	6690	4127	4127	4127	1525	1525
Energy 2	4726.9	3871.5	4303.4	5139.9	3424.2	2406.4	2427.4	1158.9	1220.5
Saved energy	1963.1	2818.5	2386.6	1550.1	702.8	1720.6	1699.6	366.1	304.5

CONCLUSION

A reconfigurable and energy-efficient manufacturing system (REMS) is a typical reconfigurable discrete event control system. It allows two kinds of dynamic system reconfigurations: local and global reconfigurations. The former ones are applied to save energy for components, whereas the latter ones are applied to change system configurations according to changed inner/outer execution environments. Meanwhile, normal events should be conditionally allowed to occur simultaneously with these system reconfigurations, such that the system can reconfigure smoothly and safely. In order to easily model conditioned concurrence of reconfiguration events and normal events and represent all interesting system behavior, this paper extends the reconfigurable timed net condition event systems (R-TNCESs) formalism. Original reconfiguration functions are newly assigned with action ranges and concurrent decision functions. Accordingly, the dynamics of R-TNCES is updated. After that, a TNCES-based implementation method for the proposed extended R-TNCES is developed such that automatic model checking can be applied. The verified properties include functional, temporal, and energy properties that are specified by Computation Tree Logic (CTL), extended Computation Tree Logic (eCTL), or Timed Computation Tree Logic (TCTL). An automatic assembly system is used to illustrate the whole work.

In the future, the authors will focus on reasonably optimal reconfigurable control systems that can save more energy and the applications of the proposed method to the crude-oil operation enterprises with huge energy consumption [48].

ACKNOWLEDGMENTS

This work was supported in part by the National Natural Science Foundation of China under Grant no. 61374068 and the Science and Technology Development Fund, MSAR, under Grant nos. 065/2013/A2 and 066/2013/A2.

REFERENCES

1. Y. Koren, U. Heisel, F. Jovane et al., "Reconfigurable manufacturing systems," CIRP Annals—Manufacturing Technology, vol. 48, no. 2, pp. 527–540, 1999.
2. M. Khalgui, O. Mosbahi, J. F. Zhang, Z. W. Li, and A. Gharbi, "Feasible dynamic reconfigurations of petri nets: application to a production systems," in Proceedings of the 6th International Conference on Software and Database Technologies (ICSOFT ‹11), pp. 105–110, Sevilla, Spain, July 2011.
3. T. Parisini and S. Sacone, "Fault diagnosis and controller re-configuration: an hybrid approach," inProceedings of the IEEE International Symposium on Intelligent Control (ISIC ‹98), pp. 163–168, September 1998.
4. P. Leitão, J. Alves, J. M. Mendes, and A. W. Colombo, "Energy aware knowledge extraction from petri nets supporting decision-making in service-oriented automation," in Proceedings of the IEEE International Symposium on Industrial Electronics (ISIE ‹10), pp. 3521–3526, Bari, Italy, July 2010.
5. S. Karnouskos, A. W. Colombo, J. L. M. Lastra, and C. Popescu, "Towards the energy efficient future factory," in Proceedings of the 7th IEEE International Conference on Industrial Informatics (INDIN ‹09), pp. 367–371, Cardiff, Wales, June 2009.
6. K. Bunse, M. Vodicka, P. Schönsleben, M. Brülhart, and F. O. Ernst, "Integrating energy efficiency performance in production management—gap analysis between industrial needs and scientific literature," Journal of Cleaner Production, vol. 19, no. 6-7, pp. 667–679, 2011.
7. S. Mechs, S. Lamparter, and J. P. Müller, "On evaluation of alternative switching strategies for energy-efficient operation of modular factory automation systems," in Proceedings of the IEEE 17th International Conference on Emerging Technologies and Factory Automation (ETFA ‹12), pp. 1–8, IEEE, Kraków, Poland, September 2012.
8. S. Mechs, J. P. Muller, S. Lamparter, and J. Peschke, "Networked priced timed automata for energy-efficient factory automation," in Proceedings of the American Control Conference (ACC ‹12), pp. 5310–5317, Montreal, Canada, June 2012.
9. Z. M. Bi and L. Wang, "Optimization of machining processes from the perspective of energy consumption: a case study," Journal of Manufacturing Systems, vol. 31, no. 4, pp. 420–428, 2012.
10. Y. Oda, Y. Kawamura, and M. Fujishima, "Energy consumption reduction by machining process improvement," Procedia CIRP, vol. 4, pp. 120–124, 2012.
11. Cannata, S. Karnouskos, and M. Taisch, "Energy efficiency driven process analysis and optimization in discrete manufacturing," in Proceedings of the 35th Annual Conference of the IEEE Industrial Electronics Society(IECON ‹09), pp. 4449–4454, Porto, Portugal, November 2009.·
12. G. Mouzon and M. B. Yildirim, "A framework to minimise total energy consumption and total tardiness on a single machine," International Journal of Sustainable Engineering, vol. 1, no. 2, pp. 105–116, 2008.
13. D. Shorin and A. Zimmermann, "Model-based development of energy-efficient automation systems," in Proceedings of the 17th IEEE Real-Time and Embedded Technology and Applications Symposium (RTAS ‹11), Chicago, Ill, USA, April 2011.
14. C.-W. Park, K.-S. Kwon, W.-B. Kim et al., "Energy consumption reduction technology in manufacturing—a selective review of policies, standards, and research," International Journal of Precision Engineering and Manufacturing, vol. 10, no. 5, pp. 151–173, 2009.
15. P. Stoffels, W. M. Boussahel, M. Vielhaber, and G. Frey, "Energy engineering in the virtual factory," inProceedings of the IEEE 18th International Conference on Emerging Technologies and Factory Automation (ETFA ‹13), pp. 1–6, Cagliari, Italy, September 2013.
16. N. Wu, M. Zhou, and Z. Li, "Resource-oriented Petri net for deadlock avoidance in flexible assembly systems," IEEE Transactions on Systems, Man, and Cybernetics Part A:Systems and Humans, vol. 38, no. 1, pp. 56–69, 2008.
17. Z. Li and M. Zhou, "Two-stage method for synthesizing liveness-enforcing supervisors for flexible manufacturing systems using Petri nets," IEEE Transactions on Industrial Informatics, vol. 2, no. 4, pp. 313–325, 2006.
18. Z. W. Li, H. S. Hu, and A. R. Wang, "Design of liveness-enforcing supervisors for flexible manufacturing systems using Petri nets," IEEE Transactions on Systems, Man and Cybernetics Part C: Applications and Reviews, vol. 37, no. 4, pp. 517–526, 2007. ·
19. Z. Li and M. Zhou, "Control of elementary and dependent siphons in Petri nets and their application,"IEEE Transactions on Systems, Man, and Cybernetics Part A: Systems and Humans, vol. 38, no. 1, pp. 133–148, 2008.
20. Z. Li, M. Zhou, and N. Wu, "A survey and comparison of Petri net-based deadlock prevention policies for flexible manufacturing systems," IEEE Transactions on Systems, Man and Cybernetics Part C: Applications and Reviews, vol. 38, no. 2, pp. 173–188, 2008. ·
21. Z. Li, N. Wu, and M. Zhou, "Deadlock control of automated manufacturing systems based on petri nets—a literature review," IEEE Transactions on Systems, Man and Cybernetics Part C: Applications and Reviews, vol. 42, no. 4, pp. 437–462, 2012.
22. Z. Y. Ma, Z. W. Li, and A. Giua, "Design of optimal Petri net controllers for disjunctive generalized mutual exclusion constraints," IEEE Transactions on Automatic Control, 2015.
23. J. H. Ye, Z. W. Li, and A. Giua, "Decentralized supervision of Petri nets with a coordinator," IEEE Transactions on Systems, Man, and Cybernetics: Systems, vol. 45, no. 6, pp. 955–966, 2015.

24. Z. W. Li, M. C. Zhou, and M. D. Jeng, "A maximally permissive deadlock prevention policy for FMS based on petri net siphon control and the theory of regions," IEEE Transactions on Automation Science and Engineering, vol. 5, no. 1, pp. 182–188, 2008.
25. Z. W. Li, G. Y. Liu, H.-M. Hanisch, and M. C. Zhou, "Deadlock prevention based on structure reuse of petri net supervisors for flexible manufacturing systems," IEEE Transactions on Systems, Man and Cybernetics, Part A: Systems and Humans, vol. 42, no. 1, pp. 178–191, 2012.
26. Y. F. Chen and Z . W. Li, "Design of a maximally permissive liveness-enforcing supervisor with a compressed supervisory structure for flexible manufacturing systems," Automatica, vol. 47, no. 5, pp. 1028–1034, 2011.View at MathSciNet ·
27. Y. F. Chen, Z. W. Li, M. Khalgui, and O. Mosbahi, "Design of a maximally permissive liveness- enforcing Petri net supervisor for flexible manufacturing systems," IEEE Transactions on Automation Science and Engineering, vol. 8, no. 2, pp. 374–393, 2011. ·
28. Y. F. Chen, Z. W. Li, K. Barkaoui, and M. Uzam, "New Petri net structure and its application to optimal supervisory control: Interval inhibitor arcs," IEEE Transactions on Systems, Man, and Cybernetics: Systems, vol. 44, no. 10, pp. 1384–1400, 2014.
29. X. Wang, I. Khemaissia, M. Khalgui, Z. Li, O. Mosbahi, and M. Zhou, "Dynamic low-power reconfiguration of real-time systems with periodic and probabilistic tasks," IEEE Transactions on Automation Science and Engineering, vol. 12, no. 1, pp. 258–271, 2015.
30. J. Zhang, M. Khalgui, Z. Li, O. Mosbahi, and A. M. Al-Ahmari, "R-TNCES: a novel formalism for reconfigurable discrete event control systems," IEEE Transactions on Systems, Man, and Cybernetics: Systems and Humans, vol. 43, no. 4, pp. 757–772, 2013. ·
31. J. F. Zhang, M. Khalgui, Z. W. Li, G. Frey, O. Mosbahi, and H. B. Salah, "Reconfigurable coordination of distributed discrete event control systems," IEEE Transactions on Control Systems Technology, vol. 23, no. 1, pp. 323–330, 2015.
32. C. Gerber, S. Preuße, and H.-M. Hanisch, "A complete framework for controller verification in manufacturing," in Proceedings of the 15th IEEE International Conference on Emerging Technologies and Factory Automation (ETFA ‹10), pp. 1–9, September 2010. ·
33. C. Gerber, Implementation and Verification of Distributed Control Systems, Logos, Berlin, Germany, 2011.
34. H.-M. Hanisch, J. Thieme, A. Lueder, and O. Wienhold, "Modeling of PLC behavior by means of timed net condition/event systems," in Proceedings of the IEEE 6th International Conference on Emerging Technologies and Factory Automation (ETFA ‹97), pp. 391–396, IEEE, Los Angeles, Calif, USA, September 1997.
35. M. Rausch and H. M. Hanisch, "Net condition/event systems with multiple condition outputs," inProceedings of the 1995 INRIA/IEEE Symposium on Emerging Technologies and Factory Automation, pp. 592–600, Paris, France, October 1995.
36. T. Murata, "Petri nets: properties, analysis and applications," Proceedings of the IEEE, vol. 77, no. 4, pp. 541–580, 1989.
37. Y. F. Chen, Z. W. Li, and M. C. Zhou, "Optimal supervisory control of flexible manufacturing systems by Petri nets: a set classification approach," IEEE Transactions on Automation Science and Engineering, vol. 11, no. 2, pp. 549–563, 2014.
38. Z. W. Li and M. C. Zhou, "Elementary siphons of Petri nets and their application to deadlock prevention in flexible manufacturing systems"," IEEE Transactions on Systems, Man, and Cybernetics Part A: Systems and Humans., vol. 34, no. 1, pp. 38–51, 2004.
39. Z. W. Li and M. C. Zhou, "Clarifications on the definitions of elementary siphons in Petri nets," IEEE Transactions on Systems, Man, and Cybernetics, Part A: Systems and Humans, vol. 36, no. 6, pp. 1227–1229, 2006.
40. Z. Li and M. Zhao, "On controllability of dependent siphons for deadlock prevention in generalized Petri nets," IEEE Transactions on Systems, Man, and Cybernetics Part A: Systems and Humans, vol. 38, no. 2, pp. 369–384, 2008.
41. P. H. Starke and S. Roch, "Analysing signal-net systems," Tech. Rep., Informatik Berichte, Humboldt-University, Berlin, Germany, 2002.
42. E. M. Clarke, O. Grumberg, and D. Peled, Model Checking, MIT Press, 1999.
43. E. A. Emerson, A. K. Mok, A. P. Sistla, and J. Srinivasan, "Quantitative temporal reasoning," inComputer-Aided Verification, vol. 531 of Lecture Notes in Computer Science, pp. 136–145, Springer, Berlin, Germany, 1991.
44. S. Roch, "Extended computation tree logic," in Proceedings of the Informatik-Bericht Workshop on Concurrency, Specification and Programming, pp. 225–234, 2000.
45. S. Preuße, D. Missal, C. Gerber, M. Hirsch, and H. M. Hanisch, "On the use of model-based IEC 61499 controller design," International Journal of Discrete Event Control Systems, vol. 1, no. 1, pp. 115–128, 2011.
46. S. Preuse, H.-C. Lapp, and H.-M. Hanisch, "Closed-loop system modeling, validation, and verification," in Proceedings of the IEEE 17th International Conference on Emerging Technologies and Factory Automation (ETFA ‹12), pp. 1–8, IEEE, Kraków, Poland, September 2012.
47. S. Preußse and H.-M. Hanisch, "Verifying functional and non-functional properties of manufacturing control systems," in Proceedings of the 3rd International Workshop on Dependable Control of Discrete Systems (DCDS ‹11), pp. 41–46, Saarbrücken, Germany, June 2011.
48. N. Q. Wu, M. C. Zhou, and Z. W. Li, "Short-term scheduling of crude-oil operations: petri net-based control-theoretic approach," IEEE Robotics & Automation Magazine, 2015.

CHAPTER 17 MODELING AND ANALYSIS OF REENTRANT MANUFACTURING SYSTEMS: MICRO- AND MACROPERSPECTIVES

Fenglan He[1], Ming Dong[1], and Dong Yang[2]

[1]Antai College of Economics & Management, Shanghai Jiao Tong University, 535 Fahua Zhen Road, Shanghai 200052, China

[2]College of Management, Donghua University, 1882 Yan-An Road, Shanghai 200051, China

ABSTRACT

In order to obtain the better analysis of the multiple reentrant manufacturing systems (MRMSs), their modeling and analysis from both micro- and macroperspectives are considered. First, this paper presents the discrete event simulation models for MRMS and the corresponding algorithms are developed. In order to describe MRMS more accurately, then a modified continuum model is proposed. This continuum model takes into account the re-entrant degree of products, and its effectiveness is verified through numerical experiments. Finally, based on the discrete event simulation and the modified continuum models, a numerical example is used to analyze the MRMS. The changes in the WIP levels and outflux are also analyzed in details for multiple re-entrant supply chain networks. Meanwhile, some interesting observations are discussed.

INTRODUCTION

In recent years, factories and production systems have become larger and more complicated. The reentrant manufacturing system is a typical example, in which the work in process (WIP) repeatedly passes through the same workstation at different stages of the process routes. Figure 1 gives the structure of a reentrant manufacturing process. Here, *B* represents the storage areas of each work center. In the large factories, no experiments can be done involving whole supply chains. Therefore, simulation models are developed, which can be used to substitute for the real systems. Especially the multiscale simulations of production flows in manufacturing systems have become a very important research topic recently. Scaling also plays a role in biomedical engineering field [1]. The production flows naturally show two time scales: a shorter one that describes single items processed through the individual workstations and a longer one that describes the response time of the whole factory [2]. In fact, two scales of flows, small and large, are well explained in fractals [3]. Furthermore, the representations of their bounds are also provided [4].

Citation: Fenglan He, Ming Dong, and Dong Yang, "Modeling and Analysis of Reentrant Manufacturing Systems: Micro- and Macroperspectives," Mathematical Problems in Engineering, vol. 2011, Article ID 139896, 17 pages, 2011. doi:10.1155/2011/139896.

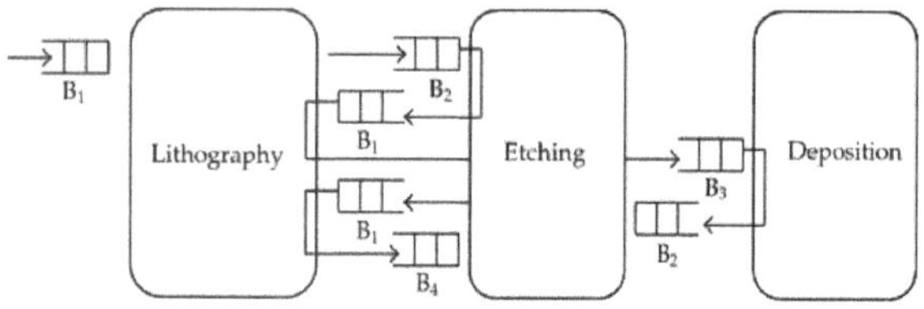

Figure 1: The structure of a reentrant manufacturing system.

Currently, most of the semiconductor manufacturing systems are described by the discrete models, the advantages of the discrete models are (1) some of the actual static problems can be directly reflected in the discrete models. (2) In reality, the methods of data collection are discrete. (3) There are many existing methods to solve the discrete models. Based on the theory of the discrete event dynamic systems, there are several discrete methods used to model the multiple reentrant semiconductor production flows.

First, queuing networks can intuitively describe the discrete production process of the wafer production line and obtain the analytical expression of the performance evaluation. A methodology for supply chain inventory analysis and optimization was presented by linking production authorization (PA) strategy to queuing models [5]. The statistics of arrival flows of the fluid models in queuing systems has been studied [6, 7]. Dong and Chen [8] proposed a network of inventory-queue models for the performance modeling and analysis of an integrated supply chain network. For a single multiple semiconductor manufacturing system, S. Kumar and P. R. Kumar [9] focused on the analysis of queuing theory of the reentrant manufacturing systems. However, with the emergence of new modes of production, especially for the development of the large quantity and multi-class order production mode, modeling and analysis of the traditional queuing network models becomes more difficult, a simplified model of wafer manufacturing systems becomes less practical. On the other hand, it is very difficult to solve the large-scale queuing network models. Most of the queuing models can only be used to evaluate the stability of some scheduling policies, while they are difficult to be used directly for the real-world supply chains. The queuing models are only used to evaluate the stability of some scheduling policies and can not be directly used for the actual production systems [10].

Second, Petri net (PN) model has been widely used in modeling, simulating, analyzing, and controlling the discrete event dynamic systems. Compared with some other description tools, PN model is especially easy to describe concurrent phenomena and simulate the parallel systems. However, with the increasing complexity and size of the manufacturing systems, the complexity of the PN model analysis is also a corresponding increase. Lin et al. [11] established a model of the reentrant semiconductor production lines using Petri-nets and studied on the stability of the system using buffer-boundless approach. Dong and Chen [12] developed a modular modeling approach based on object-oriented predicate/transition nets (OPTNs) for the analysis of supply chain process models. Liu et al. [13] proposed a model of semiconductor manufacturing systems on the basis of object-oriented colored time Petri nets and proved the validity of the model by simulating different scheduling rules.

Third, fluid networks model comes from traffic theory and was introduced by Newell [14] to approximately solve queuing problems. This method treats the discrete manufacturing process as the dynamic allocation of processing capacity of equipment. The model can greatly reduce the dimensions of the system, thereby can reduce the difficulty of analysis, while it has good robustness and can be applied to random environment. However, the expression of the model is obtained by the constraint equations, so it is difficult to apply to multi-class and high volume make-to-order production mode. Dai and Weiss [15] analyzed the relationship between the stability of the fluid model and the stability of the scheduling policies for the related queuing networks. Here, the stability of the fluid models is expressed by the boundedness of the fluid variables for a fixed influx. Göttlich [16] deduced the conservation laws under the form of ordinary differential and proved the existence of its solution based on the fluid models of supply chain networks.

On the other hand, the continuous models also have several advantages in description of the production lines. That is, they are scalable, more detailed results that can be found as compared to fluid models, and more important, they are amenable to optimization and control. Recently, the continuous models have been applied to many fields and have achieved some significant research results. Anderson [17] established the basic continuous model for supply chains and described production flow of the systems using rate equations macroscopically. Lee et al. [18] studied on the supply chain simulation with discrete-continuous combined modeling approach. This could combine the wide applicability of the discrete event simulation (DES) and fast computation of the continuum models together. Armburster and Ringhofer [19] introduced the concept of materials' density and established the continuous model of large-scale reentrant manufacturing systems and proposed new state equations so that the continuous model could be used to the real production systems. Compared with a discrete model, van den Berg et al. [20] verified the validity of the continuous model of simple serial production systems and then solved the optimal control problems with consideration of the demand growth.

The continuous models can more accurately reflect the actual situation, while the corresponding discrete models need to be discretized in time or space, then takes a constant value in each discrete node to represent the state of a period of time. This method is the approximation of the real situations and cannot fully reflect the actual situations. Using the common ground and similarities of the semiconductor manufacturing systems, the large-scale complex network can be decomposed into relatively small-scale simple problems. The multiscale methods can be more accurate modeling and analysis of the manufacturing systems. Armbruster and Ringhofer [21] extended the standard stochastic models by introducing the concept of a random phase velocity. This leads to the concepts of temperature and diffusion in the corresponding kinetic and fluid models for supply chains. Zou et al. [22] demonstrated certain features of equation free coarse-grained computation for a reentrant supply-chain model. They took the advantage of the time scale separation to directly solve coarse-grained equations through an equation-free computational approach. Unver et al. [23, 24] presented a continuum (traffic flow like) model for the flow of products through complex production networks, based on statistical information obtained from extensive observations of the system. The resulting model consists of a system of hyperbolic conservation laws, which exhibit the correct diffusive properties given by the variance of the observed data.

In this paper, in order for better analysis of the multiple reentrant manufacturing systems, their modeling and analysis from both micro- and macroperspectives are considered. First, the discrete event simulation models and their basic algorithm are proposed. In order to describe the multiple reentrant semiconductor manufacturing systems more precisely, a modified continuous model is proposed, which can reflect how the reentrant degree of a product impacts on the system performance. Finally, the changes of the WIP levels and outflux are analyzed based on the discrete event simulation models and the modified continuous models.

The structure of this paper is organized as follows: Section 2 presents the discrete event simulation models and their basic algorithm. In Section 3, in order to describe the manufacturing systems more accurately, a modified continuous model is proposed, which introduces the concept of reentrant factor, and the effectiveness is also verified through a numerical example. In Section 4, the changes of the WIP levels and outflux are analyzed between the discrete event simulation models and the modified continuous models. Meanwhile, some interesting observations are discussed. Finally, some conclusions and suggested further areas of investigation are given in Section 5.

SIMULATION MODELS

Discrete Event Simulation Models

In the discrete event simulation (DES) models, each item to be produced is treated as individual. The production process consists of M stages $(s_1,s_2,\ldots,s_M)$, a_n^m denotes the time at which the lot number n arrives at stage sm, and e_n^m denotes the time at which the lot number n exits at stage s_m (or the lot number n arrivers at stage s_{m+1}). τ_n^m denotes the processing time that the lot number n takes at stage s_m, which is known as the throughput time (TPT). The relationship between the arrival and exit times is given via the law

$$\text{(a) } e_n^m = a_n^m + \tau_n^m, \qquad \text{(b) } dP\{\tau_n^m = r\} = T_m(r, a_n^m)dr, \qquad \text{(c) } a_n^{m+1} = e_n^m. \tag{2.1}$$

Here, $T(r,a)$ denotes the time-dependent distribution of throughput times of stage s_m, and P denotes the probability distribution of the processing time. The TPT is determined by the state of the supplier s_m at time a_n^m when lot number n arrives. The throughput time distribution T_m is usually dependent on the total number lots $W(t)$ processed at stage s_m at time t, the so-called work in progress (WIP). The WIP $W(t)$ is computed as

$$W_m(t) = \sum_n H(t - a_n^m) - H\left(t - a_n^{m+1}\right), \tag{2.2}$$

where H denotes the usual Heaviside function.

In engineering practices, the most significant influence on the form of this distribution comes from the total number of items in progress, that is, the work in progress (WIP). So the TPT distribution T_m in (2.1) is written as $T_m{=}T(r,W_m($ a_n^m)). When the TPT at the beginning of the simulation is fixed, we essentially treat the whole factory as a single queue whose length at the item arrival time determines the time that the item needs to get processed. In the case of complicated supply chain networks, where each individual supplier models a whole factory, the dependence of the distribution T_m on the WIP W_m is more complex and could be given by experimental data.

For a high-volume multiple reentrant manufacturing system, the time interval between jobs becomes less important, and then the jobs can be seen as a continuous way. In these continuous production models, the WIP $W(t)$ and the influx $\lambda_m(t)$ are treated as a continuous function. This leads to a class of models often referred to as fluid models in the supply chain literature [25]. The primary variables are the WIP $W(t)$ and the fluxes $\lambda_m(t)$from stage s_{m-1} to s_m. A random throughput time function (t) can be computed from the WIP obtained before. The WIP $W(t)$ is then given by the conservation law

$$\frac{d}{dt}W_m(t) = \lambda_m(t) - \mu_m(t), \qquad \lambda_{m+1}(t) = \mu_m(t). \tag{2.3}$$

Here, λ and μ represent the influx and outflux of the lots, respectively. According to (2.1), the fluxes are given

$$\mu_m(t) = \int \delta(s + \tau(s) - t)\lambda_m(s)ds. \tag{2.4}$$

For a large number of lots, the individual lots are replaced by a continuous quantity, and the fluxes λ_m are replaced by continuous functions. The WIP $W(t)$ is evolved by the conservation law $(d/dt)W(t){=}\lambda(t){-}\mu(t)$. Using the discrete time steps, the following formulation can be obtained:

$$W(t + \Delta t) = W(t) + \Delta t\left[\lambda(t) - \mu(t)\right]. \tag{2.5}$$

So the question can be used to solve outflux (t) in terms of influx $\lambda(t)$.

However, if supplier s_m is a reentrant system itself, then the lots arriving after time a_n^m will significantly influence the TPT τ_n^m since they will compete with lot number n in the individual subqueues of the system. The set of policies, such as FIFO, PUSH, or PULL, will be used to govern the competition. This situation is treated by introducing the concepts of phase (a scaled position) and phase velocity.

In phase models, the evolution of a large ensemble of lots is modeled by describing the trajectories of each individual lot in phase space based on Newton equations. The trajectories (t) and the throughput time $\tau(t)$ are introduced.

The throughput time (t) is a random variable sampled from the distribution $T(r,t)$. Given a prescribed TPT distribution function (r,t), which is expressed in terms of WIP as $\tau(r,\text{WIP})$), the position and the throughput time after a small enough time interval Δt are given by

$$\xi(t+\Delta t) = \xi(t) + \frac{\Delta t}{\tau(t)}, \qquad \tau(t+\Delta t) = (1-k)\tau(t) + k\eta(t), \tag{2.6a}$$

$$P\{k(t)=1\} = w\Delta t, \qquad P\{k(t)=0\} = 1 - w\Delta t, \qquad dP\{\eta(t)=r\} = T(r,t)dr, \tag{2.6b}$$

$$\xi(0) = 0, \qquad dP\{\eta(0)=r\} = T(r,0)dr. \tag{2.6c}$$

A random variable k is used to decide whether to update the throughput time or not, which is either one or zero. The probability of k=1 equals $w\Delta t$ and the probability of k=0 is 1−$w\Delta t$. Here, w is the update frequency. If k=1, the throughput time is updated and the new throughput time is given by a random number (t), which is obtained from the throughput time distribution $\tau(r,\text{WIP})$. If k=0, the current throughput time remains unchanged. In general, suppose that w depends on both (t) and time t, then ((2.6b)) will be replaced by

$$P\{k(t)=1\} = \Delta t w(\eta(t),t), \qquad P\{k(t)=0\} = 1 - \Delta t w(\eta(t),t),$$
$$dP\{\eta(t)=r\} = T(r,t)dr. \tag{2.6b'}$$

According to [18], w can be chosen as

$$w(r,t) = \frac{\lambda(t)}{rT_{-1}}, \tag{2.7}$$

where $\lambda(t)$ is the influx of items and $T_{-1}=\int r^{-1}T(r,t)dr$. Based on the evolution (2.6a), (2.6b), and (2.6c), the algorithm procedure to solve the phase and TPT of an item for the discrete event and continuous production is then as follows.

Step 1. Set the initial phase to 0 as the item enters the factory, increase the WIP (t) by one and adjust the probability distribution $T(r,\text{WIP})$ accordingly.

Step 2. Compute (r,t) based on (2.7), and sample the parameters $k(t)$ and $\eta(t)$ according to their distribution ((2.6b)).

Step 3. Compute the TPT $(t+\Delta t)$ and the phase $\xi(t+\Delta t)$ according to ((2.6b)).

Step 4. Update the distribution TPT (r,WIP), according to the current WIP at the time $t+\Delta t$, and back to Step2, repeat this loop until the phase ξ=1. Now the time when ξ=1 becomes the time of the item leaving the system.

The function (t) plays a role of a position which is advanced with a randomly changing velocity $1/\eta(t)$. It is, therefore, to derive a kinetic formulation. In the context of supply chains, these models are often referred to as traffic flow models [26]. A product arrives at the first stage at time t=a and moves from x=0 to x=1 along a trajectory with velocity $1/\eta(t)$, in such a way that it reaches the end x=1 at time t=e, then the velocity has to be chosen in the following way $\int_a^e (1/\eta_a(t))dt = 1$. The total WIP (t) and the flux $F(x,t)$ at any point $x\in[0,1]$are then obtained as follows:

$$W(t) = \int [H(\xi_a(t)) - H(\xi_a(t)-1)]\lambda(a)da, \qquad F(x,t) = \int \eta_a(t)\delta(x-\xi_a(t))\lambda(a)da, \tag{2.8}$$

where $\lambda(a)$ is concentrated on the arrival times a_n for a discrete model or is a continuous function for a continuous production model.

Basic Partial Differential Equation (PDE) Models

Partial differential equation (PDE) models are actually continuum approximation of fluid models. Recently, PDE models for large-scale multiple reentrant production systems have become an important research topic. PDE models do have several advantages, that is, they might be preferred when evaluating the overall performance of large-scale manufacturing systems or solving the optimization problem for a longer time plan. Currently, the continuous models have been applied to many fields and have achieved some significant research results.

This is appropriate to describe a semiconductor manufacturing fab involving a large number of items in many stages: (x,t) is the density of the products with units (units/space) in the system, which is the conserved variable. Here, x denotes the degree of completion (DOC), x=0 describes raw products that have just entered into the factory, and x=1 denotes finished products that are ready to exit from the system. So, x is in the closed interval: $x\in[0,1]$. The total number of products in the system can be obtained by taking the integral of density (x,t) of products over the stage variable x from 0 to 1. Therefore, the total WIP (t) as a function of time can be obtained as follows:

$$W(t) = \int_0^1 \rho(x,t)dx. \tag{2.9}$$

Assuming that there is a unique entry and exit for the system and the yield is 100%, PDE models can be given bellow according to the conservation law,

$$\frac{\partial\rho(x,t)}{\partial t} + \frac{\partial(v(\rho(x,t))\rho(x,t))}{\partial x} = 0, \quad x\in[0,1],\ t\in(0,\infty), \tag{2.10}$$

where $v(\rho(x,t))$ is a velocity function that depends on the density $\rho(x,t)$ only. For a reentrant supply chain system, we assume that $(\rho(x,t))$ can be described by a state equation of the following form:

$$v(\rho(x,t)) = v_0\left(1 - \frac{W(t)}{W_{max}}\right), \tag{2.11}$$

where v_0 is the velocity for the empty supply chain system, and W_{max} is the maximal load (capacity of the supply chain system). Clearly, the velocity $(\rho(x,t))$ is determined by the total WIP. The boundary condition for the start rate (t) of products entering the supply chain system at x=0 is then defined as

$$\lambda(t) = \rho(0,t)v(t). \tag{2.12}$$

An arbitrary initial condition for the density of the products can be expressed as

$$\rho(x,0) = \rho_0(x). \tag{2.13}$$

The production process of the supply chain network can be described as an equivalent $M/M/1$ queue. The state variable ρ_{eq} denotes the equilibrium density of the supply chain system as a whole. Let $\rho=\rho_{eq}$, then $(0,t)=\rho_{eq}$ and $v=v_{eq}=1/\tau$. Correspondingly, the associated cycle time in steady state is $\tau=1/v_{eq}$. Since a job arriving at a queue with a processing rate is $\mu=1/v_{max}$, the cycle time $\tau=(1+L)$ can be obtained according to the queuing theory, the equilibrium velocity therefore becomes

$$v(\rho) = \frac{v_{\max}}{1+\int_0^1 \rho(x,t)dx} = \frac{v_{\max}}{1+w(t)}. \tag{2.14}$$

Equation (2.14) is a widely used expression between v and ρ for large-scale multiple reentrant supply chain systems, notice that the velocity v only depends upon the WIP at stage x. Therefore, (2.10) can be rewritten as

$$\frac{\partial\rho(x,t)}{\partial t} + v\frac{\partial\rho(x,t)}{\partial x} = 0, \quad x \in [0,1],\ t \in (0,\infty). \tag{2.15}$$

Hence, assume that an initial WIP distribution $\rho_0(x)$ in the factory is given, the resulting full PDE models for the single-product multiple reentrant supply chain systems are given as follows:

$$\frac{\partial\rho(x,t)}{\partial t} + v(\rho)\frac{\partial\rho(x,t)}{\partial x} = 0,$$
$$\rho(x,0) = \rho_0(x),$$
$$\rho(0,t)v(t) = \lambda(t),$$
$$v(\rho) = \frac{v_{\max}}{1+\int_0^1 \rho(x,t)dx}. \tag{2.16}$$

If the influx (t) and the initial condition $\rho_0(x)$ are nonnegative, then the density will remain nonnegative. We use an upwinding scheme [27] to discretize the PDE which is given by the following equation:

$$\rho(x_i,t_{j+1}) = \rho(x_i,t_j) - \frac{\Delta t}{\Delta x}v(t_j)\left[\rho(x_i,t_j) - \rho(x_{i-1},t_j)\right], \tag{2.17}$$

where $i=1,2,\ldots,N$, and $j=0,1,\ldots,M-1$. Δt and Δx are the step sizes in time and space, respectively. Based on the boundary condition, the propagation scheme is given by

$$\rho(x_0,t_{j+1}) = \rho(x_0,t_j) - \frac{\Delta t}{\Delta x}\left[v(t_j)\rho(x_0,t_j) - \lambda(t_j)\right]. \tag{2.18}$$

Since the Courant-Friedrich-Levy (CFL) condition is necessary for stability, the time step Δt and the space step Δx must satisfy the following formulation: $|(\Delta t/\Delta x)v_{max}(t)| \le 1$. Here, $v_{max}(t)$ is the maximum velocities in the system at time t.

Based on the above formula, the density distribution of each moment (x_i,t_j) can be obtained. Therefore, the system throughput rate (x_N,t_j) of each moment can also be computed as follows:

$$q(x_N,t_j) = \rho(x_N,t_j)v(\rho(x_N,t_j)). \tag{2.19}$$

Furthermore, the total WIP $w(t) = \int_0^1 \rho(x,t)dx$ can be obtained via the extended Simpson's rule quadrature. Finally, the density distribution (xi,tj) and throughput $q(x_N,t_j)$ of each moment can be obtained.

THE MODIFIED CONTINUUM MODEL

Numerical Experiment

Mini-Fab is a simplified model of the semiconductor production lines, which has all the important features of the reentrant semiconductor manufacturing systems, such as reentrant, different processing time, and batch production. Currently, many scholars have done a lot of research work based on the Mini-Fab. The Mini-Fab contains 5 machines grouped into 3 work centers, the product comprises of 6 processing steps and each work center has a reentrant step, which is shown in Figure 2. The system designed in this study is FIFO (First In First Out).

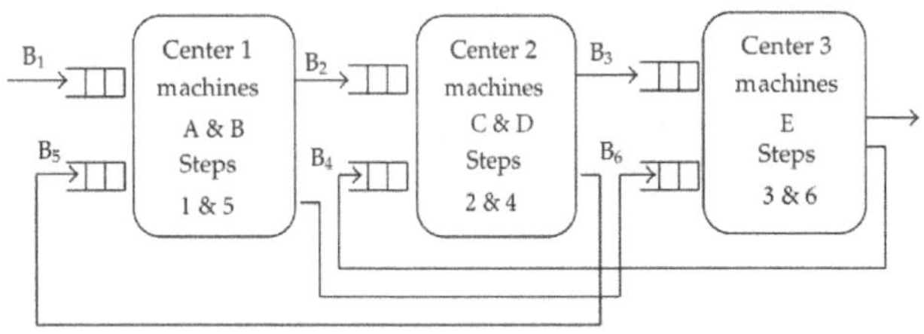

Figure 2: Process flow diagram of the Mini-Fab.

For convenience, we make the following basic assumptions on the model.(1)The product yield rate is 100%, namely, there is no rework problem.(2)The system is a continuous production process for 24 hours a day.(3)The system does not take into account the time of carrying, loading and discharging, adjusting equipment, premaintaining equipment, and downtime.

Now assuming that there is one product D in the Mini-Fab, its processing steps and processing time are shown in Table 1.

Table 1: Processing time of the product D at each step

Machining centers	Processing time (hours)	
Machines A & B	Step 1: 1.5	Step 5: 1.5
Machines C & D	Step 2: 0.5	Step 4: 1
Machine E	Step 3: 1	Step 6: 0.5

From the above table, we can see that the total processing time is P=0.25 (days). Based on the basic PDE model, the parameters are set as follows: v_{max}=4 (units/day), λ=5 (units/day). We typically start up with an empty production system, the total running time of the production lines is 10 (days). Let Δt=0.001 and Δx=0.01, then Δx and Δt that satisfy the CFL stability condition is formulated as follows: $(\Delta t/\Delta x)_{max}$<1. The system throughput can be obtained through the simulation based on the basic PDE model of (2.16). Figure 3indicates that the system throughput is about 3.2 (units/day) during the steady state.

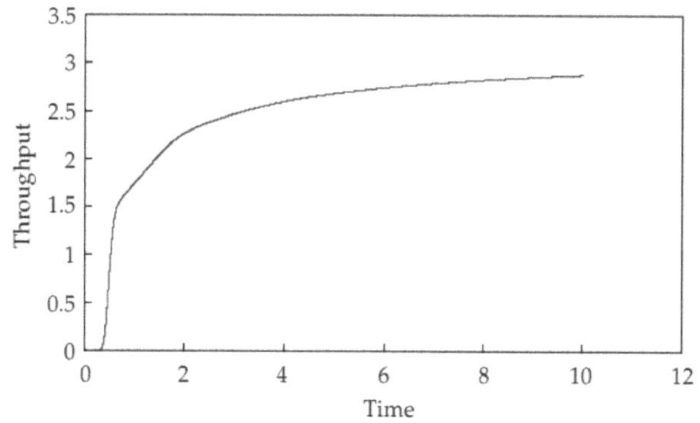

Figure 3: Throughput as a function of time for the PDE simulation.

When the throughput of the basic continuous models is obtained, the corresponding Mini-Fab simulation model can be built using simulation package ExtendSim [28] to verify the validity of the models with a period of one year. The jobs are sorted by FIFO. From Figure 4, it can be seen that the throughput during the steady state is about 5 (units/day) through ExtendSim simulation.

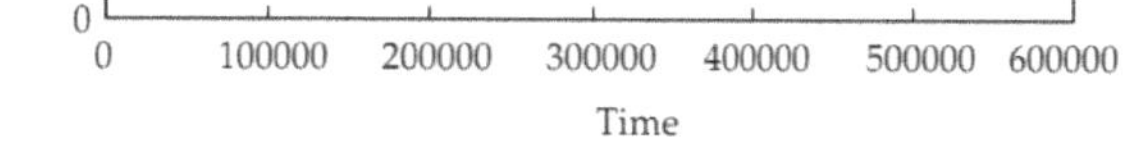

Figure 4: Throughput as a function of time for the ExtendSim simulation.

Comparing Figure 3 with Figure 4, obviously, it can be found that there are some differences between the throughput results of the two models. This is because the basic continuous model of (2.16) is built on the basis of a large number of materials and many reentrant steps in the systems and some important characteristics of semiconductor production systems are not captured. Therefore, the further investigation is needed to explore more precise models for multiple reentrant production systems.

The Modified Continuum Model

It is worth noting that the state equations could reflect the characteristics of systems—any changes of the multiple reentrant production systems may lead to a different state equation. Equation (2.14) is a quite general state equation, in order to capture some important features of multiple reentrant supply chain systems, a more specified relationship is required for computing the velocities numerically. Lefeber and Armbruster [29] presented a more sophisticated reentrant factory model through the use of integration kernels. For the general supply chain systems (nonreentrant systems), Sun and Dong [30] described several kinds of state equations and the corresponding cycle times. Although there are many kinds of continuum models currently, they do not reflect how the reentrant degree of the product impacts on the system performance. Hence, a new concept is introduced to describe the reentrant degree of production system.

Definition 3.1: Let α be a product reentrant factor, α equals to the ratio of the product's processing time of reentrant steps and the product's total processing time.

Reentrant factor α is the property of product's process flows, with the increase of reentrant factor α, the degree of reentrant of the product becomes larger and larger. Let P be the total processing time, P_1 be the reentrant processing time, and P_2 be the nonreentrant processing time. According to the above definition, the following formula can be obtained:

$$\alpha = \frac{P_1}{P_1 + P_2} = \frac{P_1}{P}. \qquad (3.1)$$

In reality, the velocity of products in the system is not only related with the WIP level, but also with the reentrant factor. Let $w(t)$ be the WIP level in the system at a given time t, $w(\Delta x,t)$ be the WIP level at interval Δx at time t, Δx be the interval of the completion of the product, and Δt be the processing time required to complete the interval Δx. Assuming that $(\Delta x,t)$ is proportional to Δt, and the system consists of two parts: reentrant processes and nonreentrant processes. Hence, $\alpha(t)$ is the WIP level of the reentrant processes at a given time t, also referred as reentrant WIP. Similarly, $(1-\alpha)(t)$ is the WIP level of nonreentrant processes at a given time t, also called as nonreentrant WIP.

According to the queuing theory, the processing cycle time of the reentrant process τ_1 can be expressed as follows:

$$\tau_1 = [1 + \alpha w(t)]P_1. \qquad (3.2)$$

As for the nonreentrant processes, assuming the total number of workstations is m, $a_1, a_2, \ldots, a_m$ are the corresponding processing times at each workstation, respectively, and τ_2 is the nonreentrant processing cycle time, then we have

$$P_2 = a_1 + a_2 + \cdots + a_m,$$

$$\tau_2 = \sum_{i=1}^{m}\left(1 + \frac{a_i}{P}w(t)\right)a_i = \sum_{i=1}^{m} a_i + \sum_{i=1}^{m}\frac{a_i^2}{P}w(t) = P_2 + \sum_{i=1}^{m}\frac{a_i^2}{P}w(t). \qquad (3.3)$$

So the total cycle time τ can be written in the following form:

$$\tau = \tau_1 + \tau_2 = [1 + \alpha w(t)]P_1 + \left[P_2 + \sum_{i=1}^{m} \frac{a_i^2}{P} w(t)\right]$$
$$= P + \left(\alpha^2 + \sum_{i=1}^{m} \frac{a_i^2}{P^2}\right) w(t) P. \quad (3.4)$$

Then, the resulting new state equation for the velocity will be given by

$$v = \frac{1}{\tau} = \frac{1}{P + (\alpha^2 + \sum_{i=1}^{m} a_i^2/P^2) w(t) P}. \quad (3.5)$$

In order to avoid computational complexity of $\sum_{i=1}^{m} a_i^2/P^2$, an approximate method is proposed to deal with the nonreentrant processes. Supposing that WIP in nonreentrant processes follows the uniform distribution in each workstation, then the mean processing time P_2/m and the mean WIP level at each workstation $((1-\alpha)/m)(t)$ can be obtained. According to the queuing theory, the mean cycle time of the nonreentrant processes at each workstation is $[1+((1-\alpha)/m) w(t)]P_2/m$, and the total cycle time of the nonreentrant processesτ_2 is $[1+((1-\alpha)/m)w(t)]P_2$. Therefore, the total cycle time τ can be expressed as follows:

$$\tau = \tau_1 + \tau_2$$
$$= [1 + \alpha w(t)]P_1 + \left[1 + \frac{(1-\alpha)}{m} w(t)\right] P_2$$
$$= P_1 + P_2 + \left[\alpha P_1 + \frac{(1-\alpha)}{m} P_2\right] w(t)$$
$$= P + \left[\alpha \cdot \alpha P + \frac{(1-\alpha)\cdot(1-\alpha)P}{m}\right] w(t)$$
$$= P + \left[\alpha^2 + \frac{(1-\alpha)^2}{m}\right] w(t) P. \quad (3.6)$$

The corresponding new state equation can be obtained as follows:

$$v = \frac{1}{\tau} = \frac{1}{P + \left[\alpha^2 + (1-\alpha)^2/m\right] w(t) P}. \quad (3.7)$$

Therefore, the resulting modified whole PDE model is given below:

$$\frac{\partial \rho(x,t)}{\partial t} + v(\rho)\frac{\partial \rho(x,t)}{\partial x} = 0,$$
$$\rho(x,0) = \rho_0(x),$$
$$\rho(0,t)v(t) = \lambda(t),$$
$$v = \frac{1}{\tau} = \frac{v_{\max}}{1 + \left[\alpha^2 + (1-\alpha)^2/m\right] w(t)}. \quad (3.8)$$

Validity of the Modified Models

Once the new state equation is obtained, the validity of the modified models can be verified by the same example described in the previous section. According to Figure 2, the total number of workstations is m=3, and the steps 4 to 6 are reentrant steps. From Table 1, the reentrant factor of products α=0.5 can be easily obtained by definition. Assuming that the influx λ of items is 5 (units/day) and the total running time is 10 (days), the system starts running from an empty factory. With Δt=0.001, Δx=0.01, then Δx and Δt satisfy the CFL stability condition: $v_{\max} \cdot \Delta t/\Delta x < 1$. Similarly, the modified continuous model (3.8) can also be solved via the upwind scheme, and the throughput of the modified continuum models can be obtained, which is plotted in Figure 5.

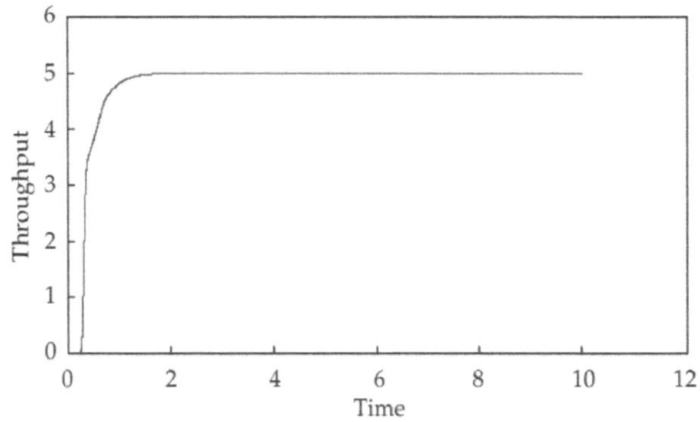

Figure 5: Throughput as a function of time for the modified PDE model.

It can be seen that the throughput is initially zero for the reentrant manufacturing systems. This is due to the time delay and the system initialization (the system starts up with an empty factory), then the system begins to have throughput at about 0.25 days and increases drastically until the throughput reaches a stable value of 5 (units/day) at about 1.5 days. Comparing Figure 4 with Figure 5, it can be obviously found that the results are basically consistent with the simulation results obtained from ExtendSim. Therefore, it can be seen that the modified continuum models are effective for multiple reentrant supply chain systems.

NUMERICAL EXPERIMENTS

Once the modified continuum models are obtained, in order to describe the multiple reentrant manufacturing systems from the micro- and macroperspectives, based on the DES and modified PDE models, a numerical experiment is carried out to demonstrate the consistency of the models for the Mini-Fab case. The WIP levels and outflux of the DES and the modified continuum models for the multiple reentrant supply chain systems are presented in this section.

In this section, a mean field assumption is made first, namely, the TPT distribution (r,t) is a given function while in reality it depends on the WIP $W(t)$. The idea underlying this assumption is that, for many lots presenting in the system, the impact of one individual lot on the distribution (r,t) is negligible. In engineering practices, the TPT distribution depends on the lot trajectories through the WIP (t). Therefore, the lots can be treated as approximately statistically independent. The prescribed TPT distribution is shown as well as the upper and the lower limits as a function of WIP in Figure 6. The mean possible TPT is a constant between the maximal possible TPT and the lower possible TPT, while the minimally possible TPT remains constant independent of the WIP.

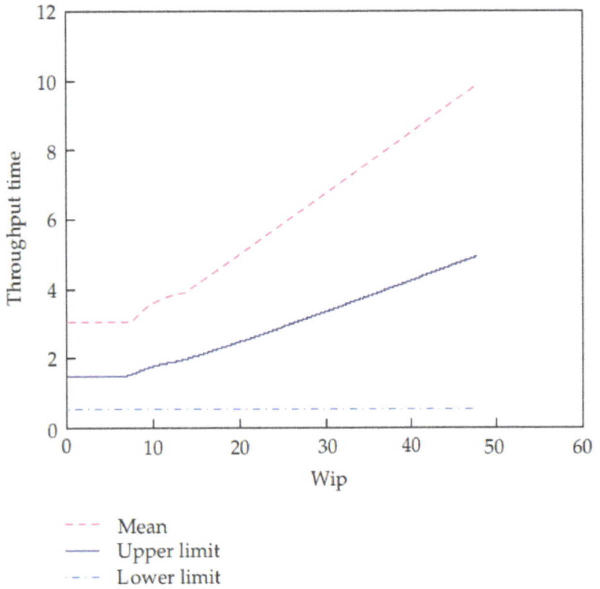

Figure 6: Probability distribution of the TPT as a function of WIP.

Same as the previous section, taking the multiple reentrant production system as an example, the system begins to run from an empty factory. The DES model is executed by generating a total of 200 lots. In order to facilitate the computation, we assume that the influx is defined as a constant value in the experiment first. The trajectories are computed by the random phase model, and the lots' arrival time a_n is given by $a_{n+1}=a_n+(1/\lambda(a_n))$, where $a_0=0$. The WIP levels and outflux of DES models are computed according to (2.8) with discrete arrival times *an*. For the modified PDE models, let $v_{max}=4$ (units/day), a uniform time step size $\Delta t=0.001$ and a uniform spatial interval length $\Delta x=0.01$, then Δx and Δt also satisfy the CFL stability condition: $(\Delta t/\Delta x)_{max}<1$. Figures 7 and 8 show the WIP levels and outflux of the DES and modified PDE models, in which dotted line indicates the WIP levels and outflux variation of the DES models and the straight line represents the WIP levels and outflux variation of the modified PDE models, respectively.

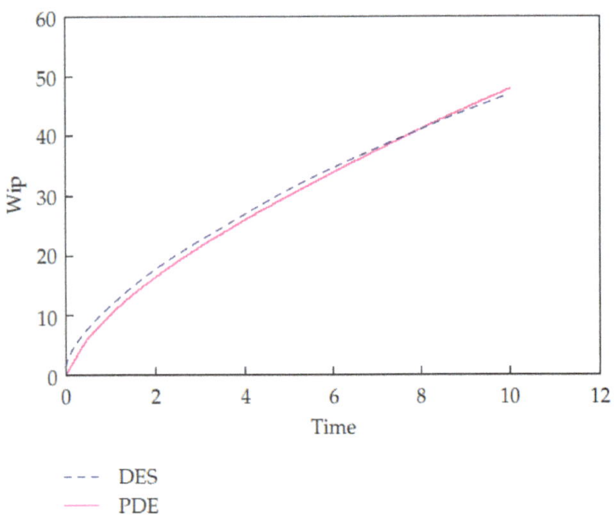

Figure 7: WIP levels of the DES and modified PDE models as a function of time.

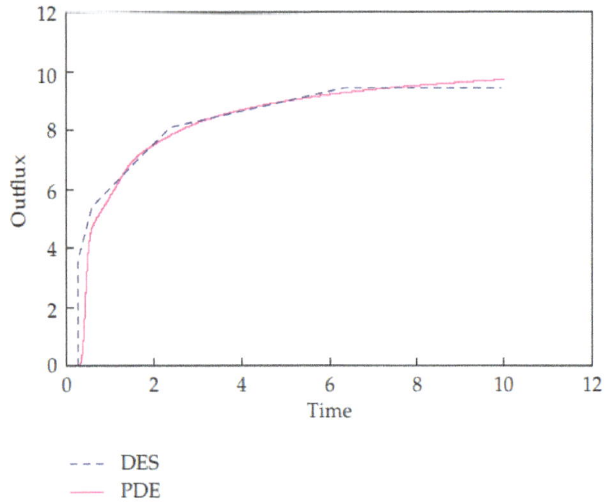

Figure 8: Outflux of the DES and the modified PDE models as a function of time.

Comparing the WIP levels and outflux computed from the DES models to the solution of the modified PDE models, it can be observed that, although the WIP levels and outflux values of the two models are not exactly equal, the results are close enough to each other to substantiate the consistency of two models. The same results of the DES and modified PDE models are given for the steady state. The WIP levels of the two models show gradual increase, while the outflux of the two models first shows an increasing trend gradually to reach the maximum and then reaches a stable value.

CONCLUSIONS

In this paper, better analysis of the multiple reentrant manufacturing systems can be obtained if both micro- and macroperspectives are adopted. The discrete event simulation models are first proposed and their basic algorithm is also presented in detail. The low accuracy of the basic PDE models for the multiple reentrant supply chain networks is explained by a numerical experiment. In order to model such complex systems more precisely, a modified PDE model that takes into account the reentrant degree of the product is presented, while the validity of the modified PDE model is also illustrated through a numerical experiment for multiple reentrant supply chain systems. Then, based on the DES and modified PDE models, a numerical experiment is provided to compare the WIP levels and outflux changes. Meanwhile, some interesting observations are discussed. Once the results of the micro- and macro simulations are obtained, some analysis for the multiple reentrant manufacturing systems based on the multiscale methods becomes possible.

ACKNOWLEDGMENTS

The work presented in this paper has been supported by a Grant from the National High-Tech Research and Development Program (863 Program) of China (2008AA04Z104) and a Grant from National Natural Science Foundation of China (70871077).

REFERENCES

1. C. Cattani, "Fractals and hidden symmetries in DNA," Mathematical Problems in Engineering, vol. 2010, Article ID 507056, 31 pages, 2010.
2. S. S. Qian and Y. H. Guo, Optimization and Control of the Re-entrant Semiconductor Manufacturing System, Publishing House of Electronics Industry, Beijing, China, 2008.
3. M. Li and S. C. Lim, "Modeling network traffic using generalized Cauchy process," Physica A, vol. 387, no. 11, pp. 2584–2594, 2008.
4. M. Li and W. Zhao, "Representation of a stochastic traffic bound," IEEE Transaction on Parallel and Distributed Systems, vol. 21, no. 9, pp. 1368–1372, 2010.
5. M. Dong, "Inventory planning of supply chains by linking production authorization strategy to queueing models," Production Planning and Control, vol. 14, no. 6, pp. 533–541, 2003.
6. M. Li, "Fractal time series—a tutorial review," Mathematical Problems in Engineering, vol. 2010, Article ID 157264, 26 pages, 2010.
7. M. Li, W. Zhao, and S. Y. Chen, "mbm-Based scalings of traffic propagated in internet," Mathematical Problems in Engineering, vol. 2011, Article ID 389803, 21 pages, 2011.
8. M. Dong and F. F. Chen, "Performance modeling and analysis of integrated logistic chains: an analytic framework," European Journal of Operational Research, vol. 162, no. 1, pp. 83–98, 2005.
9. S. Kumar and P. R. Kumar, "Queuing network models in the design and analysis of semiconductor wafer fabs," IEEE Transactions on Robotics and Automation, vol. 17, no. 5, pp. 548–561, 2001.
10. V. D. Shenal, A mathematical programming based procedure for the scheduling of lots in a wafer fab, M.S. thesis, Virginia Polytechnic Institute and State University, 2001.
11. C. Lin, M. Xu, D. C. Marinescu, F. Ren, and Z. Shan, "A sufficient condition for instability of buffer priority policies in re-entrant lines," Institute of Electrical and Electronics Engineers, vol. 48, no. 7, pp. 1235–1238, 2003.
12. M. Dong and F. F. Chen, "Process modeling and analysis of supply chain networks using object-oriented Petri nets," Robotics and Computer Integrated Manufacturing, vol. 17, no. 1, pp. 121–129, 2001.
13. H. R. Liu, Z. B. Jiang, Y. F. Lee, and R. Y. K. Fung, "Multiple-objective real-time scheduler for semiconductor wafer fab using colored timed object-oriented petri nets (CTOPN)," in Proceedings of the IEEE International Conference on Systems, Man and Cybernetics, vol. 1–5, pp. 510–515, 2003.
14. G. F. Newell, "Scheduling, location, transportation, and continuum mechanics: some simple approximations to optimization problems," SIAM Journal on Applied Mathematics, vol. 25, no. 3, pp. 346–360, 1973.
15. J. G. Dai and G. Weiss, "Stability and instability of fluid models for reentrant lines," Mathematics of Operations Research, vol. 21, no. 1, pp. 115–134, 1996.
16. S. Göttlich, M. Herty, and A. Klar, "Network models for supply chains," Communications in Mathematical Sciences, vol. 3, no. 4, pp. 545–559, 2005.
17. E. J. Anderson, "A new continuous model for job shop scheduling," International Journal of Systems Science, vol. 12, no. 12, pp. 1469–1475, 1981.
18. Y. H. Lee, M. K. Cho, S. J. Kim, and Y. B. Kim, "Supply chain simulation with discrete-continuous combined modeling," Computers and Industrial Engineering, vol. 43, no. 1-2, pp. 375–392, 2002.
19. D. Armbruster and C. Ringhofer, "Continuous models for production flows," in Proceedings of the American Control Conference, vol. 5, pp. 4589–4594, 2004.
20. R. Van Den Berg, E. Lefeber, and K. Rooda, "Modeling and control of a manufacturing flow line using partial differential equations," IEEE Transactions on Control Systems Technology, vol. 16, no. 1, pp. 130–136, 2008.
21. D. Armbruster and C. Ringhofer, "Thermalized kinetic and fluid models for reentrant supply chains,"Multiscale Modeling & Simulation, vol. 3, no. 4, pp. 782–800, 2005.
22. Y. Zou, I. G. Kevrekidis, and D. Armbruster, "Multiscale analysis of re-entrant production lines: an equation-free approach," Physica A, vol. 363, no. 1, pp. 1–13, 2006.
23. A. K. Unver, C. Ringhofer, and D. Armbruster, "A hyperbolic relaxation model for product flow in complex production networks," Discrete and Continuous Dynamical Systems, supplement 2009, pp. 791–800, 2009.
24. A. K. Unver and C. Ringhofer, "Estimation of transport coefficients in re-entrant factory models," in 15th IFAC Symposium on System Identification (SYSID '09), vol. 15, pp. 705–710, 2009.
25. J. J. Hasenbein, "Stability of fluid networks with proportional routing," Queueing Systems, vol. 38, no. 3, pp. 327–354, 2001.
26. D. Armbruster, D. Marthaler, and C. Ringhofer, "Kinetic and fluid model hierarchies for supply chains,"Multiscale Modeling & Simulation, vol. 2, no. 1, pp. 43–61, 2003.
27. R. J. LeVeque, Finite Difference Methods for Ordinary and Partial Differential Equations, Society for Industrial and Applied Mathematics (SIAM), Philadelphia, Pa, USA, 2007.
28. T. B. Qin and Y. F. Wang, Application Oriented Simulation Modeling and Analysis with ExtendSim, Tsinghua University Press, Beijing, China, 2009.
29. E. Lefeber and D. Armbruster, Aggregate Modeling of Manufacturing Systems, Systems Engineering Group, 2007.
30. S. Sun and M. Dong, "Continuum modeling of supply chain networks using discontinuous Galerkin methods," Computer Methods in Applied Mechanics and Engineering, vol. 197, no. 13–16, pp. 1204–1218, 2008.

CHAPTER 18

AN INTEGRATED MODEL FOR PRODUCTION PLANNING AND CELL FORMATION IN CELLULAR MANUFACTURING SYSTEMS

Reza Raminfar[1,2] Norzima Zulkifli[1] Mohammadreza Vasili[2] and Tang Sai Hong[1]

[1]Department of Mechanical and Manufacturing Engineering, Faculty of Engineering, Universiti Putra Malaysia, 43400 Selangor, Selangor Darul Ehsan, Malaysia

[2]Department of Industrial Engineering, Islamic Azad University, Lenjan Branch, isfahan, Iran

ABSTRACT

Cellular manufacturing (CM) is a production approach directed towards reducing costs, as well as increasing system's flexibility in today's small-to-medium lot production environment. Many structural and operational issues should be considered for a successful CM design and implementation such as cell formation (CF), production planning, and facility layout. Most researchers have addressed these issues sequentially or independently, instead of jointly optimizing a combination of these issues. In order to attain better results to ensure that the system will be capable of remaining efficient in unknown future situations, these issues should be addressed simultaneously. In this paper, a mathematical model is developed using an integrated approach for production planning and cell formation problems in a CM. A set of numerical examples are provided from existing the literature in order to test and illustrate the proposed model. In order to evaluate and verify the performance of the proposed model, it is compared with a well-known cell formation methods (rank order clustering and direct clustering analysis), using group capability index (GCI) measure. The results and comparisons indicate that the proposed model has a significantly higher and satisfactory performance and it is reliable for the design and the analysis of CM systems.

INTRODUCTION

Cellular manufacturing (CM) is a production system that involves processing a collection of similar parts (part families) on dedicated cluster of machines or manufacturing processes (cells) [1]. CM is an application of group technology (GT) which offers the advantages of both job shops (flexibility in producing a wide variety of products) and flow lines (efficient flow and high production rate) [2]. The advantages of CM include simplified material flows and reduced material handling cost, reduced production time, reduction in setup time, reduced production cost, reduction in inventory and work-in-process (WIP) inventories, reduction in scrap and waste, decentralization of responsibilities, and saving manufacturing space [3–6]. The design and implementation of an effective CM system involves many issues such as machine-part cell formation (CF), production planning, layout design, and scheduling.

Citation: Reza Raminfar, Norzima Zulkifli, Mohammadreza Vasili, and Tang Sai Hong, "An Integrated Model for Production Planning and Cell Formation in Cellular Manufacturing Systems," Journal of Applied Mathematics, vol. 2013, Article ID 487694, 10 pages, 2013. doi:10.1155/2013/487694.

In the design of a CM system, similar parts are grouped into families and associated machines into groups so that one or more part families can be processed within a single machine group [7]. The process of determining part families and machine groups in order to form the manufacturing cells is referred to as the CF problem. This problem may be solved by using different methods and considering different manufacturing features. The CF problems can be classified into binary and comprehensive problems depending on whether or not processing times and the machine capacities are considered.

The binary problem arises if the part demands are unknown when the CM system is being developed [8–10]. If the part demand can be accurately predicted, processing time and machine capacities have to be included in the analysis. This gives rise to comprehensive problems [11]. In addition, the comprehensive models can be classified considering the production requirements. They are static versus dynamic production requirements and deterministic versus stochastic production requirements [12]. A static production requirement implies a single period when designing a CM system. It assumes that product mix and part demand are constant for the entire planning horizon. The product mix refers to a set of part-types to be produced at each period, and the part demand is the quantity of each part-type to be produced [13]. Product mix and demand in such cases can be deterministic or stochastic. For static and deterministic production requirements, there is only one possible set of product mix and demand which are known [14, 15]. In contrast, static and stochastic production requirements have a set of possible product mixes and demands to occur; each has its probability of occurrence [16]. However, dynamic production requirements imply multiple periods when designing a CM system. In the dynamic environment, the entire planning horizon is divided into multiple smaller periods, and each period has different product mix and part demand. Meanwhile, as mentioned earlier, production requirements can be deterministic or stochastic in each period. For dynamic and deterministic production requirements, product mix and demand in each period are known [17, 18]. For dynamic and stochastic production requirements, the possible product mixes and demands in each period are known with certain probabilities [19, 20].

In today's world of rapidly changing product demand, small internet orders, tight delivery schedules, high competition, and high service level requirements, it will be increasingly difficult to maintain good operational performance over the time. In such a case, managing the production resources and balancing them between successive time periods with the aim of minimizing the production costs is known as "production planning" [21]. Based on the literature, numerous mathematical models and solution methods have been developed to solve the existing problems in general manufacturing or service industries [22, 23]. However, in the recent researches, critical manufacturing features such as production flexibility and manufacturing cell formations have received considerable attention in developing production planning models. Furthermore, in order to achieve more practical results, the production planning features such as machine capacity, machine cost, operations sequence, inventories holding, backorders, and subcontracting can be considered in order to form the manufacturing cells. In this regard, integrated approaches should be pursued in manufacturing system analysis, since different aspects of the system are interrelated in many ways [24]. Integrated system approaches can minimize the possibility of certain important aspects of the system being neglected, while other issues are being studied [25]. In general, integrating the concepts of the CM system design and production planning is a fundamental requirement for modeling and simulating of the real manufacturing systems. This integrated concept has been discussed, for example, in Chen [25], Safaei and Tavakkoli-Moghaddam [21], Ah kioon et al. [26], and Mahdavi et al. [27].

Although each of these researches has covered some of the features problem, or utilize special method to solve problems and their efficiencies are different too. For instance, Chen [25] solved problems with a decomposition-based heuristic algorithm or he considered unit cost to move for every part-type in batches between cells; however, it seems not practical enough. Other example is a Safaei and Tavakkoli-Moghaddam [21], although this is a very comprehensive dynamic integrated model and utilizes mathematical linear programming to solve problems, but for material handling cost feature, this research does not take into account different travel distance between cells. Ah kioon et al., [26] research is also a very well-integrated model too, but it does not take into consideration some issues that are addressed within this research paper such as machine set-up cost (with respect to the machine set-up cost for each operation) or they consider intercell material handling cost feature, but they do not take into account different travel distance between cells too. Mahdavi et al., [27] or other called researches proposed their integrated model and solved numerical examples by their models, but they do not try to evaluate and verify performance of their models, although we try to evaluate and verify the CF section performance of the proposed model within this research.

This study aims at introducing a comprehensive dynamic deterministic integrated model to offer an optimal solution for grouping the part families and machine cells, as well as an optimal production plan for minimizing inventory and machine set-up costs. The aim of proposed model is to minimize machine operating cost, intercell material handling cost (with respect to the different travel distances between cells), machine operating cost, finished-goods inventory cost, and machine set-up cost (with respect to the machine set up cost for each operation). Our model considers cell size limitation, machine duplication in one period time, and operation sequence features too.

The rest of this paper is organized as follows. Detailed description of the problem and the proposed model are presented in Section 2. Some numerical examples are presented in Section 3, in order to illustrate the proposed model. Discussions to verify the model are presented in Section 4. Summary and conclusions of the research are discussed in Section 5.

MATHEMATICAL MODEL DEVELOPMENT

Problem Description

In this section, a mixed integer nonlinear programming model is developed to simultaneously solve the cell formation and production planning problems in CM systems. The objective function of this model is to minimize intercell material handling cost, machine operating cost, production set-up cost for every operation in every part, and part inventory cost. A manufacturing system which includes a number of machines to process different part-types is considered. Each part-type has a number of operations, and that must be processed as numbered respectively, in order to take the sequence of operations into account. The manufacturing system is considered in a number of time periods t, where $t = 1, 2, \ldots, T$, with $T>1$. One time period could be a day, a week, or a month. Demands for different part-types are assumed to be known and deterministic.

Demands may be satisfied from production in the same time period or from inventory. Back orders and shortage cost are not considered in this study. This model assumes that there is a single process plan for each part-type. Since both the quantity of different part-types to be processed by the machines and the formation of the machine cells are decision variables, nonlinear terms are presented in the objective function of the mixed integer programming model.

Notations

Indices:

t: time period index: $t = 1, \ldots, T$.

- i: PART-TYPE INDEX: $i = 1, \ldots, I(t)$.
- j: INDEX OF OPERATIONS OF PART-TYPE : $j = 1, \ldots, Ji$.
- l: CELL INDEX: $l = 1, \ldots, L$.
- k: MACHINE INDEX: $k = 1, \ldots, K$.

Coefficients and Parameters:

- $H(t)$: unit inventory holding cost of part-type i for time period t.
- $D(t)$: known demand of part-type i for time period t.
- $M(t)$: unit machine operating cost for machine-type k in time period t.
- $S_{i[jk]}$: set-up cost to performing operation j on machine k from part-type i.
- $R_{ll'}$: cost of moving a unit of part-type from cell l to cell l'.
- LB_l: minimum number of machines in cell l.
- UB_l: maximum number of machines in cell l.

Binary Decision Variables:

$$\beta_i(t) = \begin{cases} 1, & \text{if part type } i \text{ is processed in period } t, \\ 0, & \text{otherwise,} \end{cases}$$

$$n_{kl}(t) = \begin{cases} 1, & \text{if one unit of type } k \text{ machine is placed in cell } l \text{ at time } t, \\ 0, & \text{otherwise,} \end{cases}$$

$$\delta_{i[jk]l}(t) = \begin{cases} 1, & \text{if operation } j \text{ of part-type } i \text{ to be processed by machine type } k, \text{ is done in cell } l \text{ during time } t, \\ 0, & \text{otherwise.} \end{cases} \tag{1}$$

Subscripts $[jk]$ of variable $\delta_{i[jk]l}(t)$ indicates that machine k is required to process operation j of part-type i. This information is known from the given part process plan.

Continuous Decision Variables:

$X(t)$: amount of part-type i to be processed in time period t,

$V(t)$: amount of part-type i in inventory at the end of time period t.

Mathematical Models Development

With the given notations, the proposed model is formulated as follows:

Minimize Z

$$= \sum_{t=1}^{T}\sum_{k=1}^{K} M_k(t) \cdot \sum_{l=1}^{L} n_{kl}(t)$$

$$+ \sum_{t=1}^{T}\sum_{i=1}^{I(t)} X_i(t)$$

$$\cdot \sum_{j=1}^{j_i-1}\sum_{k=1}^{K_{ji}}\sum_{k'=1}^{K}\sum_{l=1}^{L}\sum_{l'=1}^{L} R_{ll'}\delta_{i[jk]l}(t)\,\delta_{i[(j+1)k']l'}(t)$$

$$+ \sum_{t=1}^{T-1}\sum_{i=1}^{I(t)} H_i(t)\,V_i(t) + \sum_{t=1}^{T}\sum_{l=1}^{L}\sum_{i=1}^{I(t)}\sum_{j=1}^{J_i}\sum_{k=1}^{K} S_{i[jk]}\delta_{i[jk]l}(t). \tag{2}$$

Subject to:

$$\sum_{l=1}^{L}\delta_{i[jk]l}(t) = \beta_i(t), \quad i = 1,\ldots,I, \; j = 1,\ldots,J_i, \quad k = 1,\ldots,K_{ji} \;\forall(t), \tag{3}$$

$$\sum_{i=1}^{I}\sum_{j=1}^{J_i}\delta_{i[jk]l}(t) \geq n_{kl}(t), \quad i = 1,\ldots,I, \; j = 1,\ldots J_i, \quad k = 1,\ldots,K, \; l = 1,\ldots,L \;\forall(t), \tag{4}$$

$$\delta_{i[jk]l}(t) \leq n_{kl}(t), \quad k = 1,\ldots,K, \; l = 1,\ldots,L \;\forall(t), \tag{5}$$

$$LB_l \leq \sum_{k=1}^{K} n_{kl}(t) \leq UB_l, \quad k = 1,\ldots,K \;\forall(t), \tag{6}$$

$$v_i(t+1) = v_i(t) + X_i(t) - D_i(t), \quad i = 1,\ldots,I, \quad t = 1,\ldots,T-1, \tag{7}$$

$$\sum_{t=1}^{T} X_i(t) = \sum_{t=1}^{T} D_i(t), \quad i = 1,2,\ldots,I, \tag{8}$$

$$\beta_i(t) = \begin{cases} 1, & \text{if } X_i(t) > 0 \\ 0, & \text{if } X_i(t) = 0, \end{cases} \quad i = 1,2,\ldots,I, \;\forall(t), \tag{9}$$

$$X_i(t) \geq 0, \quad v_i(t) \geq 0, \quad \delta_{i[jk]l}(t), \quad \beta_i(t), \quad n_{kl}(t) = 0, 1, \quad \forall i, j, l, l', k, k', t. \tag{10}$$

The objective function of the proposed model has been shown in (2), and it consists of four terms. The first term of the objective function is the machine operating cost. It is assumed that the machines can be included when they are needed and can be removed from the system when they are not required.

Table 1: Part processing demand and unit inventory cost for Example 1

	Part-type									
	1	2	3	4	5	6	7	8	9	10
	Unit inventory cost									
	1.0	1.5	0.5	1.0	1.2	0.5	2.5	1.2	1.5	0.8
Time period, t	Part processing demand									
1	22	20	18	12	23	24	22	30	19	25
2	28	16	20	24	30	22	18	28	18	14
3	36	22	30	39	32	26	16	26	33	22

Table 2: Partial input data for Example 1

Machine number	Machine cost	Part-type									
		1	2	3	4	5	6	7	8	9	10
		Part operation requirement (upper number) and refixturing cost (lower number)									
1	15	2 15					2 11				1 8
2	10	1 10	1 12			2 8			2 9		
3	20		2 18	2 5		1 12			1 16		2 7
4	15			3 9	2 12			2 7		2 4	
5	12						1 14				
6	15			1 6							3 10
7	16				1 15			1 11		1 21	

Table 3: Intercell material handling cost for Example 1

Cell number	1	2	3
1	0.0	1.0	1.4
2	1.0	0.0	1.2
3	1.4	1.2	0.0

The second term of the objective function is the intercell material handling cost. This cost function is similar to those in Atmani et al. [28] and Chen [25]. In a typical cell formation problem, the number of part-type i to be produced in time t, $X(t)$, is usually considered constant. In that case, the material handling cost function will be linear. However, in this model, the term of the material handling cost is non-linear, because it has been assumed that the distances between each pair of cells are different (part-type i after completion of its operation j by machine k in cell l moves to machine k' for its next operation, j+1, in cell l'). It is further assumed that the specifications of different part-types (e.g., size or volume of different part-types) do not influence the material handling cost. The third term is the finished goods inventory cost, and the last term of the objective function is the system set-up cost with consideration of set-up cost for each operation in each part-type. The first two cost items in the objective function are related to forming manufacturing cells, while the latter two cost items are related to production and inventory control.

Constraints of the model consist of equations and inequalities (3) to (10). Equation (3) is to ensure that if operation j of part-type i will be processed by machine k in one of the cells in time period t, then the corresponding binary variable for system set-up must be 1. Inequality (4) ensures that once machine k is assigned to cell l in time period t, then the operations of part-types may be assigned to that machine. Inequality (5) ensures that sufficient machine capacity is assigned to each cell in time period t. Inequality (6) consists of two constraints for the upper and lower limits to the number of machines in each cell. The upper limit for the number of machines is due to the limitations of the physical space. Furthermore, there should be at least one machine in each cell, otherwise the cell will disappear. Equation (7) provides the relationship of storage levels at different time periods. In fact, in planning the production, demand of part-type i at time t should be deducted from the finished parts in storage at time t. Equation (8) implicates that the production in the entire planning horizon definitely meets the total demand. Equation (9) shows the relationship between the set-up variable $\beta(t)$ and the part processing quantity $X_i(t)$. Equation (10) imposes nonnegativity and integrality, respectively.

Table 4: A solution matrix in t=1 for Example 1

Cell number	Machine number	Part-type 1	2	3	5	8	10	6	7	4	9
1	1	2					1				
	2	1	1		2	2					
	3		2	2	1	1	2				
	4			3							
	6			1			3				
2	1							2			
	5							1			
3	4								2	2	2
	7								1	1	1

Table 5: Cell formation of Example 1

Time period, t	Cell 1	Cell 2	Cell 3
1	1, 2, 3, 4, 6	1, 5	4, 7
2	2, 3	7	4
3	1, 2	4, 7	2, 3

Model Linearization

The proposed model is a nonlinear mixed-integer programming model because of the existing two nonlinear parts including the second term of objective function (see (2)) and the seventh constraint (see (9)). In order to find a global solution, the nonlinear model should be transformed into the linear form. Hence, the linearization phase is implemented for these two parts. Consider the second term of the objective function:

$$\sum_{t=1}^{T}\sum_{i=1}^{I(t)} X_i(t)\cdot\sum_{j=1}^{j_i-1}\sum_{k=1}^{K_{ji}}\sum_{k'=1}^{K_{ji}}\sum_{l=1}^{L}\sum_{l'=L}^{L} R_{ll'}\delta_{i[jk]l}(t)\,\delta_{i[(j+1)k']l'}(t). \tag{11}$$

It can be modified as follows:

(12)

In order to linearize the previous expression, let us assume that

$$Y_{i[jk]lk'l'}(t) = \delta_{i[jk]l}(t)\,\delta_{i[(j+1)k']l'}(t), \tag{13}$$

$$W_{i[jk]lk'l'}(t) = X_i(t)\,Y_{i[jk]lk'l'}(t). \tag{14}$$

These variables imply that

$$Y_{i[jk]lk'l'}(t) = \begin{cases} 1, & \text{if part-type } i \text{ moves to machine } k' \text{ in cell } l' \text{ to perform operation } (j+1) \text{ after performing operation } j \text{ on machine } k \text{ in cell } l, \\ 0, & \text{otherwise,} \end{cases}$$

$$W_{i[jk]lk'l'}(t) = \begin{cases} X_i(t), & \text{if } Y_{i[jk]lk'l'}(t) = 1, \\ 0, & \text{if } Y_{i[jk]lk'l'}(t) = 0. \end{cases} \tag{15}$$

Finally, the second term of objective function can be replaced by the following linear expression:

$$\sum_{t=1}^{T}\sum_{i=1}^{I(t)}\sum_{j=1}^{J_i-1}\sum_{k=1}^{K}\sum_{k'=1}^{K}\sum_{l=1}^{L}\sum_{L'=1}^{L} R_{ll'}W_{i[jk]lk'l'}(t). \tag{16}$$

Furthermore, the following constraints should be added to this model:

$$\delta_{i[jk]l}(t) + \delta_{i[(j+1)k']l'}(t) - 2Y_{i[jk]lk'l'}(t) \geq 0, \tag{17}$$

$$\delta_{i[jk]l}(t) + \delta_{i[(j+1)k']l'}(t) - Y_{i[jk]lk'l'}(t) \leq 1, \tag{18}$$

$$W_{i[jk]lk'l'}(t) \geq X_i(t) + MY_{i[jk]lk'l'}(t) - M, \tag{19}$$

$$W_{i[jk]lk'l'}(t) \leq X_i(t), \tag{20}$$

$$W_{i[jk]lk'l'}(t) \leq MY_{i[jk]lk'l'}(t). \tag{21}$$

Inequalities (17) and (18) imply that $Y_{i[jk]lk'l'}(t)$ is equal to 1, if one unit of part-type i is moved to machine k' in cell for operation (j + 1) after performing operation j on machine k in cell l. Inequalities (19) and (20) enforce $W_{i[jk]lk'l'}(t)$ to be $X(t)$ when $Y_{[jk]lk'l'i}(t)$ is equal to 1. The last constraint (21) enforces $W_{i[jk]lk'l'}(t)$ to be 0, when $Y_{[jk]lk'l'i}(t)$ is equal to 0.

For (9) which should be linearized, the conditional 0-1 requirement for variable $\beta(t)$ can be simply converted to the following set of constraints:

$$\beta_i(t) \le X_i(t), \tag{22}$$

$$M\beta_i(t) \ge X_i(t), \tag{23}$$

where $\beta_i(t)$ are 0-1 variables and M is a large positive number. One can easily verify that (22) serve the same purpose as (9) in the model.

Table 6: Part processing sequence and batch size in Example 1

Time period, t	Part-type									
	1	2	3	5	8	10	6	7	4	9
1	50	36	68	23	30	61	72	22	36	37
2	0	0	0	30	28	0	0	18	0	0
3	36	22	0	32	26	0	0	16	39	33

Table 7: Part processing demand and unit inventory cost for Example 2

	Part-type																		
	1	2	3	4	5	6	7	8	9	10	11	12	13	14	15	16	17	18	19
	Unit inventory cost																		
	70	140	120	135	190	56	40	104	145	80	85	90	150	60	45	67	130	65	95
t	Part processing demand																		
1	500	800	1000	600	600	900	700	700	400	700	400	800	1000	900	900	500	0	700	600
2	900	500	500	700	500	600	600	0	600	700	500	700	800	900	800	600	800	800	600
3	500	800	600	100	1000	700	500	600	700	0	500	1000	800	600	1000	800	700	1000	500

A Single Model for Cell Formation Problem

The proposed model (see (2) to (10)) can be simplified through some modifications in the formulation, in order to solve the cell formation problem, without considering the production planning issues. In this model, the variable $X(t)$ is equal to $D_i(t)$. After linearization, the model can be expressed as follows:

$$\text{Minimize } Z = \sum_{t=1}^{T}\sum_{k=1}^{K} M_k(t)\cdot\sum_{l=1}^{L} n_{kl}(t) + \sum_{t=1}^{T}\sum_{i=1}^{I(t)} D_i(t)\cdot\sum_{j=1}^{j_i-1}\sum_{k=1}^{K}\sum_{k'=1}^{K}\sum_{l=1}^{L}\sum_{l'=L}^{L} R_{ll'}\cdot Y_{i[jk]lk'l'}(t). \tag{24}$$

Subject to

$$\sum_{l=1}^{L}\delta_{i[jk]l}(t) = 1, \quad i = 1,\dots,I,\ j = 1,\dots,J_i,\quad k = 1,\dots,K,\ \forall(t), \tag{25}$$

$$\sum_{i=1}^{I}\sum_{j=1}^{J_i}\delta_{i[jk]l}(t) - n_{kl}(t) \ge 0, \quad i = 1,\dots,I,\ j = 1,\dots,J_i,\quad k = 1,\dots,K,\ l = 1,\dots,L\ \forall(t), \tag{26}$$

$$-\delta_{i[jk]l}(t) + n_{kl}(t) \ge 0, \quad i = 1,\dots,I,\ j = 1,\dots J_i,\quad k = 1,\dots,K,\ l = 1,\dots,L,\ \forall(t), \tag{27}$$

$$\sum_{k=1}^{K} n_{kl}(t) \ge LB_l, \quad k = 1,\dots,K,\ \forall(t), \tag{28}$$

$$\sum_{k=1}^{K} - n_{kl}(t) \ge -UB_l, \quad k = 1,\dots,K,\ \forall(t), \tag{29}$$

$$\delta_{i[jk]l}(t) + \delta_{i[(j+1)k']l'}(t) - 2Y_{i[jk]lk'l'}(t) \ge 0,\quad i = 1,\dots,I,\ j = 1,\dots,j_i - 1,\ l,l' \in L,\ k,k' \in K,\ \forall(t), \tag{30}$$

$$-\delta_{i[jk]l}(t) - \delta_{i[(j+1)k']l'}(t) + Y_{i[jk]lk'l'}(t) \ge -1,\quad i = 1,\dots,I,\ j = 1,\dots,j_i - 1,\ l,l' \in L,\ k,k' \in K,\ \forall(t), \tag{31}$$

$$n_{kl}(t) \ge 0,\quad W_{i[jk]lk'l'}(t) \ge 0,\quad Y_{i[jk]lk'l'}(t),\quad \delta_{i[jk]l}(t), \tag{32}$$

$$n_{kl}(t) = 0, 1 \quad \forall i, j, l, l', k, k', t. \tag{33}$$

NUMERICAL EXAMPLES

Two numerical examples with different structures from the existing literature are presented in this section. The examples have been solved using LINGO 12.0, a commercially available optimization software, on a personal computer with Intel Core2 Duo T6400 @ 2.00?GHz processor and 4 GB RAM.

Example 1

The data used in this example has been adapted from Chen [25] with slight modifications. In this example, 3 cells, 3 time-periods, 10 part-types, and 7 machine-types are considered. The minimum and maximum numbers of machines in each cell are 1 and 5, respectively. Detailed production demand and unit inventory cost to hold each part-type are presented in Table 1. Table 2 shows machine operating costs, part-machine requirements, and refixturing costs. Machine operating costs are presented in the second column of this table. For example, it will cost 15 units to operate 1 unit of machine 1 in the system 10 units for machine 2, and so on. The 6th column of Table 2 shows that there are 2 operations for processing part 4. It also indicates that machine 4 and 7 are required to perform operations 2 and 1, respectively, for part 4. Meanwhile, intercell material handling costs are shown in Table 3.

Table 8: Partial input data for Example 2

Machine number	Machine cost	Part-type 1	2	3	4	5	6	7	8	9	10	11	12	13	14	15	16	17	18	19
		Part operation requirement (upper number) and refixturing cost (lower number)																		
1	1600		1 100		4 240		3 200						3 440							
2	1300						1 200					3 150	1 250				2 160			
3	1200	3 450				3 550						1 120							2 400	
4	2000	1 200									2 200					2 310		2 200		
5	1800		5 350	3 320		1 600			3 350											
6	1600		3 200		3 430					3 500					2 270			1 280		
7	1400				2 500				1 100	1 400							1 240			
8	1100			2 200	1 330			2 300			3 100			3 340		1 150				1 150
9	2000		4 250				4 220			2 600	1 300			1 110	3 420					
10	1800		2 150	1 400					2 200				2 320						1 260	
11	1600	4 300					2 300	3 440				2 230			1 250					
12	1400	2 300				2 700		1 100						2 230						2 420

Table 9: Intercell material handling cost for Example 2

Cell number	1	2	3
1	0	55	44
2	55	0	25
3	44	25	0

It is assumed that the specifications of time periods do not influence the production-related costs which were presented through Tables 1 to 3 (e.g, there is no inflation). Considering the part-operation requirements in Table 2, in order to reduce the number of variables and constraints, the variables which can be fixed to zero were removed from the model using sparse set membership filtering technique of LINGO [29]. After fixing these variables, some constraints became redundant and were subsequently removed. LINGO solver defined the model of Example 1 as an integer linear problem (MILP) and used the branch and bound (B-andB) method to solve it. The resulting formulation has a total of 1089 variables and 1526 constraints. The solution was achieved after 102 minutes of the solver running. The total cost of this problem (which appears as objective value in LINGO solution report) was 997.40 units. Table 4 shows the solution matrix for the cell formation problem (machines and part-types groups) in time-period 1. In addition, Table 5 presents machine cells for different time periods. The entries in this table are the machine numbers.

From Tables 4 and 5, it can be observed that multiple units of the same machine can be used in different cells. However, the usage of more than one unit of each machine-type in the same cell in neither necessary nor economical, since there is no machine capacity constraint in the model. For example, when t = 1, there are 2 units of machine-type 1 in cells 1 and 2, and 2 units of machine-type 4 in cells 1 and 3. Table 6 presents the solution of production planning problem. It shows the times and amounts of production to satisfy demands of all time periods. The sequence of parts in Table 6 the is same as that in the solution matrix for t =1(Table 4). From Tables 2, 5 and 6, it can be observed that, for example, 22 units of parttype 7 processed in t = 1 will flow from cell 3 to cell 1 to be processed by machines 7 and 4, respectively.

Example 2

The data set of this example have been used from Mungwattana [12] with little modifications. In this example, 3 cells, 3 time periods, 19 part-types, and 12 machine-types are considered. The minimum and maximum numbers of machines in each cell are 3 and 9, respectively. Production demands for 19 part-types in 3 time-periods and unit inventory cost to hold each part-type are shown in Table 7. Table 8 shows machine operating costs, part-machine requirements, and refixturing costs. In addition, intercell material handling costs are presented in Table 9.

The linear model of Example 2 consists of 3015 variables (including 849 integer variables) and 4421 constraints. The solver was interrupted by the authors, since it was unable to obtain the results after almost 70 hours of running. It is due to increasing the number of variables and constraints in comparison with Example 1. However, this example can be solved by the single cell formation model which was presented in Section 2.5. In this case, under the single model for the cell formation and with one time period, the model includes 663 variables and 727 constraints. The solver obtained the results after almost 6 minutes of running and objective function value was 19335. Table 10 shows the solution matrix of Example 2 (for one time period).

PERFORMANCE COMPARISON

Several different CF problem-solving methodologies have been proposed in the literature. Likewise, various performance measures have been suggested in order to evaluate and compare the efficiency of these methodologies. Hsu [30] proposed the group capability index (GCI) as a measure of goodness and claimed that this measure maybe is more consistent in predicting the suitability of a manufacturing system for CM. GCI simultaneously considers production volume and processing time of operations factors and excludes voids ("zero" entries) from the calculation of goodness. GCI is defined by

$$GCI = 1 - \frac{e_o}{e}, \tag{34}$$

where e_o is the number of exceptional elements in the machine-part matrix and e is the total number of "one" entries in the machine-component matrix.

Therefore, in this section, the CF results of the proposed model are compared with the results of rank order clustering (ROC) method [31] and direct clustering analysis (DCA) [32] methods, by means of GCI measure. Table 11 shows this comparison based on the results presented in Tables 4 and 10 and also the results of ROC and DCA methods. For the sake of concise presentation, those steps required to solve the examples with ROC and DCA methods are not described in this paper. From Table 11, it is observed that the proposed model represents significant improvements in GCI, in comparison with the ROC and DCA methods for Examples 1 and 2.

Table 10: A solution matrix for Example 2

Cell number	Machine number	Part-type																		
		6	11	12	14	1	4	5	7	15	19	2	3	8	9	10	13	16	17	18
	2	1	3	1														2		
1	3	3	1			3		3												2
	11	2	2		1	4			3											
	4					1	1	1		2						2			2	
2	8								2	1	1		2			3	3			
	12					2		2	1		2						2			
	1			3			4					1								
	5											5	3	3						
3	6				2		3					3			3				1	
	7						2							1	1			1		
	9	4			3							4			2	1	1			
	10			2								2	1	2						1

Table 11: CF results of the proposed model versus the results of ROC method

Example number	Method	e_0	e	GCI
	ROC	6	22	72.72%
1	DCA	13	22	40.90%
	Proposed model	0	22	**100%**
	ROC	33	57	42.11%
2	DCA	33	57	42.11%
	Proposed model	20	57	**64.91%**

CONCLUSIONS

In this paper, a comprehensive dynamic deterministic integrated mathematical model was developed to simultaneously solve the production planning and cell formation problems in CM systems. The overall objective function of the proposed model minimizes the intercell material handlings (with respect to the different travel distances between cells), machine operating cost finished goods inventory, and machine set-up costs (considering the machine set up cost for each operation) for CM systems. The constraints of proposed model define the relationships between variables, such as the relationship of storage levels at different time periods, and upper machine and minimum machine limits for each cell. Some numerical examples with different sizes from the existing literature were considered to test and illustrate the proposed model. The examples were solved by means of LINGO optimization software. In order to evaluate the performance of the proposed model, it was compared with the rank order clustering (ROC) method and direct clustering analysis (DCA) method, using group capability index (GCI) measure. It was shown that the proposed CF model has a higher and satisfactory performance.

The proposed mathematical model offers the advantage of solving CF problem with high performance, while it simultaneously considers the production planning issues with sequence data. However, it can be observed that the proposed model is not suitable for solving the large scale problems. Hence, the use of heuristic methods to deal with such problems deserves further study.

REFERENCES

1. N. Singh, "Design of cellular manufacturing systems: an invited review," European Journal of Operational Research, vol. 69, no. 3, pp. 284–291, 1993.
2. A. Tariq, I. Hussain, and A. Ghafoor, "A hybrid genetic algorithm for machine-part grouping," Computers & Industrial Engineering, vol. 56, no. 1, pp. 347–356, 2009.
3. U. . Wemmerlöv and N. L. Hyer, "Cellular manufacturing in the U.S. Industry: a survey of users,"International Journal of Production Research, vol. 27, no. 9, pp. 1511–1530, 1989.
4. S. Kalpakjian and S. R. Schmid, Manufaturing Enginnering and Technology, Prentice Hall, 6th edition, 2010.
5. S. S. . Heragu, "Group technology and cellular manufacturing," IEEE Transactions on Systems Man and Cybernetics, vol. 24, no. 2, pp. 203–215, 1994.
6. R. . Shanker and P. Vrat, "Some design issues in cellular manufacturing using the fuzzy programming approach," International Journal of Production Research, vol. 37, no. 11, pp. 2545–2563, 1999.
7. M. Murugan and V. Selladurai, "Optimization and implementation of cellular manufacturing system in a pump industry using three cell formation algorithms," International Journal of Advanced Manufacturing Technology, vol. 35, no. 1, pp. 135–149, 2007.
8. J. McAuley, "Machine grouping for efficient production," Production Engineering, vol. 51, no. 2, pp. 53–57, 1972.
9. R. G. Askin, et al., "A Hamiltonian path approach to reordering the part-machine matrix for cellular manufacturing," International Journal of Production Research, vol. 29, no. 6, pp. 1081–1100, 1991.
10. S. J. . Chen and C. S. Cheng, "A neural network-based cell formation algorithm in cellular manufacturing," International Journal of Production Research, vol. 33, no. 2, pp. 293–318, 1995.
11. S. Zolfaghari, "Design and plannng for cellular manufacturing: application of neural networks and advanced search techniques," in Mechanical Engineering, Universty of Ottawa, Ottawa, Canada, 1997.
12. A. Mungwattana, Design of Cellular Manufacturing Systems for Dynamic and Uncertain Production Requirements with Presence of Routing Flexibility, Virginia Polytechnic Institute and State University, Blacksburg, Va, USA, 2000.
13. R. Tavakkoli-Moghaddam, M. B. Aryanezhad, N. Safaei, and A. Azaron, "Solving a dynamic cell formation problem using metaheuristics," Applied Mathematics and Computation, vol. 170, no. 2, pp. 761–780, 2005.
14. R. Logendran, "A binary integer programming approach for simultaneous machine-part grouping in cellular manufacturing systems," Computers & Industrial Engineering, vol. 24, no. 3, pp. 329–336, 1993.
15. S. . Zolfaghari and M. Liang, "Comprehensive machine cell/part family formation using genetic algorithms," International Journal of Manufacturing Technology and Management, vol. 15, no. 6, pp. 433–444, 2004.
16. H. Seifoddini, "A probabilistic model for machine cell formation," Journal of Manufacturing Systems, vol. 9, no. 1, pp. 69–75, 1990.
17. M. Chen, "A mathematical programming model for system reconfiguration in a dynamic cellular manufacturing environment," Annals of Operations Research, vol. 77, no. 0, pp. 109–128, 1998.
18. E. M. Wicks, Designing Cellular Manufacturing Systems with Time Varying Product Mix and Resource Availability, Virginia Polytechnic Institute and State University, Blacksburg, Va, USA, 1995.
19. G. Harhalakis, R. Nagi, and J. M. Proth, "An efficient heuristic in manufacturing cell formation for group technology applications," International Journal of Production Research, vol. 28, no. 1, pp. 185–198, 1990.
20. G. Harhalakis, G. Ioannou, I. Minis, R. Nagi, et al., "Manufacturing cell formation under random product demand," International Journal of Production Research, vol. 32, no. 1, pp. 47–64, 1994.
21. N. Safaei and R. Tavakkoli-Moghaddam, "Integrated multi-period cell formation and subcontracting production planning in dynamic cellular manufacturing systems," International Journal of Production Economics, vol. 120, no. 2, pp. 301–314, 2009.
22. Yu. N. Sotskov and T.-C. Lai, "Minimizing total weighted flow time under uncertainty using dominance and a stability box," Computers & Operations Research, vol. 39, no. 6, pp. 1271–1289, 2012.
23. N. M. Matsveichuk, Yu. N. Sotskov, N. G. Egorova, and T.-C. Lai, "Schedule execution for two-machine flow-shop with interval processing times," Mathematical and Computer Modelling, vol. 49, no. 5-6, pp. 991–1011, 2009.
24. B. . Arvindh and S. A. Irani, "Cell formation the need for an integrated solution of the subproblems,"International Journal of Production Research, vol. 32, no. 5, pp. 1197–1218, 1994.
25. M. Chen, "A model for integrated production planning in cellular manufacturing systems," Integrated Manufacturing System, vol. 12, no. 4, pp. 275–284, 2001.

26. S. Ah kioon, A. A. Bulgak, and T. Bektas, "Integrated cellular manufacturing systems design with production planning and dynamic system reconfiguration," European Journal of Operational Research, vol. 192, no. 2, pp. 414–428, 2009.

27. I. Mahdavi, A. Aalaei, M. M. Paydar, and M. Solimanpur, "Designing a mathematical model for dynamic cellular manufacturing systems considering production planning and worker assignment," Computers & Mathematics with Applications, vol. 60, no. 4, pp. 1014–1025, 2010.

28. A. Atmani, R. S. Lashkari, and R. J. Caron, "A mathematical programming approach to joint cell formation and operation allocation in cellular manufacturing," International Journal of Production Research, vol. 33, no. 1, pp. 1–15, 1995.

29. LINDO Systems Inc, LINGO User's Guide, LINDO System, Chicago, Ill, USA, 2010.

30. C. P. Hsu, The Similarity Coefficient Approaches of Machine-Component Cell Formation in Cellular Manufacturing: A Comparative Study, in Department of Industrial and Systems Engineering, of Wisconsin-Milwaukee, Milwaukee, Wis, USA, 1990.

31. M. P. Groover, Automation, Production Systems, and Computer-Integrated Manufacturing, Prentice Hall Press, 2007.

32. U. Wemmerlov and N. Hyer, "Comments on direct clustering algorithm for group formation in cellular manufacture," Journal of Manufacturing Systems, vol. 3, no. 1, 1984.

INDEX